Advances in precision livestock farming

It is widely recognised that agriculture is a significant contributor to global warming and climate change. Agriculture needs to reduce its environmental impact and adapt to current climate change whilst still feeding a growing population, i.e. become more 'climate-smart'. Burleigh Dodds Science Publishing is playing its part in achieving this by bringing together key research on making the production of the world's most important crops and livestock products more sustainable.

Based on extensive research, our publications specifically target the challenge of climate-smart agriculture. In this way we are using 'smart publishing' to help achieve climate-smart agriculture.

Burleigh Dodds Science Publishing is an independent and innovative publisher delivering high quality customer-focused agricultural science content in both print and online formats for the academic and research communities. Our aim is to build a foundation of knowledge on which researchers can build to meet the challenge of climate-smart agriculture.

For more information about Burleigh Dodds Science Publishing simply call us on +44 (0) 1223 839365, email info@bdspublishing.com or alternatively please visit our website at www.bdspublishing.com.

Related titles:

Improving organic animal farming
Print (ISBN 978-1-78676-180-4) Online (ISBN 978-1-78676-183-5; 978-1-78676-182-8)

Robotics and automation for improving agriculture
Print (ISBN 978-1-78676-272-6) Online (ISBN 978-1-78676-274-0; 978-1-78676-275-7)

Improving dairy herd health
Print (ISBN 978-1-78676-467-6) Online (ISBN 978-1-78676-469-0; 978-1-78676-470-6)

Improving data management and decision support systems in agriculture
Print (ISBN 978-1-78676-340-2) Online (ISBN 978-1-78676-342-6; 978-1-78676-343-3)

Chapters are available individually from our online bookshop:
https://shop.bdspublishing.com

climate•SMART•publishing
IN AGRICULTURAL SCIENCE

BURLEIGH DODDS SERIES IN AGRICULTURAL SCIENCE
NUMBER 105

Advances in precision livestock farming

Edited by Professor Daniel Berckmans, Katholieke University of Leuven, Belgium

burleigh dodds
SCIENCE PUBLISHING

Published by Burleigh Dodds Science Publishing Limited
82 High Street, Sawston, Cambridge CB22 3HJ, UK
www.bdspublishing.com

Burleigh Dodds Science Publishing, 1518 Walnut Street, Suite 900, Philadelphia, PA 19102-3406, USA

First published 2022 by Burleigh Dodds Science Publishing Limited

Library of Congress Control Number: 2021947614

British Library Cataloguing in Publication Data
A catalogue record for this book is available from the British Library

ISBN 978-1-78676-471-3 (Print)
ISBN 978-1-78676-474-4 (PDF)
ISBN 978-1-78676-473-7 (ePub)
ISSN 2059-6936 (print)
ISSN 2059-6944 (online)

DOI 10.19103/AS.2021.0090

Typeset by Deanta Global Publishing Services, Dublin, Ireland

Contents

Series list

Title	Series number
Achieving sustainable cultivation of maize - Vol 1 From improved varieties to local applications Edited by: Dr Dave Watson, CGIAR Maize Research Program Manager, CIMMYT, Mexico	001
Achieving sustainable cultivation of maize - Vol 2 Cultivation techniques, pest and disease control Edited by: Dr Dave Watson, CGIAR Maize Research Program Manager, CIMMYT, Mexico	002
Achieving sustainable cultivation of rice - Vol 1 Breeding for higher yield and quality Edited by: Prof. Takuji Sasaki, Tokyo University of Agriculture, Japan	003
Achieving sustainable cultivation of rice - Vol 2 Cultivation, pest and disease management Edited by: Prof. Takuji Sasaki, Tokyo University of Agriculture, Japan	004
Achieving sustainable cultivation of wheat - Vol 1 Breeding, quality traits, pests and diseases Edited by: Prof. Peter Langridge, The University of Adelaide, Australia	005
Achieving sustainable cultivation of wheat - Vol 2 Cultivation techniques Edited by: Prof. Peter Langridge, The University of Adelaide, Australia	006
Achieving sustainable cultivation of tomatoes Edited by: Dr Autar Mattoo, USDA-ARS, USA and Prof. Avtar Handa, Purdue University, USA	007
Achieving sustainable production of milk - Vol 1 Milk composition, genetics and breeding Edited by: Dr Nico van Belzen, International Dairy Federation (IDF), Belgium	008
Achieving sustainable production of milk - Vol 2 Safety, quality and sustainability Edited by: Dr Nico van Belzen, International Dairy Federation (IDF), Belgium	009
Achieving sustainable production of milk - Vol 3 Dairy herd management and welfare Edited by: Prof. John Webster, University of Bristol, UK	010
Ensuring safety and quality in the production of beef - Vol 1 Safety Edited by: Prof. Gary Acuff, Texas A&M University, USA and Prof. James Dickson, Iowa State University, USA	011
Ensuring safety and quality in the production of beef - Vol 2 Quality Edited by: Prof. Michael Dikeman, Kansas State University, USA	012
Achieving sustainable production of poultry meat - Vol 1 Safety, quality and sustainability Edited by: Prof. Steven C. Ricke, University of Arkansas, USA	013
Achieving sustainable production of poultry meat - Vol 2 Breeding and nutrition Edited by: Prof. Todd Applegate, University of Georgia, USA	014

Introduction

The livestock sector is facing increasing pressure to develop more 'climate-smart' methods that can be used to prevent the onset of major diseases, whilst also monitoring the efficiency and environmental impact of livestock production.

This volume provides a comprehensive review of recent advances in the development of precision livestock technologies to monitor the health and welfare of animals as well as key areas of production such as housing and feed efficiency. The collection is split into two parts: Part 1 includes chapters on data collection and analysis using different techniques to monitor livestock, such as on-animal sensors, thermal imaging and acoustic techniques. Chapters also focus on machine vision techniques and developments in data analysis for decision making. Part 2 chapters review applications in precision livestock farming for monitoring and controlling housing conditions, developments in individual-animal feed efficiency monitoring systems and using automated systems for monitoring mastitis and lameness. Chapters also examine the developments in automated monitoring of livestock fertility and pregnancy, automatic milking systems and monitoring grazing behaviour and automatic grazing management.

Part 1 Data collection and analysis

The book opens with a chapter that reviews the developments in on-animal sensors for monitoring livestock. Chapter 1 begins by exploring the different components of an on-animal sensor system and the challenges and limitations faced by developers trying to make them operational on commercial livestock farms. The chapter then goes on to discuss form factor and the deployment of various sensors such as leg, collar, headstall or halter sensors and ear tag sensors. It also draws attention to tail or tail-head sensors, in-animal sensors as well as novel deployment modes. The chapter moves on discuss four sensor types that provide data on location, activity, movement and physiological characteristics of livestock. Energy management for on-animal sensors is also addressed, followed by sections on communication and data transfer, data management, reduction and analysis as well as the application of on-animal sensors.

The next chapter focuses on developments in thermal imaging techniques to assess livestock health. Chapter 2 begins by examining the use of these techniques to monitor animal health then goes on to discuss how these detection systems must account for key challenges such as ambient environmental conditions. From this, the chapter draws attention to the development of infrared thermography technologies to address these challenges. A section on improving diagnostic accuracy is also provided, which is then followed by a case study highlighting the use of infrared thermography to identify bovine respiratory disease.

The subject of Chapter 3 is developments in acoustic techniques to assess livestock health. The chapter first provides an overview of the different animal vocalisations and sounds produced and how these can be used to identify certain behaviours as well the health status of the animals being monitored. The chapter moves on to analyse how these sounds can be recorded and analysed. Sections on sound applications for monitoring pigs, poultry and cattle are also provided. A critical review on the status of the current technological developments in acoustic techniques is also included in the chapter.

Chapter 4 examines the developments in machine vision techniques to monitor livestock behaviour and health in precision livestock farming. The chapter begins by reviewing advances in computer vision-based technologies for precision livestock farming. It also reviews how automation in image analysis can promote smart management of livestock to improve health and welfare. The chapter discusses the main devices for data acquisition in computer-vision based systems and the range of tasks computer vision (CV) techniques can perform. It reviews key steps such as initialisation, tracking, pose detection and recognition. The chapter includes illustrative case studies of precision livestock farming applications based on existing CV techniques. The chapter concludes by reviewing recent advances in CV techniques based on artificial neural network techniques, as well as future challenges.

The next chapter focuses on the developments in activity and location technologies for monitoring cattle movement and behaviour. Chapter 5 begins by examining the range of options for monitoring cattle location and activity by highlighting the different techniques currently available. The chapter then moves on to discuss the application of these technologies in livestock and how they can also be integrated into the farm system and industry. It concludes by providing an overview on the future development of these technologies and how machine learning techniques are so important in identifying potential issues in livestock.

The final chapter of Part 1 draws attention to the developments in data analysis for decision making in precision livestock farming systems. Chapter 6 first discusses the fundamental concepts in data science and reviews the various techniques that can be used in the data mining process, such as classification, regression and clustering. For each technique, an example has been added for livestock data. The chapter also examines the use of machine learning for supervised learning and unsupervised segmentation. Fitting the model to data and the phenomenon of overfitting are also described. The chapter concludes by providing an overview of why data mining is a useful tool in precision livstock farming and highlights future research areas.

Part 2 Applications

The first chapter of Part 2 focuses on the monitoring and control of livestock housing conditions using precision livestock farming techniques. Chapter 7 discusses the effects of variables such as temperature and humidity on animal productivity, health and welfare. It also shows how environmental conditions affect production of greenhouse gases such as methane (CH_4) and nitrous oxide (N_2O) as well as other damaging emissions such as ammonia (NH_3), The chapter reviews ways of measuring and managing these environmental variables in dairy, poultry and pig production to make livestock production more sustainable.

The next chapter examines the developments in individual-animal feed efficiency monitoring systems for livestock. Chapter 8 begins by first reviewing three key literature databases that provided a basis for the chapter's main discussion on the individual feed-intake measurement system. It then goes on to examine how wearable sensors and electronic scales are used to monitor livestock in various settings, which is then followed by a section on the use of camera sensors for the same purpose. The chapter also looks at image analysis algorithms and deep learning, then moves on to discuss how feed-intake measurement systems can be evaluated.

Chapter 9 analyses developments in automated systems for monitoring livestock health, focusing specifically on mastitis in dairy cattle. The chapter describes the current state of automated mastitis detection systems in dairy cattle and potential new developments in mastitis detection. It first reviews the various components of a mastitis sensor system, then goes on to review the current commercially available sensors. Sections on developing biosensors and combing data sources and improving algorithms are also provided.

Touching on topics previously discussed in Chapter 9, Chapter 10 moves on to review the use of automated systems for monitoring lameness in dairy cattle. The chapter examines existing methods for manual and automated detection of lameness, including approaches that detect changes and abnormalities in the gait or stance of the animal, and methods that directly or indirectly detect changes in individual and social behaviour. It also highlights approaches that use automated technology such as video, accelerometers and spatial positioning systems, and discusses methods to analyse trends and signals in these data.

Chapter 11 focuses on developments in automated monitoring of livestock fertility and pregnancy. The chapter first describes the oestrous cycle in dairy cows, focusing specifically on oestrus-associated changes in dairy cows' behaviour as well as the physiological parameters that can be affected. A section on oestrus detection, pregnancy diagnosis and reproductive performance is also provided, which is then followed by an overview of the various methods currently in use for oestrus detection in cows. The chapter concludes by emphasising the importance of developing oestrus detection methods in future research in order to ensure that any potential issues are detected early on.

The subject of Chapter 12 is advances in automatic milking (AM) systems. The chapter provides comprehensive information on the different aspects of AM to ensure that the potential benefits of the system are available to dairy farmers. It begins by highlighting the importance of identifying which breeds of dairy cow are most suited to automatic milking systems. The chapter moves on to review the effects of AM systems on milk yield and quality, feeding concentrate supplementation in AM systems as well as grazing and grassland management. A section on benchmarking and optimising performance using key performance indicators is also provided. Cow behaviour, training of cows and the transition to AM is also addressed, which is followed by an analysis of health and welfare management. The chapter also analyses how people are involved in AM systems, the impact of these systems on labour requirements, the economics of AM and the possibility of mobile AM systems. The chapter concludes by providing an overview of the current and future focus of AM research.

The final chapter of the book provides an overview of the development and use of information technology to improve the management and welfare of grazing livestock under extensive systems. Chapter 13 discusses the evolution of grazing management through the animal/pasture interface, and the development and application of grazing management technologies, including, on-animal managements though virtual fencing technology and on-animal monitoring of pasture intake, location, health and activity. Technological examples will focus on the Australian beef industry as a potential early adopter of PLF solutions. A case-study is also presented on the commercial implementation of the eShepherd® virtual fencing system for protection of an environmentally sensitive area. The chapter concludes with a summary of developments to date, discussion of future research trends, and where to look for additional information.

Acknowledgement

We wish to acknowledge the following for their help in reviewing particular chapters:

- Chapter 1: Chris Knight, Professor Emeritus, University of Copenhagen, Denmark
- Chapter 5: Prof. Marian Dawkins, University of Oxford, UK; Dr Heiner Lehr, Syntesa, Faroe Islands; and Prof. Erik Vranken, Katholieke University of Leuven, Belgium
- Chapter 9: Professor David Kelton, University of Guelph, Canada
- Chapter 12: Professor Reiner Brunsch, Leibniz Institute for Agricultural Engineering and Bioeconomy, Germany

Part 1

Data collection and analysis

Chapter 1

Developments in on-animal sensors for monitoring livestock

Mark Trotter, CQUniversity Institute for Future Farming Systems, Australia; Derek Bailey, New Mexico State University, USA; and Jaime Manning, Caitlin Evans, Diogo Costa, Elle Fogarty and Anita Chang, CQ University Institute for Future Farming Systems, Australia

1 Introduction

Since the very beginning of animal domestication, humans have sought to understand several key attributes of the livestock under their care. Very simply, these questions can be summarised as: Where are they (location)? What are they doing or experiencing (behaviour)? Are they in a biological state that will meet my needs in terms of the purpose for which I manage them (state)?

Traditionally, these attributes have been discerned by closely watching the animals, in some cases through almost constant human association (e.g. early shepherding practices) or through observations undertaken at critical times (e.g. dairy farmers observing their cows as they are moved into the milking parlour). This resolution of monitoring, aligned with the highly trained eye of the manager, meant that any problems or challenges to the health and productivity of individual animals could be quickly identified and ameliorated.

Unfortunately, the same degree of intensity in monitoring is no longer achievable in modern livestock production systems, particularly in remote and extensive areas, as the costs of labour preclude the ability to have trained staff

http://dx.doi.org/10.19103/AS.2021.0090.01

monitoring animals at similar temporal resolutions. While much has changed in terms of the characteristics of modern animal production compared to early subsistence husbandry, modern livestock managers still seek the same information as their historic counterparts: location, behaviour and state. Given the ever-increasing pressure to reduce costs (particularly time) and the decreasing skill of farm labourers to interpret animal behaviour, producers are increasingly turning to automated remote monitoring systems to provide this key information. Sensors have been developed for a variety of purposes and come in many shapes, sizes and means of interaction with an animal. Sensors can be broadly divided into two categories: on-animal sensors (OASs) - which are attached to, or inserted into, the animal in some manner and which are the focus of this chapter - and off-animal sensors, which collect data by observing the animal without maintaining a permanent attachment. Off-animal sensors include systems that automatically weigh, collect imagery or record vocalisations of livestock. Off-animal sensors provide valuable information, and a brief discussion of the benefits of integration between the two platforms is provided in the applications section (Section 8).

One clear benefit of OASs over off-animal systems is that the former collect information wherever the animal is located. While it is true that some off-animal imaging systems can monitor animals almost constantly in confined areas (e.g. cattle feedlots and pig barns), OASs provide the unique capability of being able to monitor individual livestock in larger areas and more extensive landscapes. As a result, the animal is monitored 24 hours a day and 7 days a week. It is this feature that has proven to be one of the key strengths of these systems, as this degree of intense monitoring could never be achieved using human labour. Most livestock are essentially herbivorous prey animals, which have evolved to hide signs of pain and weakness when observed. It is this tendency for resilience that makes detecting health and welfare problems, through occasional human inspection, a challenge (McLennan, 2018). OASs provide an opportunity to overcome this challenge as they allow for changes in location and movement to be detected while the animal goes about its daily activity without interference. This provides insights that have never been available to animal managers. Remotely detecting the location of an animal also provides an additional key benefit of enabling the animal to be easily found. This simple application is valuable, especially in extensive rangeland systems, where simply finding livestock is a time-consuming and expensive process.

OASs are not the product of the digital age; they have been used, at least in a rudimentary form, for centuries. The simple cow bell collar (Fig. 1) could be described as the earliest version of a remote animal monitoring device. The gentle ringing of the bell as the animal walks enables a manager to remotely locate the animal in difficult landscapes (usually terrain or vegetation related). The sound provided by the bell also provides a clue as to the behaviour of

Figure 1 The 'original' on-animal sensor, a bell on a cow in the Italian Alps North of Torino and a current commercially developed GNSS tracking collar, the Smart Paddock Blue Bell™. The original cow bell provides the livestock manager with the ability to remotely detect the location of the animal using an audible signal. The new technology enables the producer to remotely locate the cow using their smart phone. While digital technologies are transforming the livestock sector, the principle of on-animal sensing has been used for centuries (photo: Bailey and Wolchyn).

an animal; a rapidly ringing bell indicates that the cow is running, either from a predator or from the manager as they attempt to round up the herd. While the cow bell can be considered the original OASs, the digital revolution has delivered a range of new sensor platforms that have significantly advanced this concept. The development of sensors has primarily been driven by other industries, particularly the mobile phone and human wearable platforms. At first glance, it appears a simple translation to take the technology in a FitBit™ and attach it to a cow; however, in reality, there are a number of challenges that make on-animal sensing a far more complex problem.

The objective of this chapter is to provide the reader with a broad understanding of how OASs work, how they are deployed, how data from them are managed and what the data can be used for. It will also provide insights into the challenges of developing and deploying these devices in commercial livestock management operations.

2 Components of an on-animal sensor system

OASs systems have four broad functional components: (1) the device, which is attached to the animal and collects data; (2) the communication system, which transfers the data off the device; (3) the data handling and analytical system, which turns raw data into meaningful information; and (4) the data visualisation platform, which presents the information to the end user to enable decision making (Fig. 2). However, not all on-animal systems follow this simple linear

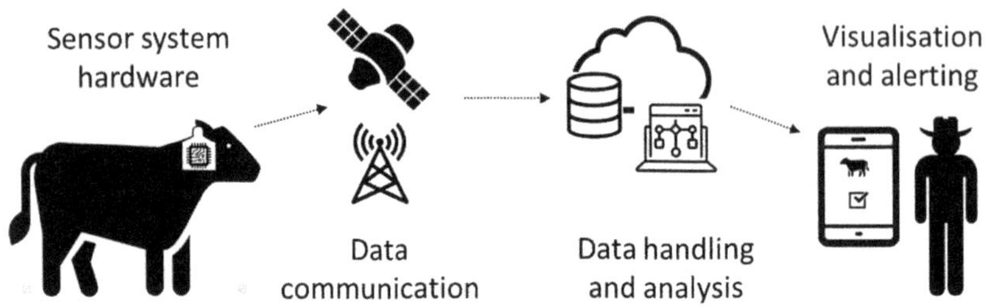

Figure 2 Components of an on-animal sensor system. It is worth noting that not all systems follow this simple design. Many undertake data reduction or preliminary analysis on the device, some even providing alerts (e.g. a coloured light may flash to indicate a problem).

model. Some systems undertake preliminary processing or data reduction on the device while others incorporate a warning system on the device itself, such as a flashing light, to signal an issue with an individual animal. These design specifications are dictated by the specific application and environment in which the system is deployed.

The on-animal device usually consists of several key components: the sensor technology that detects the desired attribute (common sensors include accelerometers, Global Navigation Satellite Systems (GNSS) and proximity loggers); a power source that may include either an energy storage system or energy harvesting system or more commonly both; a transmitting and potentially receiving radio to communicate the data with a receiver; data storage and/or processing capability; and the packaging that contains all the elements, protects them from the external environment and attaches the entire system to the animal.

3 Form factor and deployment mode

Key challenges in the deployment of OASs include attaching the device to the animal (i.e. form factor) and keeping it attached (i.e. device retention). For some species this is relatively easy, and for others the development of an industry-acceptable form factor remains a considerable challenge. There have been numerous options explored over the years, from the traditional collar-borne sensor systems to more novel subcutaneous implantable sensors. A full discussion of the various deployment modes follows.

3.1 Leg sensors

Sensors have been deployed on the legs of dairy cattle for over 40 years, initially as a research tool (Kiddy, 1977) and subsequently as commercially available systems. The earliest leg deployed sensors were simple pedometers, and although modern leg-borne systems mostly use accelerometer technology they provide similar data as these original devices. The deployment of leg sensors on commercial dairy cows is made relatively easy by several key features in how these animals are managed, including the regular movement of animals through the milking parlour, safe access to the leg of an animal by the manager (often at working height), familiarity of animals with being handled around this area (as milking apparatus are put on and taken off regularly) and the generally quiet temperament of dairy cattle. Similar accessibility and temperament features of horses have enabled their use in a research context (DuBois et al., 2015). However, the deployment of leg-borne sensors on other large livestock species (e.g. beef cattle) is more problematic in a commercial context as animals are not as well habituated to humans in handling, although they have been used for research purposes (Ungar et al., 2018). The deployment of leg-borne sensors on smaller production animals (sheep and goats) has been reported in a research context (Champion et al., 1997; Barwick et al., 2018b); there are no examples of commercial systems using this mode of deployment.

3.2 Collar sensors

Collars provide a relatively simple attachment mode and, as discussed previously, have been used by producers for centuries with bells attached. One of the key benefits of a collar sensor is that the animal can support a much larger device weight compared to other deployment modes (particularly when compared to ear tags). Wildlife researchers suggest a maximum collar mass of no more than 5% of the animal's body weight (Cuthill, 1991), which for most production animals equates to a large potential collar weight. In practice, most collar-based livestock monitoring systems weigh much less than this, and research suggests that this deployment mode has little impact on the natural behaviour of grazing cattle (Manning et al., 2017a,b) and sheep (Hulbert et al., 1998). There are numerous collar-borne sensors currently available on the market, with most of them being targeted at the dairy industry for the detection of oestrus (e.g. Allflex SCR, Cow Monitor and Moocall). Most of these systems involve an accelerometer sensor (Kamphuis et al., 2012), although the integration of audio sensing has been evaluated (Reith and Hoy, 2012). At present, collars remain the only viable means of deploying virtual fencing systems that are currently being developed by numerous commercial entities.

3.3 Headstall or halter sensors

The headstall or halter is a slightly more complex version of a collar-based attachment enabling sensors to be placed in locations that are not accessible by other means, particularly over the nose and under the jaw. This allows for more specific head movements to be captured, including bite rate or rumination. While these systems have been used in a research context (Rahman et al., 2018), they are not common in commercial animal production systems (Benaissa et al., 2019). The headstall or halter sensors are the only widely commercialised applications that have been developed for monitoring horses with the primary aim of detecting parturition (e.g. Smart Halter).

3.4 Ear tag sensors

Since researchers first began using OASs there has been significant interest in the development of an ear tag-based form factor. This has been based on the perceived likelihood that industry acceptance will be greater using this type of device, and although there is no published evidence for this, it is a reasonable assumption (Schleppe et al., 2009; Barwick et al., 2018b). Most livestock managers are familiar with the application of ear tags as they have been used for individual animal identification purposes for decades. In some countries (e.g. Australia and New Zealand) electronic Radio Frequency Identification Device (RFID) ear tags are used as part of regulated animal traceability schemes. In addition, the application of ear tags to large livestock is a relatively safe process compared to collars and leg tags, which require animals to be restrained and/or for the manager to work closely around the animal. While there is still some risk to the handler, the application of an ear tag is generally considered safer than other options.

One of the key challenges with ear tag monitoring devices is the weight limit that is imposed by the combination of the small pin attachment and the relatively soft tissue of the animal's ear. This area of research remains one that has largely been unexplored in any systematic fashion, although several case studies have reported the results of various ear tag weights. Currently, most commercial ear tag sensors weigh less than 40 g; for example the Smartbow weighs 35 g (Schweinzer et al., 2019), and the CowManager Sensor weighs 32 g (Zambelis et al., 2019). However, ear tag sensors as heavy as 227 g have been deployed on cattle (Schleppe et al., 2009) but have had to be removed after a short period of time. Heavier tags likely have a higher risk of negatively impacting animal welfare as a result of reduced wound healing and/or a lower tag retention. Commercial developers have explored several solutions to this problem including dual/multiple pin attachments and modified tag designs. While the long-term goal must be to reduce the weight of the tag, these design modifications may provide a reliable solution in the short term. A key gap in the literature exists around the maximum weight that can be sustained by different livestock species in different conditions and using different pin configurations and tag designs. This information would be of value to commercial developers seeking to balance functionality against retention.

3.5 Tail or tail-head sensors

Tail or tail-head-mounted sensors have been developed both as research tools (Krieger et al., 2018; Miller et al., 2020) and for commercial use (e.g. Moocall). These sensors are primarily designed for short-term deployment to detect calving behaviours (Mac, 2017). While the focus of tail deployed sensors has predominantly been on dairy cattle, this form of attachment could be valuable in other large ruminants (e.g. beef cattle), as the tail is usually readily and safely accessible. The challenge with long-term use of tail-head-mounted sensors would be in finding a means of attachment that both was reliable and did not eventually restrict blood flow around the tail itself.

3.6 'In-animal' sensors

In addition to the range of wearable technologies (OASs) reviewed earlier, there are a number of devices designed to be deployed internally, such as within the digestive system or subcutaneously.

3.6.1 Intra-ruminal devices

Rumen bolus sensors or intra-ruminal devices have been used in a research context for many years (Mottram et al., 2008). Once administered, these devices sit in the reticulum or adjacent to the junction between the rumen and the reticulum (Koltes et al., 2018). Most rumen sensors record pH (Dijkstra et al., 2020), temperature (Ipema et al., 2008) and/or more recently, movements through accelerometers (Hamilton et al., 2019). Despite being widely reported in research, the uptake of these sensors in a commercial context remains limited in comparison to other OASs (Knight, 2020a,b). Rumen bolus sensors are highly influenced by the feed and water ingested by the animal (Bewley and Schutz, 2010), and while this may be a challenge in terms of detecting some characteristics (e.g. seeking to measure core body temperature), it may provide valuable insights into other applications (e.g. understanding fodder characteristics or detecting drinking behaviours). Recent research has suggested that simple analytical processes can be developed to enable data from a rumen bolus to be related to core body temperature (Lees et al., 2019).

3.6.2 Intra-vaginal implants

These sensors are placed in the reproductive tract, most commonly in the vagina, and have been primarily used for detecting body temperature (Burdick et al., 2012). This form factor provides one of the few opportunities for reliably measuring core body temperature. The challenge with the use of vaginal implants is that they are usually expelled during parturition, a regular process for most production animals. This process has been leveraged to provide alerts to this event, with several devices having been developed that transmit radio signals after expulsion to alert to calving events (Chung et al., 2020).

3.6.3 Tympanic sensors

Tympanic or ear canal sensors have been developed to provide data on livestock body temperature. They consist of an ear tag with a probe that extends down the ear canal with a temperature sensor attached to the end. Conceptually, this is an ideal place to source data related to the core body temperature of livestock; however, several issues have been reported, including difficulty in correctly fitting the tag and subsequent displacement of the probe out of the ear canal (McCorkell et al., 2014).

3.6.4 Implantable sensors

Some small sensors have been developed that can be surgically implanted (Lee et al., 2016) or more simply injected (Chung et al., 2020). Most of these sensors have been used to collect animal body temperature data; however, there is an increasing interest in their use to monitor a range of other characteristics such as heart rate and metabolites (Neethirajan, 2017). These systems hold obvious value in measuring actual body temperature, which may aid in the detection of diseases. As with other in-animal sensors, implantable sensors have the advantage of higher retention as there is no external component that can be pulled out by being caught on structures in the field. The downside of implantable sensors is the invasiveness of implantation and the need to have small devices capable of long-term deployment. There is also some concern about the technology entering the food chain, if not successfully removed during the processing of meat-producing animals.

3.7 Novel deployment modes

In addition to the more commonly reported device types described above, some more novel strategies have been explored for sensor deployment. Glue on patches has been tested for monitoring bull mating activity (Abell et al., 2017). This particular device could prove valuable for monitoring bulls during the breeding season as other methods (collars and ear tags) are likely to be damaged if bulls interact aggressively. Other novel deployment options include horn attachments (where the animals have horns) and hair clip attachments (where an animal has a fleece or hair coat long enough); however, there is no evidence of these being formally evaluated in the literature.

4 Sensors

There is an obvious and inextricable relationship between the shape and attachment of the device and the type of sensor being used to collect the data of interest. While there have been many different sensors explored for use in monitoring animals, there are a handful that are now regularly used and have undergone considerable refinement in their application to livestock. They can be very broadly categorised into the different types of data they provide: location, activity and movement, and physiological.

4.1 Location sensors

These sensors provide information on either the absolute or relative location of an animal. Absolute location relates to an animal's geographic position and is most commonly referenced as latitude or longitude or may also be represented as X and Y (and sometimes Z) coordinates. Relative location data provide information on an animal's position relative to features of interest (e.g. water point) or other animals.

4.1.1 Global Navigation Satellite Systems

The geo-location technologies that use the network of space-borne satellites to provide positional information are collectively known as Global Navigation Satellite Systems (GNSSs). There are many national and collaborative multinational versions of these technologies; the most commonly referenced is the USA-controlled Global Positioning System (GPS). GNSS uses the radio signals broadcasted by the network of satellites to geolocate the receiver in a process known as trilateration. GNSS has been widely used in livestock research (Swain et al., 2011) and is increasingly being used as part of commercial animal monitoring systems. While GNSS can provide valuable data in most terrestrial environments, one of its key limitations is its current level of power consumption, which is relatively high compared to other sensors.

4.1.2 Local radio signal-based positioning

Another common form of radio-based location involves the use of OASs which emit a signal which is then detected by receivers enabling the location of the animal to be calculated. In these systems, the receivers are usually stationary antennas positioned in such a way as to optimise the signal reception and enable the calculation of the animal's location through either time of flight or signal strength (Hindermann et al., 2020). There are several examples in the literature of these systems applied to both intensive animal production systems (Wolfger et al., 2017) and extensive grazing environments (Menzies et al., 2016).

4.1.3 Location by proximity

While some systems are designed to provide an exact and absolute estimate of location, simpler technologies have been developed that provide the position of an animal relative to another animal or landscape feature. Most of these systems have been developed using a radio signal, with the majority of modern devices employing the Bluetooth radio protocol. The sensors usually consist of both an emitter and receiver and log a 'contact' when they come into range of each other. The number of contacts logged between individual animals is considered to relate to the amount of time they spend with each other. This technology has been used to explore social relationships between individual animals for a variety of applications, from parturition detection to more complex assessments of social structure (Swain and Bishop-Hurley, 2007; Paganoni et al., 2020).

4.2 Activity and movement sensors

Some of the earliest sensors used in livestock research consisted of simple activity meters or vibration recording devices (Stobbs, 1970). These simple devices enabled researchers to autonomously record and understand the behavioural patterns of animals. Since this time, the ability to collect high-resolution movement and activity data has been refined considerably with the development of digital systems that can detect a range of movement types and record large amounts of high-frequency data.

4.2.1 Inertia Monitoring Units

Inertia Monitoring Units (IMUs) are a combination of different individual sensor systems (an accelerometer, gyroscope and magnetometer). Each sensor provides a unique piece of data that enables the calculation of an animal's movement patterns and intensity, and when used in conjunction with one another, the animal's movement path and trajectory can be determined through a process of dead reckoning. The accelerometer provides a measure of the moving force in three directions (for triaxial accelerometers with X, Y and Z axes). The gyrometer measures the sensor's angular velocity by using the earth's gravitational force to determine orientation. Finally, the magnetometer provides the direction or bearing of the sensor orientation in relation to the earth's magnetic fields.

4.2.2 Accelerometer

There is only limited use of IMUs within current OAS systems. Most systems simply use data from the accelerometers (without the gyroscope and magnetometer) to monitor movement and then translate this into activities of interest. Accelerometer data can be collected at a very high temporal resolution and is considered to have a relatively low power consumption. These devices are now commonly available at affordable prices and are probably the most universally applied sensor within livestock sensor systems. One of the key challenges with accelerometers is dealing with the large volume of data. Even simple off-the-shelf systems can record data at over 500 Hz (500 times a second) across three perpendicular axes (X, Y and Z). In most cases, researchers and commercial developers reduce this data either by sampling at lower frequencies, combining multiple axes into a single measure or creating features that summarise data across a defined period of time (e.g. 5 s to 5 min), referred to as an epoch.

The data from accelerometer sensors can be used to detect a range of common or aberrant behaviours. Accelerometers have been used to distinguish between grazing or feeding, standing, walking and lying behaviours (Diosdado et al., 2015; Alvarenga et al., 2016; Barwick et al., 2018b; Fogarty et al., 2020). A more subtle behaviour, in terms of the movement of the animal, is rumination, a highly valuable behaviour in terms of understanding an animal's health and welfare. Many commercial systems in the dairy industry can detect and quantify rumination with variable results in independent tests (Wolfger et al., 2015; Pereira et al., 2018). Reports of scientific instruments successfully detecting rumination in the literature have been largely isolated to headstall or collar deployments (Shen et al., 2020; Tamura et al., 2020), although it has been reported that rumen-based accelerometers have also been successful (Hamilton et al., 2019).

Probably the most widely used applications for such sensor systems in a commercial context are collars and ear tags for the detection of oestrus events to improve the timing of artificial insemination. This application relies on accelerometer sensors to detect the increased activity associated with an oestrus event in dairy cows (Roelofs and Van Erp-van der Kooij, 2018). Accelerometers have also been used to detect specific behaviours associated with unusual behavioural activities that might provide opportunities for management intervention such as parturition events (Smith et al., 2020), disease-related behaviours (Tobin et al., 2020) and heat stress events (Davison et al., 2020).

4.3 Physiological sensors

There are a range of sensors that have been developed to monitor the physiological characteristics of livestock. Perhaps the simplest characteristic to measure, and one of most interest to the livestock industry to date, is body temperature. There are numerous examples of internal sensors such as rumen boluses, rectal or vaginal implants being used to provide measures of body temperature (Koltes et al., 2018; Lees et al., 2019; Baida et al., 2021). There is some conjecture around the interpretation of each deployment mode's data in relation to what it actually measures, whether core body or peripheral temperature (Godyń et al., 2019). Two types of ear tag sensors have been developed for commercial use. The first involves an ear-tag-mounted sensor with a probe that is placed in the ear canal (McCorkell et al., 2014), and the second is mounted on the ear and measures the ear surface temperature (Koltes et al., 2018). While the collection of ear surface temperature is likely to be the easiest and less intrusive of all systems, the relationship between ear surface temperature and core body temperature remains poorly understood. Subcutaneous temperature sensors have been used in some animals (Koltes et al., 2018); however, their widespread application remains limited most likely due to concerns around meat contamination should the device inadvertently enter the food chain.

Rumen pH is a physiological characteristic that has been extensively explored using reticulorumen devices. The primary target for these devices has been the detection of acidosis in cattle, a metabolic disorder commonly associated with high concentrate diets. While many systems have been developed within this research domain, their extension as widespread commercial tools remains limited as the pH sensor itself has a limited lifespan within the rumen (Hamilton et al., 2019).

One other key physiological sensor type of interest is that which measures the heart rate. Several systems have been explored to provide heart rate data ranging from electrocardiography (ECG) to photoplethysmography (PPG) techniques that are used in modern fitness trackers (Nie et al., 2020). For animals in extensive grazing environments, the adaption of these systems to an on-animal sensor will likely be necessary. In comparison, intensive livestock industries may benefit from heart rate monitor systems adapted to off-animal sensor systems (e.g. cameras). Heart rate data is of significant interest to livestock researchers for a range of reasons, from monitoring welfare to understanding the energetics of grazing animals (Laister et al., 2011; Kovács et al., 2014). However, its development for commercial situations remains isolated to applications within the equine industry where temporary fitment of electrodes can be achieved.

4.4 Audio sensors

Audio sensing systems have been explored in the context of off-animal deployment modes primarily for the intensive animal industries. In this situation, they have been applied to detect a range of sounds associated with abnormal or disease states (Chung et al., 2013; Huang et al., 2019). The development of audio sensing in an on-animal form factor remains limited. This is likely due to higher levels of processing power required to handle high volumes of audio data and the infrequent vocalisations of interest (Bishop et al., 2019). Audio sensing could potentially enable the measurement of ingestive behaviours and feed intake at pasture (Galli et al., 2006; Sheng et al., 2020), predation detection (from alert vocalisations) and mother/offspring interactions. However, a significant body of development work remains before these can be realised in a commercial context.

4.5 Future sensors that warrant further investigation

There are a handful of sensor systems in development in other industries that could have applications within the livestock sector and are worth considering. Implantable devices that monitor blood chemistry and metabolites are rapidly evolving and have primarily been used in human medicine (Yang, 2018). While this technology could provide key insights into stress-related responses of animals as a means to improve animal management, the need for surgical implantation, high initial costs and challenges around energy management may limit short-term adoption in a commercial context. However, it is likely that these sensor systems will provide key insights for animal scientists in a research context.

5 Energy management for on-animal sensors

Powering on-animal sensor systems remains one of the key challenges for developers. The general increase in energy density of storage systems, improvements in energy harvesting technologies, along with the reduction in power utilisation of sensor systems have resulted in smaller form factor systems being achieved. However, there is still room for improvement in this area, including two key issues that need to be considered in energy management, the storage of energy and where relevant the recharging of this storage system.

5.1 Energy storage

Most on-animal sensor systems integrate an energy storage system. This commonly takes the form of a battery that can hold energy (in some form of chemistry) for a long period of time and gradually supply it to the sensor system. There are some sensor systems reportedly being developed using capacitors (where the energy is stored as an electric field as opposed to the chemical energy in batteries). Systems reliant on batteries are of two types: rechargeable (secondary) and non-rechargeable (primary). Many research devices (Trotter et al., 2010) and some commercial systems reportedly use primary batteries due to their high energy density, where recharging is considered unviable. Rechargeable batteries are probably the most widely used system across all current commercial and research-grade OAS systems. Within this category, the most commonly used chemistry is that based on Lithium, the lightest weight, most energy-dense and affordable option.

5.2 Energy harvesting

Recharging batteries or harvesting energy for the direct operation of OASs remains another key challenge for the livestock industry. A handful of different technologies have been evaluated, including kinetic motion harvesting and inductive coupling. However, the only realistic option at this stage appears to be the collection of solar radiation using photo-voltaic (PV) technology. All current commercial on-animal sensor systems that have an energy harvesting system use PV technology (e.g. CeresTag), with small solar panels (PV cells) placed on the tag or collar. The use of PV systems on ear tags remains challenging as the available area is necessarily smaller and the placement of the tag on the ear can cause problems with an orientation towards the sun. Most early ear tag sensor systems were developed with the idea of having the device facing forward (inside the ear); however, this can result in shading of the tag, which reduces energy harvest. PV energy harvest on collars appears a viable option as a larger surface area of PV cells can be deployed when compared to ear tags.

A key issue with the use of PV energy harvesting is that animals will actively seek shade during the hottest parts of the day, which compromises the amount of solar energy that can be harvested during these optimal radiation times. One alternative for energy harvesting, as previously mentioned, is inductive coupling which is thought to provide a feasible option for intensively managed animals that can be restrained in a small space (Minnaert et al., 2018).

6 Communication and data transfer

There are several means of retrieving the data from the on-animal sensor, ranging from simple manual downloading to real-time radio telecommunications. Manual downloading of sensors has been used for many years in the context of research and is still used today for the collection of very high-resolution data where the transfer would be problematic due to the volume of information collected. Data transfer systems can be broadly classified by the distance over which they transmit, from short-range local to longer-range terrestrial and satellite systems.

6.1 Short-range terrestrial radio options

There are a number of commercially available systems that use short-range radio connections, particularly the Bluetooth, Wi-Fi and Zigbee standards. The limitation with these systems is that the animal must be located within the range of the receiving antenna. This varies according to the individual platform but is generally limited to distances less than 150 m. While useful in some intensive livestock industries, these systems pose limitations for use in extensively grazed livestock. Despite this, many systems still use short-range terrestrial radio and rely either on the animal returning to a common point (e.g. water trough) or as part of a network of connected nodes distributed across the landscape. The latter is known as a wireless sensor network and consists of a series of nodes that initially collect and then 'hop' the data using a series of short-range receivers/transmitters (Huircán et al., 2010). Short-range radio systems are likely to remain valuable options for the intensive livestock industries as the lower power consumption and reduced infrastructure costs make them more feasible in these situations.

6.2 Longer-range terrestrial radio options

For animals in extensive grazing systems, longer-range radio connectivity is considered more ideal. Generally, these systems work up to a range of 10–20 km, but this varies considerably depending on the specific radio system and the terrain in which it is deployed. There are a number of radio protocols that have been used in this context, but the most common is the Low-power Long-range Wide Area Network (LoRaWAN, commonly shortened to LoRa). LoRa radio connectivity has been used in many of the current commercially available on-animal sensor systems (e.g. Moovement, Smart Paddock and Cattle Watch); however, the objective assessment of these systems remains lacking in the literature. Other similar radio protocols such as the proprietary Sigfox platform have also been used for communications, but with less wide application.

6.3 Satellite-based radio options

While satellite communication has been frequently used in a research context, its application in commercial systems has been limited. Some systems use satellites to transfer data to the cloud once it is collected from the field (using a LoRa system). Recent trials with direct to low earth orbit satellite communication from an ear tag have been successful (Bishop-Hurley, 2021) and are currently used in at least one commercial system (CeresTag).

6.4 Hybrid communication systems

Some systems have been proposed and developed which integrate different components of various communication platforms to form an optimal solution.

Perhaps the most common is the use of short-range communication ear tags (deployed on many animals) with a single animal carrying a collar that acts as a mobile gateway. This system allows for the deployment of a large number of low-cost and low-energy usage devices with short-range data transfer, combined with a small number of higher-cost, higher-energy usage devices that handle data transfer over a long range.

7 Data management, reduction and analysis

The management of data and analysis is dealt with in detail elsewhere in this book; however, it is worthwhile discussing the relationship between sensor type, energy management, communications and the challenge of data processing and analysis. The volume of data produced by sensors varies greatly and is sometimes limited by the energy demand. For example, GNSS has relatively large power consumption and as a result many devices are programmed to collect location data sparingly. Conversely, accelerometer sensors have a relatively low power usage and, consequently, are frequently run continuously and at a sub-second rate. In terms of data production, GNSS consequently produces only a small amount of data while the accelerometer produces large volumes. For those sensors that produce large volumes of data, a common strategy is to reduce it by summarising the data over a period of time. In the case of accelerometers, a number of 'features' are often calculated. These features range from simple summations of the three axes over time (e.g. signal magnitude area) to more complex algorithms (e.g. energy and entropy). This on-device algorithm computation is commonly referred to as edge computing (Hu et al., 2016), and the summarised output can be more efficiently communicated as the volume of data is reduced compared to the original raw data.

8 Applications of on-animal sensors

Although other chapters in this book deal with specific applications of OASs, it is worthwhile exploring and summarising the broad range of uses reported within the various livestock industries. When OASs were first applied, it was primarily undertaken in the context of research with the purpose of providing data to answer scientific questions. As the focus shifted to application on-farm, many new applications of OASs have been both reported and proposed.

8.1 Applications in the intensive small animal industries (pigs and poultry)

For the intensive industries, particularly those where animals are raised in concentrated housed systems, the potential for OASs might not be as great as in other industries. Given the number of animals per unit area, other monitoring systems, particularly off-animal systems (cameras, image analysis and audio sensing), may be a more efficient means of collecting the required information.

Accelerometer sensors have been used to classify a range of pig behaviours including parturition-related activities (Thompson et al., 2019). While collars have been used to detect different behaviours (Escalante et al., 2013), their practical use outside of research remains questionable. Applications in the pig industry have focussed on disease and welfare detection (Chapa et al., 2020). Although reports of on-animal sensors deployed in commercial pig systems can be found in the literature (e.g. ear tag, Chapa et al., 2020), there are no known industry-ready systems available for this species.

The use of OASs in chicken, turkey, duck and other avian species has been limited to more novel research applications such as using small RFID-based monitoring systems to detect nest use, feeding and range use (Li et al., 2020). Much of the sensor-based research in this sector has used off-animal sensors with a particular focus on image analysis and audio monitoring (Rowe et al., 2019).

8.2 Applications in the intensive large animal industries (housed dairy and beef feedlots)

The use of OASs in intensive large animal industries represents the most mature commercial market of on-animal sensor technology. There are several key drivers of this: first, the individual animals are often higher in value, making it more economically feasible; second, the data transfer requires less power as the reading antennas can be placed close to the animals. This means a smaller device can be developed as large power sources are not required.

Intensive dairy producers currently use OASs, including ear tag, collar and leg-borne sensors to detect oestrus and improve the timing of artificial insemination (Roelofs and Van Erp-van der Kooij, 2018). Maltz (2020) provides a good review and discussion of the success of sensors used for oestrus detection in the dairy industry. Other applications in the dairy industry include the detection of diseases such as mastitis, lameness, metritis and metabolic disorders (Rutten et al., 2013). The current challenge is that these issues are generally not exclusively detected and identified but rather a general alert to aberrant behaviour is provided, requiring investigation by the manager to determine a final diagnosis (Eckelkamp, 2019). Further discussion around the use of OASs in the dairy industry can be found in Michie et al. (2020) and Knight (2020a,b).

While the use of OASs in intensive beef feedlots is less developed, several studies have demonstrated the ability of OASs to detect key health issues. One of the main targets has been the detection of Bovine Respiratory Disease

(BRD). Although the results of studies using sensors to detect BRD have been mixed (McCorkell et al., 2014), there is a growing belief that this disease can be effectively detected using OASs (Marchesini et al., 2018). As of yet, there has not been widespread uptake of sensors in this industry, possibly due to the cost of the systems available.

8.3 On-animal sensors in the extensive animal industries

The development of OASs in large-scale extensive grazing systems is in its infancy. One of the key limitations is the challenge of achieving low power radio communication across the large landscapes common in grazing production systems. These large landscapes often come with other challenges, including variation in topography and environment. Despite these challenges, there are many systems currently in development or in the early stages of commercialisation.

There have been numerous applications proposed for OASs in the grazing sector in both research and commercial contexts. It is worth noting that there are differences between the applications developed for research purposes and those being developed for use by producers in the day-to-day management of their livestock. This often causes confusion amongst commercial developers and producers as the promotion of research applications appears irrelevant to commercial livestock managers.

8.3.1 Cattle

In the context of research, there have been ample studies using OASs in extensive beef cattle research (Swain et al., 2011; Bailey et al., 2018). These have focussed on quantifying the various behaviours of grazing cattle (Guo et al., 2009; Hamilton et al., 2019; Sprinkle et al., 2020) and then using this data to answer broader research questions around: landscape utilisation (Ganskopp and Bohnert, 2009), the impact of imposed treatments (Bailey et al., 2008), effect of pasture availability (Manning et al., 2017a,b), behavioural changes caused by naturally occurring events (Laporte et al., 2010) or even complex relationships between genetics and animal behaviour (Bailey et al., 2015).

The application of sensors as a management tool for producers has been less well explored. There are examples of specific applications being developed to monitor some key animal behaviours that would prove useful for bull-mounting activity (Abell et al., 2017), infectious disease detection (Tobin et al., 2020), drinking behaviour (Williams et al., 2020) and feeding behaviour as animals become nutritionally challenged (Roberts et al., 2015). However, producer surveys (Trotter et al., 2018) suggest that there may be many more ways in which these systems could bring value to the beef sector including the location of animals to improve the efficiency of mustering, detection of calving, prevention of stock theft, predation detection, monitoring for plant toxicity issues, assisting in feed supplementation decisions and genetic matching of cows and calves.

8.3.2 Sheep and goats

The use of OASs for smaller ruminants has largely focussed on research applications with the development of commercial systems for sheep and goats limited. Like cattle, there are numerous reports of sensors being used to classify the normal behaviours (e.g. grazing, lying, standing, walking and ruminating) of small ruminants (Umstatter et al., 2008; Giovanetti et al., 2017; Fogarty et al., 2018; Mansbridge et al., 2018; Sakai et al., 2019). More complex research applications have included the exploration of energy expenditure while grazing (Animut et al., 2005), parent and offspring interactions (Broster et al., 2012), spatial landscape utilisation (Freire et al., 2012; Akasbi et al., 2013) and response to climatic challenges (Thomas et al., 2008). Further insights into the development of sensor systems for small ruminants can be found in the review by Caja et al. (2020).

In the context of developing sensor systems for producers to use, there has been a number of studies exploring key issues such as the detection of lambing events (Dobos et al., 2014), infectious disease detection (Falzon et al., 2013), oestrus detection (Fogarty et al., 2015), predation detection (Manning et al., 2014), identification of lameness (Barwick et al., 2018a) and prediction of intake at pasture (Giovanetti et al., 2020). The survey undertaken by Trotter et al. (2018) indicated a range of other applications useful to sheep producers including general welfare monitoring, detection of stock theft, genetic matching of ewes and lambs, timing of grazing rotations and ensuring the animals have access to drinking water. As these technologies become commercially viable, it is almost certain that producers will find a range of other applications and uses that will bring value to extensively grazed sheep and goats.

9 Future trends

At a technical level, there are several innovations in the development of specific sensors and the way in which data are managed that will likely bring significant value to both researchers and ultimately the industry as OASs evolve.

The development of sensors that focus on more specific and measurable behaviours, physiological characteristics and/or metabolites will obviously continue to leverage the development in human-related research. Some of the emerging technologies of particular interest include the direct measurement of the heart rate, which has many applications from energetics research for improved production efficiency (Brosh et al., 2006) to stress and welfare assessment (Nie et al., 2020). Another key suite of sensors that will continue to evolve are those capable of directly monitoring the metabolic state of animals either directly or by inference. While implantable technologies remain a challenge for the livestock industries because of potential food chain contamination, the value of the data provided by these systems will keep them in consideration in the future.

One of the key future trends of livestock sensing systems will be the integration of data from multiple sensor systems to provide the key information being sought by both researchers and producers. There are some early examples of different types of on-animal sensor data being integrated including multiple site deployments of accelerometers (Thompson et al., 2019), integration of GNSS and magnetometer sensors (Guo et al., 2009), GNSS with accelerometer data (Brennan et al., 2021) and the integration of location data and accelerometer sensors for dairy cattle (Benaissa et al., 2020). In terms of detecting and diagnosing key behaviours or events in livestock, the integration of data from OASs with data from off-animal sensors will likely expand in the future. Some early examples of this research include where: OASs have been integrated with a range of milk characteristics, feed and water intake and body weight to detect mastitis and lameness in dairy cattle (Post et al., 2020); and GNSS, accelerometer and weather data were integrated to detect lambing events in sheep (Fogarty et al., 2021).

As the technical development of OASs will undoubtedly continue, their use will also expand. There are several key applications that are emerging drivers of this sector. The use of these devices is being explored by regulatory agencies seeking to monitor the location, behaviour and state of livestock. There are several government agencies already using RFID systems to monitor the movement of animals for biosecurity and market integrity. The potential to improve the functionality of these systems to more complex systems including GNSS and/or accelerometer is obvious; however, the path to realising widespread adoption remains challenging. Another key driver of this technology is product authentication. There are now numerous livestock products sold under a marketing claim ranging from 'grass-fed' to enhanced welfare management. On-animal sensor systems provide an opportunity to validate these market claims, particularly for those relating to animal welfare (Fogarty et al., 2019; Chapa et al., 2020; Maroto Molina et al., 2020; Manning et al., 2021) and are likely to see significant research and development into the future.

As OASs transform from research tools into readily available and affordable systems for livestock producers, there is likely to be a significant increase in the adoption of these systems. Already, there is a widespread utilisation in the intensive and high-value animal industries, particularly dairy. As these systems become cheaper and able to work across larger and more varied landscapes, they will almost certainly become more attractive to producers in all animal industries, firstly for beef cattle and then smaller ruminants. However, the timelines for likely adoption will always be limited by the technical feasibility of systems and the benefits brought to the producer. The economic benefits of OASs remain largely unexplored. Systems may have several applications improving financial benefits through a number of cost savings and/or productivity increases, making the economic analysis of benefits complex. In addition, there are many non-financial benefits of these systems. Numerous producers, having been exposed to the concept of OASs and the real-time data they provide, commented that there is a significant component of increased 'peace of mind' of knowing where their animals are and that they are in a satisfactory state.

The potential for OASs to transform the livestock industries is enormous. However, the research community and commercial technology developers need to concentrate their efforts on understanding what data and information are required by commercial producers to improve farm decision making and provide the benefits that these systems promise. The industry has evolved in many ways since we first started using cow bells, and this new generation technologies are the key to further revolution.

10 References

Abell, K. M., Theurer, M. E., Larson, R. L., White, B. J., Hardin, D. K. and Randle, R. F. 2017. Predicting bull behavior events in a multiple-sire pasture with video analysis, accelerometers, and classification algorithms. *Computers and Electronics in Agriculture* 136:221-227.

Akasbi, Z., Oldeland, J., Dengler, J. and Finckh, M. 2013. Analysis of GPS trajectories to assess goat grazing pattern and intensity in Southern Morocco. *The Rangeland Journal* 34(4):415-427.

Alvarenga, F. A. P., Borges, I., Palkovič, L., Rodina, J., Oddy, V. H. and Dobos, R. C. 2016. Using a three-axis accelerometer to identify and classify sheep behaviour at pasture. *Applied Animal Behaviour Science* 181:91-99.

Animut, G., Goetsch, A. L., Aiken, G. E., Puchala, R., Detweiler, G., Krehbiel, C. R., Merkel, R. C., Sahlu, T., Dawson, L. J., Johnson, Z. B. and Gipson, T. A. 2005. Grazing behavior and energy expenditure by sheep and goats co-grazing grass/forb pastures at three stocking rates. *Small Ruminant Research* 59(2-3):191-201.

Baida, B. E. L., Swinbourne, A. M., Barwick, J., Leu, S. T. and van Wettere, W. H. 2021. Technologies for the automated collection of heat stress data in sheep. *Animal Biotelemetry* 9(1):1-15.

Bailey, D. W., Lunt, S., Lipka, A., Thomas, M. G., Medrano, J. F., Cánovas, A., Rincon, G., Stephenson, M. B. and Jensen, D. 2015. Genetic influences on cattle grazing distribution: association of genetic markers with terrain use in cattle. *Rangeland Ecology and Management* 68(2):142-149.

Bailey, D. W., Trotter, M. G., Knight, C. W. and Thomas, M. G. 2018. Use of GPS tracking collars and accelerometers for rangeland livestock production research. *Translational Animal Science* 2(1):81–88.

Bailey, D. W., VanWagoner, H. C., Weinmeister, R. and Jensen, D. 2008. Comparison of low moisture blocks and salt for manipulating grazing patterns of beef cows. *Journal of Animal Science* 86(5):1271–1277.

Barwick, J., Lamb, D., Dobos, R., Schneider, D., Welch, M. and Trotter, M. 2018a. Predicting lameness in sheep activity using tri-axial acceleration signals. *Animals: An Open Access Journal from MDPI* 8(1):12.

Barwick, J., Lamb, D. W., Dobos, R., Welch, M. and Trotter, M. 2018b. Categorising sheep activity using a tri-axial accelerometer. *Computers and Electronics in Agriculture* 145:289–297.

Benaissa, S., Tuyttens, F. A. M., Plets, D., Cattrysse, H., Martens, L., Vandaele, L., Joseph, W. and Sonck, B. 2019. Classification of ingestive-related cow behaviours using RumiWatch halter and neck-mounted accelerometers. *Applied Animal Behaviour Science* 211:9–16.

Benaissa, S., Tuyttens, F. A. M., Plets, D., Trogh, J., Martens, L., Vandaele, L., Joseph, W. and Sonck, B. 2020. Calving and estrus detection in dairy cattle using a combination of indoor localization and accelerometer sensors. *Computers and Electronics in Agriculture* 168:105153.

Bewley, J. M. and Schutz, M. M. 2010. Recent studies using a reticular bolus system for monitoring dairy cattle core body temperature. In: *Proceedings of the First North Annuaire Conference*, Precision Dairy Management, Toronto, Canada, pp. 218–219.

Bishop, J. C., Falzon, G., Trotter, M., Kwan, P. and Meek, P. D. 2019. Livestock vocalisation classification in farm soundscapes. *Computers and Electronics in Agriculture* 162:531–542.

Bishop-Hurley, G. 2021. Preliminary test results of a sensor ear tag with satellite radio communication - CSIRO Australia. In: Trotter, M. (Ed.). Rockhampton.

Brennan, J., Johnson, P. and Olson, K. 2021. Classifying season long livestock grazing behavior with the use of a low-cost GPS and accelerometer. *Computers and Electronics in Agriculture* 181:105957.

Brosh, A., Henkin, Z., Ungar, E. D., Dolev, A., Orlov, A., Yehuda, Y. and Aharoni, Y. 2006. Energy cost of cows' grazing activity: Use of the heart rate method and the Global Positioning System for direct field estimation. *Journal of Animal Science* 84(7):1951–1967.

Broster, J. C., Rathbone, D. P., Robertson, S. M., King, B. J. and Friend, M. A. 2012. Ewe movement and ewe-lamb contact levels in shelter are greater at higher stocking rates. *Animal Production Science* 52(7):502–506.

Burdick, N. C., Carroll, J. A., Dailey, J. W., Randel, R. D., Falkenberg, S. M. and Schmidt, T. B. 2012. Development of a self-contained, indwelling vaginal temperature probe for use in cattle research. *Journal of Thermal Biology* 37(4):339–343.

Caja, G., Castro-Costa, A., Salama, A. A. K., Oliver, J., Baratta, M., Ferrer, C. and Knight, C. H. 2020. Sensing solutions for improving the performance, health and wellbeing of small ruminants. *Journal of Dairy Research* 87(S1):34–46.

Champion, R. A., Rutter, S. M. and Penning, P. D. 1997. An automatic system to monitor lying, standing and walking behaviour of grazing animals. *Applied Animal Behaviour Science* 54(4):291-305. doi: 10.1016/S0168-1591(96)01210-5.

Chapa, J. M., Maschat, K., Iwersen, M., Baumgartner, J. and Drillich, M. 2020. Accelerometer systems as tools for health and welfare assessment in cattle and pigs-A review. *Behavioural Processes* 181:104262.
Chung, H., Li, J., Kim, Y., Van Os, J. M. C., Brounts, S. H. and Choi, C. Y. 2020. Using implantable biosensors and wearable scanners to monitor dairy cattle's core body temperature in real-time. *Computers and Electronics in Agriculture* 174:105453.
Chung, Y., Oh, S., Lee, J., Park, D., Chang, H. H. and Kim, S. 2013. Automatic detection and recognition of pig wasting diseases using sound data in audio surveillance systems. *Sensors* 13(10):12929-12942.
Cuthill, I. 1991. Field experiments in animal behaviour: Methods and ethics. *Animal Behaviour* 42(6):1007-1014.
Davison, C., Michie, C., Hamilton, A., Tachtatzis, C., Andonovic, I. and Gilroy, M. 2020. Detecting heat stress in dairy cattle using neck-mounted activity collars. *Agriculture* 10(6):210.
Dijkstra, J., Van Gastelen, S., Dieho, K., Nichols, K. and Bannink, A. 2020. Review: Rumen sensors: Data and interpretation for key rumen metabolic processes. *Animal: An International Journal of Animal Bioscience* 14(S1):s176-s186.
Diosdado, J. A. V., Barker, Z. E., Hodges, H. R., Amory, J. R., Croft, D. P., Bell, N. J. and Codling, E. A. 2015. Classification of behaviour in housed dairy cows using an accelerometer-based activity monitoring system. *Animal Biotelemetry* 3(1):15.
Dobos, R. C., Dickson, S., Bailey, D. W. and Trotter, M. G. 2014. The use of GNSS technology to identify lambing behaviour in pregnant grazing Merino ewes. *Animal Production Science* 54(10):1722-1727.
DuBois, C., Zakrajsek, E., Haley, D. B. and Merkies, K. 2015. Validation of triaxial accelerometers to measure the lying behaviour of adult domestic horses. *Animal: An International Journal of Animal Bioscience* 9(1):110-114. doi: 10.1017/S175173111400247X.
Eckelkamp, E. A. 2019. Invited review: Current state of wearable precision dairy technologies in disease detection. *Applied Animal Science* 35(2):209-220.
Escalante, H. J., Rodriguez, S. V., Cordero, J., Kristensen, A. R. and Cornou, C. 2013. Sow-activity classification from acceleration patterns: A machine learning approach. *Computers and Electronics in Agriculture* 93:17-26.
Falzon, G., Schneider, D., Trotter, M. and Lamb, D. W. 2013. Correlating movement patterns of merino sheep to faecal egg counts using global positioning system tracking collars and functional data analysis. *Small Ruminant Research* 111(1):171-174.
Fogarty, E. S., Manning, J. K., Trotter, M. G., Schneider, D. A., Thomson, P. C., Bush, R. D. and Cronin, G. M. 2015. GNSS technology and its application for improved reproductive management in extensive sheep systems. *Animal Production Science* 55(10):1272-1280. doi: 10.1071/AN14032.
Fogarty, E., Swain, D., Cronin, G. and Trotter, M. 2019. A systematic review of the potential uses of on-animal sensors to monitor the welfare of sheep evaluated using the Five Domains Model as a framework. *Animal Welfare* 28(4):407-420.
Fogarty, E. S., Swain, D. L., Cronin, G. and Trotter, M. 2018. Autonomous on-animal sensors in sheep research: A systematic review. *Computers and Electronics in Agriculture* 150:245-256.
Fogarty, E. S., Swain, D. L., Cronin, G. M., Moraes, L. E., Bailey, D. W. and Trotter, M. 2021. Developing a simulated online model that integrates GNSS, accelerometer and

weather data to detect parturition events in grazing sheep: A machine learning approach. *Animals: An Open Access Journal from MDPI* 11(2):303.

Fogarty, E. S., Swain, D. L., Cronin, G. M., Moraes, L. E. and Trotter, M. 2020. Behaviour classification of extensively grazed sheep using machine learning. *Computers and Electronics in Agriculture* 169:105175.

Freire, R., Swain, D. L. and Friend, M. A. 2012. Spatial distribution patterns of sheep following manipulation of feeding motivation and food availability. *Animal: An International Journal of Animal Bioscience* 6(5):846-851.

Galli, J. R., Cangiano, C. A., Demment, M. W. and Laca, E. A. 2006. Acoustic monitoring of chewing and intake of fresh and dry forages in steers. *Animal Feed Science and Technology* 128(1-2):14-30.

Ganskopp, D. C. and Bohnert, D. W. 2009. Landscape nutritional patterns and cattle distribution in rangeland pastures. *Applied Animal Behaviour Science* 116(2-4):110-119.

Giovanetti, V., Cossu, R., Molle, G., Acciaro, M., Mameli, M., Cabiddu, A., Serra, M. G., Manca, C., Rassu, S. P. G., Decandia, M. and Dimauro, C. 2020. Prediction of bite number and herbage intake by an accelerometer-based system in dairy sheep exposed to different forages during short-term grazing tests. *Computers and Electronics in Agriculture* 175:105582.

Giovanetti, V., Decandia, M., Molle, G., Acciaro, M., Mameli, M., Cabiddu, A., Cossu, R., Serra, M. G., Manca, C., Rassu, S. P. G. and Dimauro, C. 2017. Automatic classification system for grazing, ruminating and resting behaviour of dairy sheep using a tri-axial accelerometer. *Livestock Science* 196:42-48.

Godyń, D., Herbut, P. and Angrecka, S. 2019. Measurements of peripheral and deep body temperature in cattle-A review. *Journal of Thermal Biology* 79:42-49.

Guo, Y., Poulton, G., Corke, P., Bishop-Hurley, G. J., Wark, T. and Swain, D. L. 2009. Using accelerometer, high sample rate GPS and magnetometer data to develop a cattle movement and behaviour model. *Ecological Modelling* 220(17):2068-2075.

Hamilton, A. W., Davison, C., Tachtatzis, C., Andonovic, I., Michie, C., Ferguson, H. J., Somerville, L. and Jonsson, N. N. 2019. Identification of the rumination in cattle using support vector machines with motion-sensitive bolus sensors. *Sensors* 19(5):1165.

Hindermann, P., Nüesch, S., Früh, D., Rüst, A. and Gygax, L. 2020. High precision real-time location estimates in a real-life barn environment using a commercial ultra wideband chip. *Computers and Electronics in Agriculture* 170:105250.

Hu, W., Gao, Y., Ha, K., Wang, J., Amos, B., Chen, Z., Pillai, P. and Satyanarayanan, M. 2016. Quantifying the impact of edge computing on mobile applications. In: *Proceedings of the 7th ACM SIGOPS Asia-Pacific Workshop on Systems*, p. 1-8.

Huang, J., Wang, W. and Zhang, T. 2019. Method for detecting avian influenza disease of chickens based on sound analysis. *Biosystems Engineering* 180:16-24.

Huircán, J. I., Muñoz, C., Young, H., Von Dossow, L., Bustos, J., Vivallo, G. and Toneatti, M. 2010. ZigBee-based wireless sensor network localization for cattle monitoring in grazing fields. *Computers and Electronics in Agriculture* 74(2):258-264. doi: 10.1016/j.compag.2010.08.014.

Hulbert, I. A. R., Wyllie, J. T. B., Waterhouse, A., French, J. and McNulty, D. 1998. A note on the circadian rhythm and feeding behaviour of sheep fitted with a lightweight GPS collar. *Applied Animal Behaviour Science* 60(4):359-364.

Ipema, A. H., Goense, D., Hogewerf, P. H., Houwers, H. W. J. and Van Roest, H. 2008. Pilot study to monitor body temperature of dairy cows with a rumen bolus. *Computers and Electronics in Agriculture* 64(1):49-52.

Kamphuis, C., DelaRue, B., Burke, C. R. and Jago, J. 2012. Field evaluation of 2 collar-mounted activity meters for detecting cows in estrus on a large pasture-grazed dairy farm. *Journal of Dairy Science* 95(6):3045-3056.

Kiddy, C. A. 1977. Variation in physical activity as an indication of estrus in dairy cows. *Journal of Dairy Science* 60(2):235-243.

Knight, C. H. 2020a. DairyCare 'blueprint for action': Husbandry for wellbeing. *Journal of Dairy Research* 87(S1):1-8.

Knight, C. H. 2020b. Review: Sensor techniques in ruminants: More than fitness trackers. *Animal: An International Journal of Animal Bioscience* 14(S1):s187-s195.

Koltes, J. E., Koltes, D. A., Mote, B. E., Tucker, J. and Hubbell, D. S. III. 2018. Automated collection of heat stress data in livestock: New technologies and opportunities. *Translational Animal Science* 2(3):319-323.

Kovács, L., Jurkovich, V., Bakony, M., Szenci, O., Póti, P. and Tozser, J. 2014. Welfare implication of measuring heart rate and heart rate variability in dairy cattle: Literature review and conclusions for future research. *Animal: An International Journal of Animal Bioscience* 8(2):316-330.

Krieger, S., Sattlecker, G., Kickinger, F., Auer, W., Drillich, M. and Iwersen, M. 2018. Prediction of calving in dairy cows using a tail-mounted tri-axial accelerometer: A pilot study. *Biosystems Engineering* 173:79-84.

Laister, S., Stockinger, B., Regner, A.-M., Zenger, K., Knierim, U. and Winckler, C. 2011. Social licking in dairy cattle–Effects on heart rate in performers and receivers. *Applied Animal Behaviour Science* 130(3-4):81-90.

Laporte, I., Muhly, T. B., Pitt, J. A., Alexander, M. and Musiani, M. 2010. Effects of wolves on elk and cattle behaviors: Implications for livestock production and wolf conservation. *PLoS ONE* 5(8):e11954.

Lee, Y., Bok, J. D., Lee, H. J., Lee, H. G., Kim, D., Lee, I., Kang, S. K. and Choi, Y. J. 2016. Body temperature monitoring using subcutaneously implanted thermo-loggers from holstein steers. *Asian-Australasian Journal of Animal Sciences* 29(2):299-306.

Lees, A. M., Sejian, V., Lees, J. C., Sullivan, M. L., Lisle, A. T. and Gaughan, J. B. 2019. Evaluating rumen temperature as an estimate of core body temperature in Angus feedlot cattle during summer. *International Journal of Biometeorology* 63(7):939-947.

Li, N., Ren, Z., Li, D. and Zeng, L. 2020. Review: Automated techniques for monitoring the behaviour and welfare of broilers and laying hens: Towards the goal of precision livestock farming. *Animal: An International Journal of Animal Bioscience* 14(3):617-625.

Mac, S. 2017. *Evaluating the Ability to Detect Calving Time in Dairy Cattle Using a Precision Technology That Monitors Tail Movement.*Available at: https://digitalcommons.murraystate.edu/postersatthecapitol/2018/UK/5/,

Maltz, E. 2020. Individual dairy cow management: Achievements, obstacles and prospects. *Journal of Dairy Research* 87(2):145-157.

Manning, J., Cronin, G., González, L., Hall, E., Merchant, A. and Ingram, L. 2017a. The behavioural responses of beef cattle (*Bos taurus*) to declining pasture availability and the use of GNSS technology to determine grazing preference. *Agriculture* 7(5):45.

Manning, J., Power, D. and Cosby, A. 2021. Legal complexities of animal welfare in Australia: do on-animal sensors offer a future option? *Animals: An Open Access Journal from MDPI* 11(1):91.

Manning, J. K., Cronin, G. M., González, L. A., Hall, E. J. S., Merchant, A. and Ingram, L. J. 2017b. The effects of global navigation satellite system (GNSS) collars on cattle (Bos taurus) behaviour. *Applied Animal Behaviour Science* 187:54-59.

Manning, J. K., Fogarty, E. S., Trotter, M. G., Schneider, D. A., Thomson, P. C., Bush, R. D. and Cronin, G. M. 2014. A pilot study into the use of global navigation satellite system technology to quantify the behavioural responses of sheep during simulated dog predation events. *Animal Production Science* 54(10):1676-1681. doi: 10.1071/AN14221.

Mansbridge, N., Mitsch, J., Bollard, N., Ellis, K., Miguel-Pacheco, G. G., Dottorini, T. and Kaler, J. 2018. Feature selection and comparison of machine learning algorithms in classification of grazing and rumination behaviour in sheep. *Sensors* 18(10):3532.

Marchesini, G., Mottaran, D., Contiero, B., Schiavon, E., Segato, S., Garbin, E., Tenti, S. and Andrighetto, I. 2018. Use of rumination and activity data as health status and performance indicators in beef cattle during the early fattening period. *Veterinary Journal* 231:41-47.

Maroto Molina, F., Pérez Marín, C. C., Molina Moreno, L., Agüera Buendía, E. I. and Pérez Marín, D. C. 2020. Welfare Quality® for dairy cows: Towards a sensor-based assessment. *Journal of Dairy Research* 87(S1):28-33.

McCorkell, R., Wynne-Edwards, K., Windeyer, C., Schaefer, A. and UCVM Class of 2013 2014. Limited efficacy of Fever Tag® temperature sensing ear tags in calves with naturally occurring bovine respiratory disease or induced bovine viral diarrhea virus infection. *The Canadian Veterinary Journal* 55(7):688-690.

McLennan, K. M. 2018. Why pain is still a welfare issue for farm animals, and how facial expression could be the answer. *Agriculture* 8(8):127.

Menzies, D., Patison, K. P., Fox, D. R. and Swain, D. L. 2016. A scoping study to assess the precision of an automated radiolocation animal tracking system. *Computers and Electronics in Agriculture* 124:175-183.

Michie, C., Andonovic, I., Davison, C., Hamilton, A., Tachtatzis, C., Jonsson, N., Duthie, C. A., Bowen, J. and Gilroy, M. 2020. The Internet of Things enhancing animal welfare and farm operational efficiency. *Journal of Dairy Research* 87(S1):20-27.

Miller, G. A., Mitchell, M., Barker, Z. E., Giebel, K., Codling, E. A., Amory, J. R., Michie, C., Davison, C., Tachtatzis, C., Andonovic, I. and Duthie, C. A. 2020. Using animal-mounted sensor technology and machine learning to predict time-to-calving in beef and dairy cows. *Animal: An International Journal of Animal Bioscience* 14(6):1304-1312.

Minnaert, B., Thoen, B., Plets, D., Joseph, W. and Stevens, N. 2018. Wireless energy transfer by means of inductive coupling for dairy cow health monitoring. *Computers and Electronics in Agriculture* 152:101-108.

Mottram, T., Lowe, J., McGowan, M. and Phillips, N. 2008. Technical note: A wireless telemetric method of monitoring clinical acidosis in dairy cows. *Computers and Electronics in Agriculture* 64(1):45-48.

Neethirajan, S. 2017. Recent advances in wearable sensors for animal health management. *Sensing and Bio-Sensing Research* 12:15-29.

Nie, L., Berckmans, D., Wang, C. and Li, B. 2020. Is continuous heart rate monitoring of livestock a dream or is it realistic? A review. *Sensors* 20(8):2291.

Paganoni, B., Macleay, C., van Burgel, A. and Thompson, A. 2020. Proximity sensors fitted to ewes and rams during joining can indicate the birth date of lambs. *Computers and Electronics in Agriculture* 170:105249.

Pereira, G. M., Heins, B. J. and Endres, M. I. 2018. Technical note: Validation of an ear-tag accelerometer sensor to determine rumination, eating, and activity behaviors of grazing dairy cattle. *Journal of Dairy Science* 101(3):2492–2495.

Post, C., Rietz, C., Büscher, W. and Müller, U. 2020. Using sensor data to detect lameness and mastitis treatment events in dairy cows: A comparison of classification models. *Sensors* 20(14):3863.

Rahman, A., Smith, D. V., Little, B., Ingham, A. B., Greenwood, P. L. and Bishop-Hurley, G. J. 2018. Cattle behaviour classification from collar, halter, and ear tag sensors. *Information Processing in Agriculture* 5(1):124–133.

Reith, S. and Hoy, S. 2012. Relationship between daily rumination time and estrus of dairy cows. *Journal of Dairy Science* 95(11):6416–6420.

Roberts, J., Trotter, M., Schneider, D., Lamb, D., Hinch, G. and Dobos, R. 2015. Daily grazing time of free-ranging cattle as an indicator of available feed. In: *Proceedings of the 7th European Conference on Precision Livestock Farming*.

Roelofs, J. and Van Erp-van der Kooij, E. 2018. Estrus detection tools and their applicability in cattle: Recent and perspectival situation. *Animal Reproduction* 12(3):498–504.

Rowe, E., Dawkins, M. S. and Gebhardt-Henrich, S. G. 2019. A systematic review of precision livestock farming in the poultry sector: is technology focussed on improving bird welfare? *Animals: An Open Access Journal from MDPI* 9(9):614.

Rutten, C. J., Velthuis, A. G. J., Steeneveld, W. and Hogeveen, H. 2013. Invited review: sensors to support health management on dairy farms. *Journal of Dairy Science* 96(4):1928–1952.

Sakai, K., Oishi, K., Miwa, M., Kumagai, H. and Hirooka, H. 2019. Behavior classification of goats using 9-axis multi sensors: The effect of imbalanced datasets on classification performance. *Computers and Electronics in Agriculture* 166:105027.

Schleppe, J. B., Lachapelle, G., Booker, C. W. and Pittman, T. 2009. Challenges in the design of a GNSS ear tag for feedlot cattle. *Computers and Electronics in Agriculture* 70(1):84–95.

Schweinzer, V., Gusterer, E., Kanz, P., Krieger, S., Süss, D., Lidauer, L., Berger, A., Kickinger, F., Öhlschuster, M., Auer, W., Drillich, M. and Iwersen, M. 2019. Evaluation of an ear-attached accelerometer for detecting estrus events in indoor housed dairy cows. *Theriogenology* 130:19–25.

Shen, W., Cheng, F., Zhang, Y., Wei, X., Fu, Q. and Zhang, Y. 2020. Automatic recognition of ingestive-related behaviors of dairy cows based on triaxial acceleration. *Information Processing in Agriculture* 7(3):427–443.

Sheng, H., Zhang, S., Zuo, L., Duan, G., Zhang, H., Okinda, C., Shen, M., Chen, K., Lu, M. and Norton, T. 2020. Construction of sheep forage intake estimation models based on sound analysis. *Biosystems Engineering* 192:144–158.

Smith, D., McNally, J., Little, B., Ingham, A. and Schmoelzl, S. 2020. Automatic detection of parturition in pregnant ewes using a three-axis accelerometer. *Computers and Electronics in Agriculture* 173:105392.

Sprinkle, J. E., Sagers, J. K., Hall, J. B., Ellison, M. J., Yelich, J. V., Brennan, J. R., Taylor, J. B. and Lamb, J. B. 2020. Predicting cattle grazing behavior on rangeland using accelerometers. *Rangeland Ecology and Management* 76:157–170.

Stobbs, T. 1970. Automatic measurement of grazing time by dairy cows on tropical grass and legume pastures. *Tropical Grasslands* 4(3):237–244.

Swain, D. L. and Bishop-Hurley, G. J. 2007. Using contact logging devices to explore animal affiliations: Quantifying cow–calf interactions. *Applied Animal Behaviour Science* 102(1–2):1–11.

Swain, D. L., Friend, M. A., Bishop-Hurley, G. J., Handcock, R. N. and Wark, T. 2011. Tracking livestock using global positioning systems-Are we still lost? *Animal Production Science* 51(3):167–175.

Tamura, T., Chida, Y. and Okada, K. 2020. Short communication: Detection of mastication speed during rumination in cattle using 3-axis, neck-mounted accelerometers and fast Fourier transfer algorithm. *Journal of Dairy Science* 103(8):7180–7187.

Thomas, D. T., Wilmot, M. G., Alchin, M. and Masters, D. G. 2008. Preliminary indications that Merino sheep graze different areas on cooler days in the Southern Rangelands of Western Australia. *Australian Journal of Experimental Agriculture* 48(7):889–892.

Thompson, R. J., Matthews, S., Plötz, T. and Kyriazakis, I. 2019. Freedom to lie: how farrowing environment affects sow lying behaviour assessment using inertial sensors. *Computers and Electronics in Agriculture* 157:549–557.

Tobin, C., Bailey, D. W., Trotter, M. G. and O'Connor, L. 2020. Sensor based disease detection: A case study using accelerometers to recognize symptoms of Bovine Ephemeral Fever. *Computers and Electronics in Agriculture* 175:105605.

Trotter, M., Cosby, A., Manning, J., Thomson, M., Trotter, T., Graz, P., Fogarty, E., Lobb, A. and Smart, A. 2018. *Demonstrating the Value of Animal Location and Behaviour Data in the Red Meat Value Chain*. Meat and Livestock Australia Limited. Available at: https://www.mla.com.

Trotter, M. G., Lamb, D. W., Hinch, G. N. and Guppy, C. N. 2010. Global navigation satellite systems (GNSS) livestock tracking: System development and data interpretation. *Animal Production Science* 50(5):616–623.

Umstatter, C., Waterhouse, A. and Holland, J. P. 2008. An automated sensor-based method of simple behavioural classification of sheep in extensive systems. *Computers and Electronics in Agriculture* 64(1):19–26.

Ungar, E. D., Nevo, Y., Baram, H. and Arieli, A. 2018. Evaluation of the IceTag leg sensor and its derivative models to predict behaviour, using beef cattle on rangeland. *Journal of Neuroscience Methods* 300:127–137.

Williams, L. R., Moore, S. T., Bishop-Hurley, G. J. and Swain, D. L. 2020. A sensor-based solution to monitor grazing cattle drinking behaviour and water intake. *Computers and Electronics in Agriculture* 168:105141.

Wolfger, B., Jones, B. W., Orsel, K. and Bewley, J. M. 2017. Technical note: Evaluation of an ear-attached real-time location monitoring system. *Journal of Dairy Science* 100(3):2219–2224.

Wolfger, B., Timsit, E., Pajor, E. A., Cook, N., Barkema, H. W. and Orsel, K. 2015. Technical note: Accuracy of an ear tag-attached accelerometer to monitor rumination and feeding behavior in feedlot cattle. *Journal of Animal Science* 93(6):3164–3168.

Yang, G.-Z. 2018. *Implantable Sensors and Systems: From Theory to Practice*. Springer, Cham.

Zambelis, A., Wolfe, T. and Vasseur, E. 2019. Technical note: Validation of an ear-tag accelerometer to identify feeding and activity behaviors of tiestall-housed dairy cattle. *Journal of Dairy Science* 102(5):4536–4540.

Chapter 2

Developments in thermal imaging techniques to assess livestock health

A. L. Schaefer and N. J. Cook, University of Alberta, Canada

1 Introduction

As described by O'Brian et al. (2019), precision livestock farming (PLF) is the use of advanced technologies to optimize the contribution of each animal in achieving production and other objectives. Astill et al. (2020) define PLF as 'a system of farming that optimizes the care and attention that farmed animals receive by monitoring animals in real time, ideally resulting in individual animal monitoring or at the smallest unit that can be managed'. However, it is important to be aware of the importance of monitoring the wider environment in which livestock are farmed. Some livestock species such as poultry are very sensitive to environmental conditions, particularly ambient temperature (AT) and humidity. Infrared (IR) cameras can measure the radiated temperature of a poultry flock which can reflect environmental conditions in the barn, as well as potential signs of disease in one or more birds or could indicate behavioural changes such as clustering activity in pigs that are indicative of disease. It is possible, for example, to monitor a whole flock of broiler poultry with a few cameras placed on the ceiling of the barn and from the distribution of animals,

http://dx.doi.org/10.19103/AS.2021.0090.02

determine the position of malfunctioning waterers (Kashiha et al., 2013). A broader view, therefore, is that PLF encompasses the close monitoring not just of individual animals but groups of animals, the environment and housing infrastructure, as well as the integration of these data sources into real-time, predictive models of animal health and well-being that continually evolve to meet changing conditions (Aerts et al., 2003; Wouters et al 1990).

Using PLF techniques allows farmers or producers to deliver quantitative or qualitative improvements in livestock productivity through improved health. Larger herd sizes and intensive livestock production practices indicate a greater need for automated, high throughput, accurate, real-time, non-invasive and cost-effective technologies for monitoring livestock performance and health (Barkema et al., 2015). PLF requires a regular flow of high-quality data for effective bio-surveillance (Gates et al., 2015; Nasirahmadi et al., 2017). As described by Halachmi and Guarino (2016), there are many examples of precision farming tools utilized in animal agriculture, including electronic milkers, oestrus detectors, rumination tags, robot milkers and feed intake monitors. These tools include the use of technologies such as infrared thermography (IRT) (Kastberger and Stachl, 2003).

The panel responsible for animal welfare at the European Food Safety Authority (AHAW, 2012) has highlighted the importance of on-farm animal-based measures (ABM) to assess animal health and welfare. In the case of poultry, for example, they stated: 'There is a need for more systematic flock monitoring and surveillance programmes in the broiler industry'. In terms of the pig industry they noted that 'in monitoring and surveillance systems some ABM may be useful, not only because they can indicate current welfare problems in the herd, but because they can also serve as a tool for early detection of findings that may indicate a potential, future, negative situation'. In response to these concerns, there has been significant research on developing PLF technologies specifically for monitoring animal health and welfare. Landmark precision farming work in image and auditory analysis has been reported by Sankur et al. (1992), Jahns (2006) and Song et al. (2008). Behavioural ethograms and micro-behaviour signals as well as devices such as accelerometers have been used to identify animals with health issues (Bench and Schaefer, 2012; Stewart et al., 2017). The use of automated facial aspect, kinematic assessment and visible wave cameras has proven effective in the identification of health and emotional states (Proctor and Carder, 2015; Lambert and Carder, 2019; de Olivier and Keeling, 2018; Guesgen and Bench, 2017).

2 The use of thermal imaging techniques to monitor animal health

The treatment of a health condition in an animal is almost always preceded by a diagnostic assessment. While the technology used for that assessment has changed over time, the use of a thermal biometric measurement has remained one of the cornerstones for assessing health in both humans and livestock. As described by Hodge (1987), the use of heat signals to identify medical problems dates back to at least 400 BC during the time of Hippocrates and the Greeks. Measurement of core body temperature is fundamental to the veterinary practice for the identification of febrile diseases in animals. Body temperature is particularly sensitive to change and closely related to functions such as nutrition, reproduction, activity, stress and disease (Sellier et al., 2014). The measurement of body surface temperature by techniques such as infrared thermography can be used in the detection of disease. There may be a febrile response to a systemic infection such as PRRS in pigs (Hu et al., 2013) or a localized response to inflammation such as footpad dermatitis in laying hens (Jacob et al., 2016) and turkeys (Moe et al., 2018a,b). Feather loss in laying hens has several causes, including disease, and is easily detected by IRT due to the loss of insulation indicated by temperature (Cook et al., 2006; Zhao et al., 2013).

Sellier et al. (2014) provide an excellent review of contact and non-contact methods for measuring animal temperature. Contact sensors have been developed to minimize the amount of interaction with human handlers, including sensors that can be placed in orifices (e.g. rectum, vagina, ear or gastrointestinal tract (GIT)) or implanted subcutaneously or intramuscularly. There are, however, serious limitations to the use of contact methods of measurement. These devices require additional hardware such as an antenna to obtain or transmit data. Retrieval of such devices is also a major problem in meat-producing animals because they must either be free of such devices prior to slaughter or such devices must be retrieved in the slaughter line which, with 1200 pig carcasses per hour in a slaughter line, for example, is potentially unsafe and impractical.

An alternative is to measure body surface temperature by non-contact methods such as IRT. The surface temperature can provide vital information about the physiological state of the animal and, together with the environmental context, can be predictive of important biological states such as disease, stress and pain. Many papers have compared radiated surface temperature to core body temperature, usually in an attempt to validate surface temperature as a proxy for core temperature. However, it is a mistake to conflate surface and core body temperatures since the surface temperature is at least partly a function of thermoregulatory processes to regulate core temperature and is influenced by micro-environmental variables such as air velocity as a vector. The radiated temperature often exhibits a characteristic biphasic response to infection that potentially provides an early warning of infection prior to the diagnostic limitations imposed by measuring core body temperature.

Thermal imaging cameras detect radiation in the long-IR range (approx. 9000–14 000 nm) of the electromagnetic spectrum. They map the temperature of each of the camera's pixels to produce a thermogram, that is, a temperature map of the surface of an object. There are many parameters and other factors that need to be defined in order to obtain thermal images from which useful data can be obtained. Typical examples include emissivity, distance and angle of the camera to object, exposure time, full or partial frame images, background IR radiation, AT and humidity and solar loading. In PLF, the exact parameters will depend on the species and housing environment. Other factors that affect IRT measurements are the performance characteristics of the camera, for example, accuracy, sensitivity, resolution and sampling frequency. The more expensive IR cameras have a greater resolution but to what extent this improves decision making based on temperature measurements will depend on the circumstances. The latest multispectral cameras are capable of imaging in 3D and can express surface temperature as a function of surface area. This promises to be a significant advance in the interpretation of thermal responses to febrile disease.

Monitoring a potentially large number of individual animals using techniques such as thermal imaging technology is a challenge. Though based on a relatively small number of animals, Bloch et al. (2019) have demonstrated that relatively inexpensive, low-resolution IR cameras can be placed in a feed box that can capture thermal images of the heads of poultry broiler birds when they take a meal. Cook et al. (2014) have shown that monitoring the temperature of an individual pig is possible from cameras placed in an automated feed station. Given the current costs of IRT cameras, sentinel feed-boxes equipped with IR cameras could be distributed around a barn to obtain a representative assessment of the entire flock or herd.

In terms of accuracy of measurement, it is relatively easy to account for constraints such as emissivity, reflected IR radiation or camera performance. These factors can be accurately measured and algorithms in imaging analysis software used to account for them. Much more difficult to control are the physiological processes and micro-environmental influences that confound the interpretation of surface temperature of living animals. Of the many factors that affect thermoregulation, the most common are environment, feed consumption, stress, pain, disease, oestrus, pregnancy, parturition and lactation. Measurement of radiated temperature can be used to detect and monitor each of these biological states.

Homeothermic animals attempt through various physiological processes to maintain a relatively constant core body temperature in the face of often widely fluctuating ATs. Since animals thermoregulate in response to their environment, changes in environmental conditions often result in changes in animal temperature. They regulate their body temperature via physiological processes that are very specific to different livestock species. Perhaps the most important process is by regulating blood flow to the periphery or to specific anatomical sites, for example, ears or legs. Factors that affect metabolism also have an impact on an animal's surface temperature. Consumption of food, in terms of amount and composition, results in a feeding heat increment. Body temperature also follows a circadian rhythm. Responses to one physiological variable can also confound interpretation of the reaction to a different variable. As an example, it is sometimes necessary to restrain an animal in order to obtain an IR image. However, the act of restraint is likely to be a stressful experience that initiates a stress response, including an increase in radiated temperature. The act of obtaining the image thus confounds the interpretation of the temperature from the image. Much of the research on IRT has addressed ways of accounting for these variables to achieve an accurate assessment of an animal's state of health.

3 Accounting for ambient conditions

As noted previously, an automated disease detection system based on radiated temperature measurement by IRT must take into account the animal's thermoregulatory response to changing environmental conditions. It is possible to input data on AT in either the camera settings or the image analysis software to correct the accuracy of the camera for AT. However, this only corrects for the camera performance. It does not account for thermoregulatory responses of the animals to the environment. Ambient air temperature is known, for example, to affect pig surface radiated temperature, and reported correlations vary between $r = 0.44$ and 0.91 depending mainly on three factors: the surface area within the image used for extracting temperature measurements, the size of the animals, and the ambient conditions (Soerensen and Pedersen, 2015). The first two factors can be the same if the entire surface area of the animal is used to obtain temperature measurements.

Some of the earliest work on this relationship reported a correlation between AT and the mean body surface temperature (MBST) of pigs of $r = 0.97$. This relationship is described by a linear regression equation (MBST = 0.4 × AT + 24.82°C) used to adjust the pig's temperature for ambient conditions (Loughmiller et al., 2001). Based on this formula, subsequent work by Loughmiller et al. (2005) adjusted the pig temperature by ± 0.4°C for every 1°C in AT above or below 20°C. This equation was based on the average MBST derived from four pigs housed in an environmental chamber. The relationship for individual animals was not presented, and no explanation was given for using 20°C as the pivotal temperature to make the adjustments.

The precise relationship between the AT and the pig's radiated temperature should be calculated for individual animals since each animal responds differently to environmental conditions. In addition, this relationship depends not only on the factors noted previously but also on the physiological state of the animal, that is, healthy, diseased, stress, pain, etc. Consequently, the relationship of the animal's radiated temperature to the ambient conditions is unique for each animal in each situation. Moreover, an animal is a time-varying system; the response to a specific situation will not always be the same. That is why we call living organisms 'CITD systems': complex, individually different, time-varying and dynamic (Berckmans, 2006).

It is not possible to obtain the relationship between the ambient and radiated temperature of individual animals from group imaging. Nevertheless, in a group setting, conditions that impact the relationship between the animal's radiated temperature and environmental conditions are continually changing, for example, time of day and spatial distribution and the same relationship can be described for groups. It is, therefore, important to establish the relationship between AT and the animal's radiated temperature for the specific conditions under which the thermal images were recorded, that is, it is not sufficient to simply take the equation described by Loughmiller et al. (2001, 2005) and apply to any group of pigs.

The relationship between AT and radiated temperature of the animals is linear within the thermal neutral zone of the animals and recording time-matched ambient and radiated temperatures is, therefore, important to establish this relationship. Regression analysis provides a simple linear equation that describes the relationship between ambient and radiated temperatures, and, for any given AT, it is, therefore, simple to calculate the expected radiated temperature. The difference between the expected radiated temperature and the observed radiated temperature is termed the 'residual radiated temperature.' The residual radiated temperature is a novel variable that has been shown to be a better indicator of a febrile reaction to stimulation than the uncorrected observed radiated temperature (Cook et al., CJAS, 2021).

Ambient temperatures are recorded for the purposes of correcting camera inaccuracy and consequently, there are usually far fewer recordings of AT than there are of images. A possible alternative for recording matching ambient and animal temperatures is to use the background radiated temperature in the image as a proxy for AT. Based on a segmentation algorithm that sets a threshold value separating the animals from the background (Fig. 7), the average temperature below the threshold temperature is a measure of the background radiated temperature and correlates well with AT. Thus, the background radiated temperature in the image could be used to define the relationship between the observed animal temperature with the background radiated temperature to obtain the residual radiated temperature. It may, therefore, be possible to adjust for thermoregulation to environmental conditions from the IR image rather than establishing a separate means of obtaining the AT. The calculation of residual radiated temperature, whether for individual animals or groups, is an example of an empirical, linear model of the relationship between radiated surface temperature and ambient air temperature. More sophisticated modelling of this relationship using machine learning algorithms gives a much better prediction of animal temperature (Gorczyca et al., 2018). However it is achieved, accounting for environmental conditions is an essential component of an automated disease detection system.

4 The use of infrared thermography for animal health monitoring

As noted previously, one of the primary reasons IRT has proven useful in the identification of inflammation and disease states is that the development of fever is almost always a concomitant result of the disease. As described by Schaefer and Cook (2013), IRT can be effective at identifying the presence of a fever or other temperature variation related to the disease. There are many good reviews describing the use of IRT technology in the identification of health risks. The original congress proceedings on thermography were published in 1975 by Aerts et al. (1975) describing numerous clinical applications in both humans and animals. More recent reviews relating to beef and dairy animals include the use of IRT for the non-invasive identification of health issues in cattle as well as other ruminant species such as sheep and goats (Kunc and Knizkova, 2012; Schaefer and Cook, 2013; Cook and Schaefer, 2013; Rekant et al., 2017).

Health issues relating to beef and dairy animals identified using IRT include:

- microbial infections such as bovine respiratory disease complex (BRDC) (Johnson and Dunbar, 2011; Schaefer and Cook, 2013; Hoffmann et al., 2016; Rekant et al., 2017);
- lameness (Alsaaod and Buescher, 2012; Alsaaod et al., 2015; Rekant et al., 2017; Byrne et al., 2019; LokeshBabu et al., 2018; Orman and Endres, 2016; Novotna et al., 2019); and
- mastitis (Hovinen et al., 2008; Polat et al., 2010; Poikalainen et al., 2012 ; Metzner et al., 2014, 2015; Pampariene et al., 2016; Sathiyabarathi et al., 2016; Rekant et al., 2017).

In addition to these applications, IRT has also been demonstrated to identify other conditions of interest in dairy and beef animals including:

- neonatal calf diarrhoea (NCD) (Lowe et al., 2018);
- foot and mouth infection (Rainwater-Lovett et al., 2009; Gloster et al., 2011; Rekant et al., 2017);
- rabies (Rekant et al., 2017);
- bluetongue (Rekant et al., 2017);
- metabolic diseases (MacMillan et al., 2019);

- injection site lesions (Maharjan et al., 2018);
- local injuries (Nikkhah, 2015; Ferreira et al., 2019; Rekant et al., 2017);
- stress (Montanholi et al., 2013; Jerem et al., 2019; Cuthbertson et al., 2020);
- heat stress (Unruh et al., 2017) and cold stress(Collier et al., 2003);
- fever (Schaefer et al., 2018a); and
- pain (Casa-Alvaradoa et al., 2019).

A more recent development reported by Baes and Schenkel (2020) is the use of PLF data on animal health status to establish these phenotypic health traits for use by geneticists in the development of estimated breeding values (EBVs).

5 The development of infrared thermography technologies

To be effective, thermal technology used in PLF must adhere to standardized operating procedures (SOPs). These include defining parameters such as image distance, angle, focus, image number (Okada et al., 2013; Church et al., 2013; Cook et al., 2016; Racewicz et al., 2018; Scoley et al., 2019), image site contamination (Golzarian et al., 2017) and, in particular, the normalization of environmental conditions (Schaefer et al., 2018b). Early research and development were based on hand-held thermal systems with varying degrees of camera resolution and sophistication of data management systems. There has been significant evolution in technology from platforms such as the AGEMA 782 (Fig. 1) to greater use of automation for more efficient precision farming tools (Stewart et al. 2015). Examples of the development of IRT into automated precision farming tools in beef and dairy animals can be seen in Figs 1–3. As an example, Schaefer et al. (2012) utilized an automated radio frequency identification (RFID) tag-driven IR system to assist in the identification of beef calves at risk of BRDC (Figs 2 and 3). Likewise, the company Agricam (www.agricam.SE) located in Sweden has developed an automated scanning system to assist with the identification of mastitis in dairy cattle.

Further development involved the integration of thermography with other technologies to improve sensitivity and accuracy (Galen and Gambino, 1975). Examples include combining IRT with macro and micro behaviour measurements to improve prediction of BRD (Table 1) (Bench and Schaefer, 2012, 2017; Hoffmann et al., 2016) and the combined use of movement pedometers and IRT for improved identification of animals with respiratory and GIT infections (Stewart et al., 2017).

One area of improvement has been the shift from spot radiometers to high-resolution thermal anatomical profiles to account for the fact that a disease state often demonstrates a different thermal complexity or pattern on an anatomical area such as a face. A thermal anatomical profile can better

Figure 1 Public Works Canada engineer (Edmonton) Garry Desroches collecting non-automated hand-held infrared (IR) image on a beef animal with an AGA FLIR Agema model 782. Circa 1985. R and D studies in transport stress (Schaefer et al., 1988).

identify facial hot spots linked to a disease and is easier to combine with other technologies (Jaddoa et al., 2019; Scoley et al., 2019; Lowe et al., 2020). A thermal pattern profile is also more accurate for the identification of animals at risk of a disease compared to using a single-pixel or low-resolution scan device. The early identification of infected individuals (several days before clinical signs become apparent) using these techniques enables early quarantine and reduces subsequent infection of other animals (Fig. 4 and Table 2) (Schaefer et al., 2004, 2007, 2012).

Further development in the past few years is the emergence of multispectral 3D and 4D camera systems (Guarino et al. 2018). Whilst a 2D IR image can show a maximum temperature or mean temperature for an animal profile, a 3D thermal profile can identify biological signals such as vasodilation or vasoconstriction in specific areas of interest such as nasal or

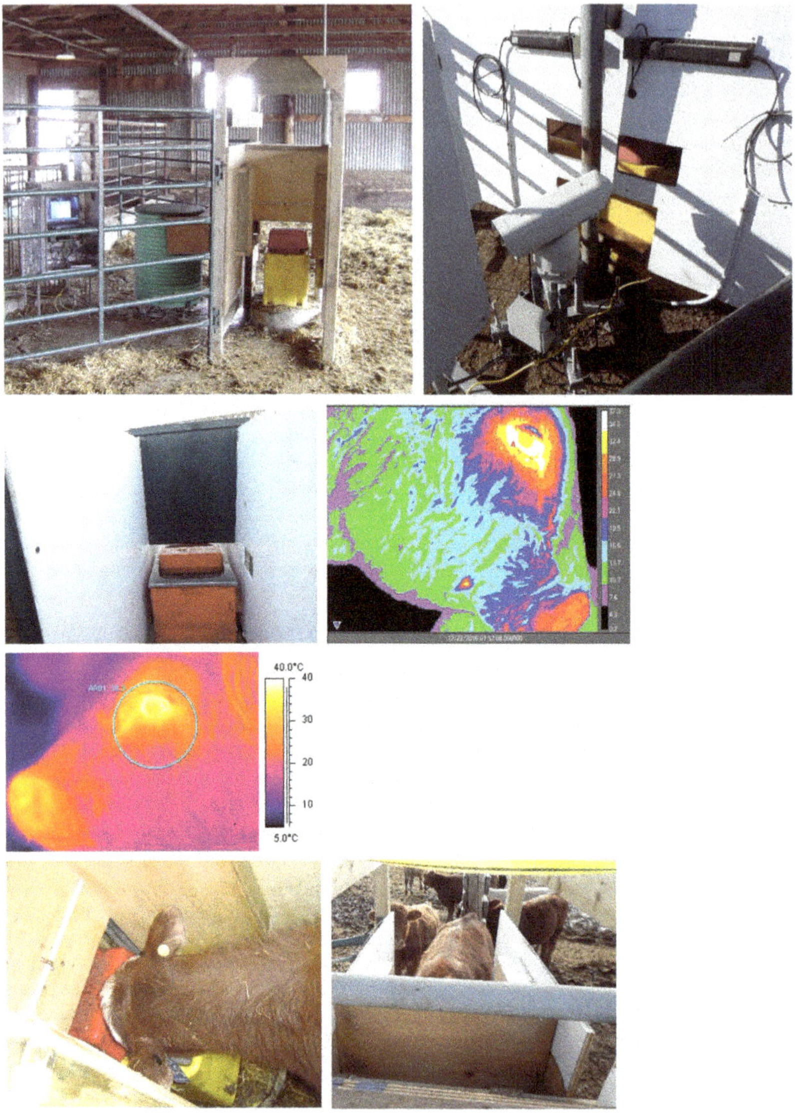

Figure 2 Automated capture of cattle infrared (IR) images for bovine respiratory disease (BRD) study. The system utilized an radio frequency identification (RFID) panel reader and a forward-looking infrared (FLIR) S60 camera interconnected to a laptop computer. Data storage was on a site hard drive. The camera rotated between view windows on water bowl (Schaefer et al., 2007, 2012). Circa 2005. Image from Lacombe Research Centre, Alberta, Canada.

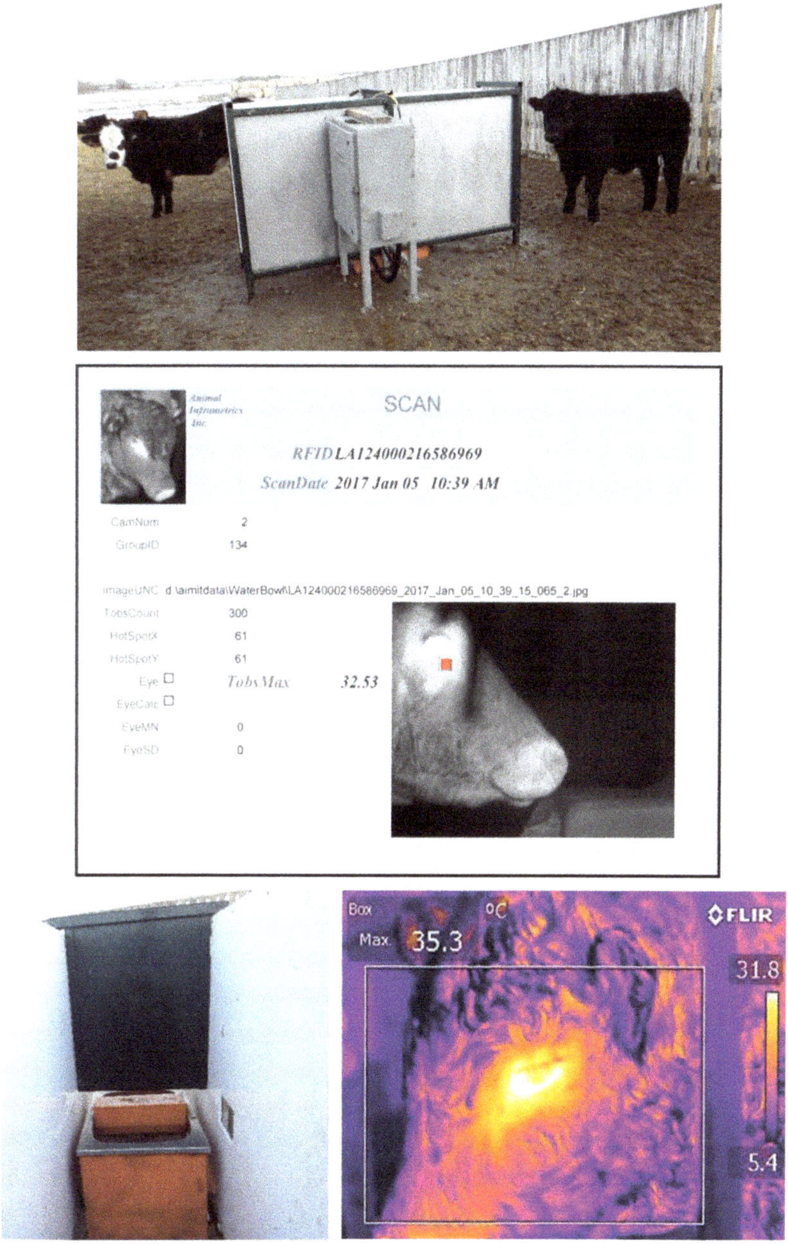

Figure 3 Automated capture of cattle infrared (IR) images for bovine respiratory disease (BRD) study. The system utilized an radio frequency identification (RFID) panel reader and a forward-looking infrared (FLIR) A315 GIGE camera interconnected to a high speed laptop with CAT5 Ethernet link to internet. Cloud data storage (Schaefer et al., 2018b). Circa 2015. Image from Olds College, Olds Alberta, Canada.

Table 1 Positive predictive value (PPV), negative predictive value (NPV) and test efficiency (E) for bovine respiratory disease complex (BRDC) detection when combining infrared and micro-behavioural scan systems. Data represents values 4-6 days pre-clinical BRD (From Schaefer et al., 2012)

Test System	PPV (%)	NPV (%)	E (%)
Clinical score	70.4	44.6	55
IRT + behaviour	85	83	84

Table 2 Correlation between IRT values and verified health status for true positive (TP) animals. Comparison of a thermal pattern profile including kinematic and micro-behavioural detail compared to single point scans for typical anatomical sites such as an eye or skin sight

Day	Parameter	R-value
Symptom and treatment day	Thermal profile	0.93
	Eye Tmax	0.86
	Skin Tmax	0.59
Symptom and treatment day-1	Thermal profile	0.74
	Eye Tmax	0.58
	Skin Tmax	0.56
Symptom and treatment day-2	Thermal profile	0.92
	Eye Tmax	0.57
	Skin Tmax	0.02
Symptom and treatment day-3	Thermal profile	0.84
	Eye Tmax	0.51
	Skin Tmax	0.14

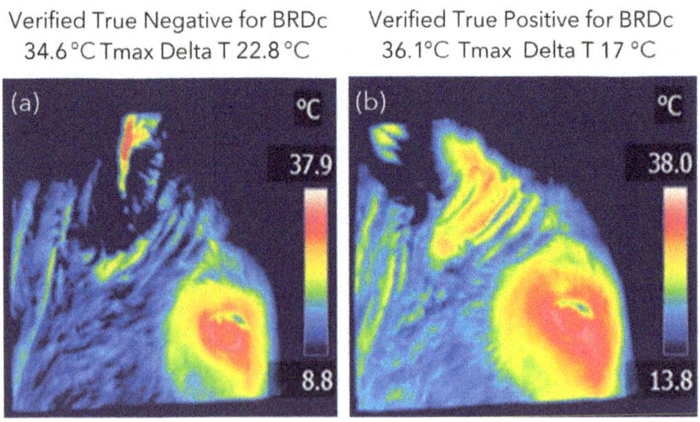

Figure 4 (a) Verified true negative (TN) for bovine respiratory disease complex (BRDC). 34.6°C Tmax Delta T 22.8°C. (b) Verified true positive (TP) for BRDC. 36.1°C Tmax Delta T 17°C.

udders. Incorporating a 4D image system adds a time series component to the assessment, providing information, for example, on 'fidgeting' behaviour that increases during a bout of respiratory disease (Bench and Schaefer, 2012, 2017). The addition of 4D time components could previously only be added by incorporating movement analysis with technologies such as Noldus Ethovision XT® (info@noldus.nl) or Tracktor® (a python-based program: https://github.com/vivekhsridhar/tractor).

Whilst these technologies have been available in other sectors for some years, it has only been recently known that certain technological and economic preconditions have existed to allow its application in the livestock sector. Preconditions include the following:

1 Until recently, IRT cameras were too expensive for image capture and too slow to provide a real-time image stream.
2 The computer processing required for real-time analysis has only become cost-effective in the last 5 years (thanks in part to advances in the gaming industry).
3 Whilst 1 GB Ethernet speed was available in 1998, 1 GB routers and switches were not commonplace until about 2010.
4 For many rural farms in North America and elsewhere, high-speed internet access is becoming a reality only recently. High-speed internet is important because it solves the problem of data transfer to remote servers for rapid/real-time analysis and storage.
5 Computer applications to solve animal motion problems taking advantage of an image stream of up to 60 images per sec are just being developed. For example, forward-looking infrared (FLIR) cameras released their GigE software development kit only 5 years ago. This has enabled the analysis of real-time problems.
6 Until recently, veterinary practitioners have not been familiar with IRT technology.

A further set of conditions required for the development of IRT as a precision farming tool is the integration of IRT technology into farm management practices. For this to happen, the following must occur:

1 IRT applications must be integrated into existing data management systems and software at the farm. For example, it may not be enough to know the immediate temperature of the animal and to make an evaluation of whether it is normal. It would be more valuable to have a system that evaluates the temperature, compares it to a threshold value or a herd average or, better yet, the animal's recent past temperature

profile. This requires real-time access to the database storing the animal's temperature records, including environmental climatic information.

2 Applications must be commercially viable and show a positive benefit/cost ratio.
3 IRT applications must fit in with existing precision farming applications such as milking robots, weighing stations and bio-surveillance systems.
4 Storing data on the 'cloud' provides access to sophisticated analysis techniques such as expert systems, statistical trend analysis and other data mining applications.

The future potential of IRT used in precision farming is excellent. Computers are now powerful enough to allow cost-efficient processing of IRT images. Small image acquisition devices are becoming more economical. Hard-wired Ethernet networks are also more than adequate for speed, and performance and will be complemented by the advent of 5G Wi-Fi networks and satellite-based high-speed internet services. The emergence of commercial farm processing systems and databases will allow for IRT integration. The introduction of multi-spectrum cameras (see e.g. Fig. 5) means that data can be processed locally and integrated into cloud-based data systems. Such systems can, for example, already provide 3D images of animals and incorporate other time motion and micro-behaviour observations. This allows for the development of algorithms that can determine whether the animal may be at risk of a health issue. Developments in big data in PLF are discussed by Wolfert et al. (2017) and Astill et al. (2020).

6 Improving diagnostic accuracy

There are essentially two methods of obtaining the radiated temperature of animals, either as individuals or as a group of animals in a pen. It is technically easier but less diagnostically helpful to obtain radiated temperature measurements on groups of animals such as pigs in a pen or broiler birds on the floor of a barn. This type of image provides temperature information on the group as a whole and is able to detect the presence of a febrile disease when only a few of the animals exhibit a febrile response. Unfortunately, group imaging is not yet able to inform the producer which of those animals are exhibiting a febrile response, but only informs that there is a potential problem in the pen or barn. Advances in imaging technology, particularly 3D imaging, may be able to identify individual animals from their body size and shape. The accuracy of identification using 3D imaging has yet to be reported. Currently, diagnostics based on group imaging is best achieved using the maximum animal temperature as the diagnostic variable because it would only require

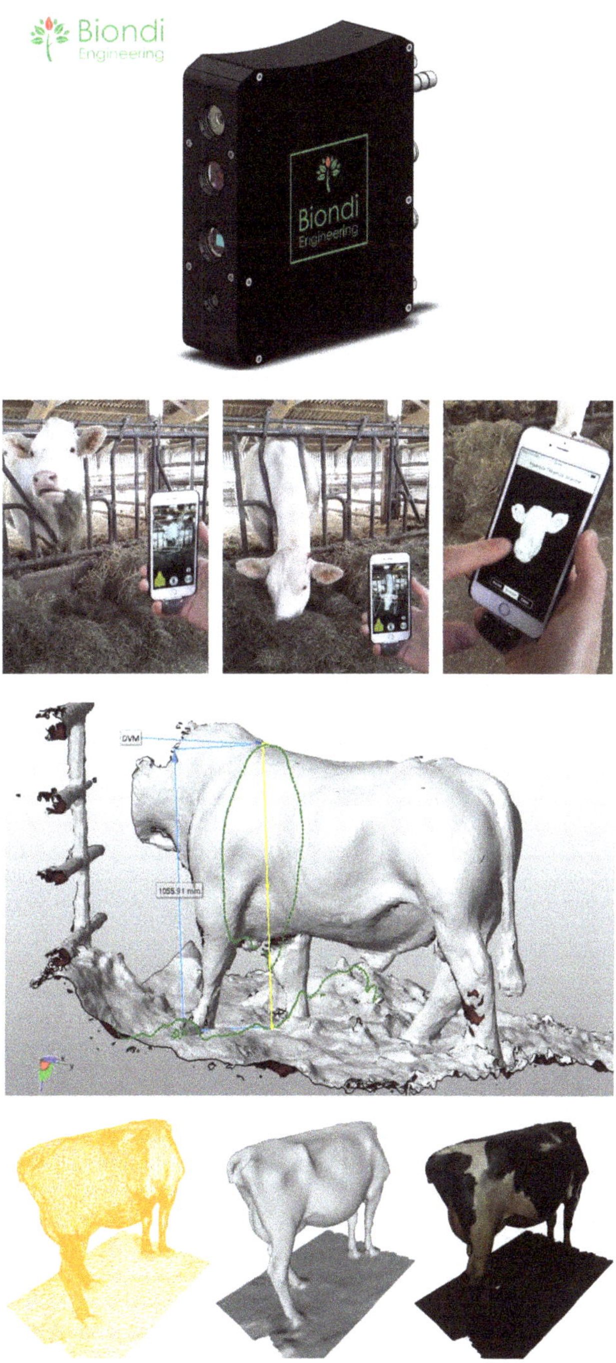

Figure 5 Images of the multi-spectral cameras (figure provided by Biondi Engineering).

one animal in a group to develop a febrile response for it to be the warmest animal, that is, the highest maximum temperature in the group.

Cook et al. (2014) tested this hypothesis using vaccination to elicit a febrile response and an IR camera fitted with a wide-angle lens to cover a pen of 26 weaner pigs, with differing numbers of animals vaccinated in each replication of the experiment to represent a range of prevalence of symptoms. An increase in the mean maximum temperature of 0.14°C for the group was noted when < 10% of the pigs were vaccinated (Fig. 6). At a prevalence of approximately 70%, the average difference in the group mean for the maximum pig temperature between pre- and post-vaccination days reached 0.8°C. Some animals exhibited a much greater response than others. The average temperature of a group of pigs is much less diagnostically relevant because the febrile reaction of one pig is subsumed by the average temperature of the group. Of course, the more animals within the group that exhibit a febrile response, the more diagnostically relevant the average group temperature becomes.

A problem with group imaging is that much of the temperature information from the image is obtained from everything in the image that is not the animals, that is, walls, floor, etc. (Fig. 7a and b). In order to overcome it and obtain temperature information on the animals separate from the background, it is necessary to apply a segmentation algorithm to the images. Segmentation algorithms decompose the IR images based on various temperature parameters. The choice of the algorithm determines the temperature range and this can include the entire temperature range of the animal or a more narrow range. In group imaging of pigs (Cook et al., 2014, 2018), an algorithm by Parker (1997) identified the peaks in temperature distribution within the image and from these selected the lowest temperature between the two warmest peaks. Applying the lowest temperature between the warmest peaks as a threshold in

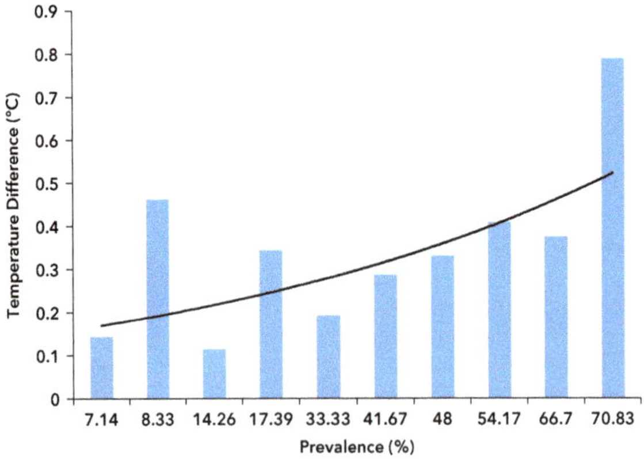

Figure 6 The effect of vaccination prevalence on the differences in daily average maximum pig temperature between control and treatment days 24 h apart.

the image analysis software isolates the animals from the background (Fig. 7c and d). A similar effect using a different algorithm is described by Zaninelli et al. (2017) to identify individual laying hens in an aviary system.

The spatial organization and (by extension) activity of weaner pigs based on a passive IR sensor is reported by Besteiro et al. (2018). Isolating the pigs from the background can be used to count clustering activity which is important because group temperature depends to some extent on clustering activity (Fig. 8). Clustering activity can be a behavioural component of the febrile response. The more the clustering activity, the higher the radiated temperature. There are two consequences to this observation. Firstly, it would be relatively easy to reject images in which there was more than one cluster, thereby reducing the variation in the group temperature measurements. Secondly, clustering behaviour is in itself indicative of disease since it is an adaptive response to infection. Another aspect of group imaging is that it is relatively easy to observe

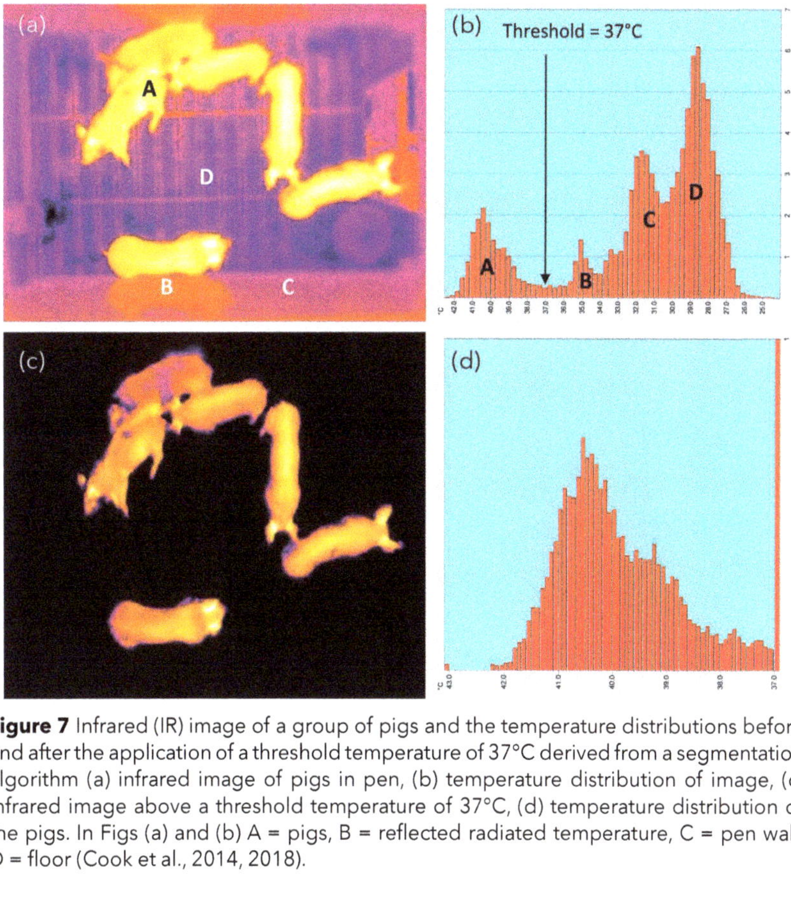

Figure 7 Infrared (IR) image of a group of pigs and the temperature distributions before and after the application of a threshold temperature of 37°C derived from a segmentation algorithm (a) infrared image of pigs in pen, (b) temperature distribution of image, (c) infrared image above a threshold temperature of 37°C, (d) temperature distribution of the pigs. In Figs (a) and (b) A = pigs, B = reflected radiated temperature, C = pen wall, D = floor (Cook et al., 2014, 2018).

the growth of the group in terms of the area of the image occupied by the animals. Figure 9 shows the growth curves of two groups of piglets fed diets with and without in-feed antibiotics. Those animals receiving in-feed antibiotics exhibit higher radiated temperatures than those on an antibiotic-free diet. IR imaging can thus be used to obtain thermal (radiated temperature), behavioural (clustering activity) and physiological (growth rate) data simultaneously on the same animals in the same images.

The more common approach to disease detection is to obtain radiometric images on individual animals. Individual animal measurements are technically more difficult than group imaging by virtue of the fact that animals must be identifiable when the images are recorded. The most common method is to install a camera at a water or feed station and capture images when the animal visits the station (Schofield et al., 2002). Such a system requires a method of automatically identifying an animal. This is usually achieved using RFID tags or transponders. In a swine unit, for example, pigs carry half-duplex RFID tags in their ear, often associated with a commercial feeding system. When a pig enters the feeding system, a unique 15-digit identifier is detected by a transceiver and transmits the ID number to a computer. Manufacturers of automated feeder or waterer systems should permit access of the IR camera system to the unique

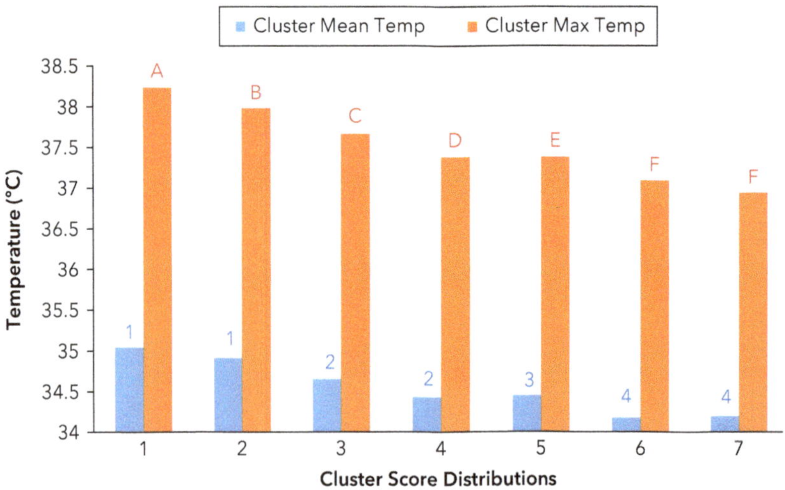

NB: Daily average mean image temperatures between Clusters not sharing the same numbers, and daily average maximum image temperatures between Clusters not sharing the same letters were statistically different ($p < 0.05$).

Figure 8 Mean average and maximum pig temperatures by cluster scores. NB: daily average mean image temperatures between clusters not sharing the same numbers, and daily average maximum image temperatures between clusters not sharing the same letters were statistically different ($P < 0.05$).

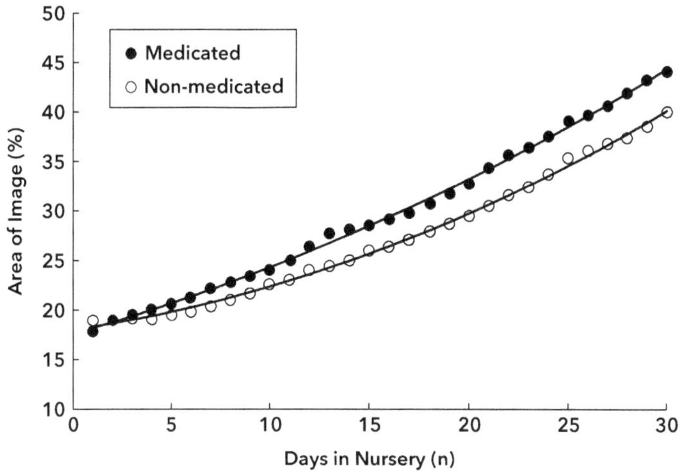

Figure 9 Relationships between the percentage areas of image occupied by the pigs and the number of days in the nursery pen for medicated and non-medicated groups (Cook et al., 2018).

identifiers of the RFID system. However, this is not yet the case as it requires installing a separate antenna to detect the ID of an animal in the feeding system and to signal the image capture software to start recording images.

Animals can lose RFID tags in interactions with the infrastructure of the pen or during aggressive interactions with cohorts. RFID tags come at a cost that some producers find difficult to manage given the vagaries in the market prices of animals such as pigs. Fitting animals with RFID tags is labour-intensive and often painful for the animals. For these reasons, the identification of animals such as pigs using facial recognition software has gained attention in recent years (Hansen et al., 2018). The application of facial recognition stems from software aimed at categorizing facial expression in relation to emotional states. McLennan et al. (2019) for example in a review on the use of facial recognition software to identify pain in mammals. Facial recognition can not only identify emotional states in animals but can also identify the animals themselves. However, the technology is far from perfect and the accuracy of the facial recognition software can vary considerably. Marsot et al. (2020) reported an accuracy of 83% out of 320 images, whereas Hansen et al. (2018) reported an accuracy of 96.7% out of 1553 images. In either case, the accuracy of the facial recognition software is still inadequate for routine application in a commercial barn. However, it not only shows the potential of combining facial recognition software with high-resolution IR cameras to measure facial temperature but also identifies the animal from the same IR image.

Feeding systems are often linked to a weighing scale, and the weight of the animal is recorded each time it enters the feeding system. Several variables

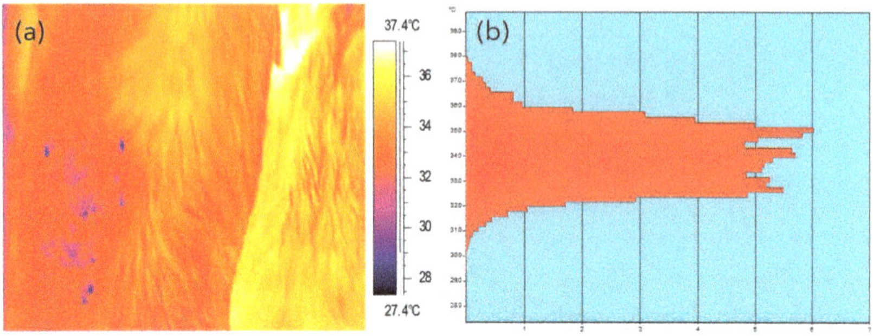

Figure 10 (a) Infrared image of two pigs in the automated feeding system. (b) Temperature distribution of two pigs.

are thus measured each time the animal enters the feeding system, including feed consumption, feeding duration, feeding frequency, animal weight and radiant temperature. The positioning of the camera dictates the anatomic location of the imaging. For pigs, the easiest location is to set the camera above the feeder and obtain images of the dorsal surface. As demonstrated by Bloch et al. (2019), a thermal camera could be placed in the feed box of poultry to obtain images of the head. Spurious images must be automatically identified and eliminated from further analysis, including getting two animals in the system at the same time (Fig. 10). Methods of automatically eliminating spurious images are generally based on the size and distribution of the animal's temperature. Thus, if only the head is captured in the image, this will represent a relatively small number of pixels compared to a full dorsal image and can be automatically excluded from further analysis based on size. It is more difficult to automatically identify images in which there are two animals. One approach is to look at the distribution of temperature within the image and, if there is a bi-phasic distribution in temperature above the threshold, there is likely to be two pigs in the image and it can be rejected.

7 Case study: using infrared thermography to identify bovine respiratory disease

The technical development and engineering of 3D cameras have been progressing for some years. A number of companies have recently produced a stereo vision camera that combines a broad spectrum IR camera with a red-green-blue (RGB) visual range camera. In 2018, for example, Alpha Phenomics produced a multi-spectral camera which includes a visual range camera, a multi-point laser data cloud camera and an IR broad range camera (Fig. 5) (www.AlphaPhenomics.com). In addition, because of a comparatively rapid image capture speed, this scan system can also collect time-sensitive behavioural images (adding the fourth dimension of time). The following example illustrates the potential of this technology in monitoring animal health.

Bovine respiratory disease is an important disease in cattle and is mostly caused by a virus infection, especially of the upper respiratory tract, affecting significant numbers of animals globally. By the time conventional clinical scores (symptoms) are observed, the animal is often well into the course of the disease, making it more difficult to treat and potentially spreading the infection to the rest of the herd. If producers could accurately identify animals as true positive (TP) or true negative (TN) at an earlier stage, they could target subsequent treatment and quarantine measures more effectively.

In a recent study, multiple sourced, comingled and recently transported calves were brought into the animal facility at the Lacombe Research Centre operated by Agriculture and Agri-Food Canada in Alberta where the animals were studied daily with a non-invasive automated IRT camera system (Fig. 2). The cattle were also monitored daily for clinical scores as well as via lab tests including serology, haematology, cortisol, acute phase proteins and core body temperature (Schaefer et al., 2012). IRT images of the animal's face along the cranial-median plane (side of the face) were collected at 50 Hz or 50 images per second. The animals were classed as TP or TN based on a 'gold standard' scoring system developed in concert with veterinary pathologists. Statistical analysis was undertaken using a conventional stats package (MedCalc, Statistics for Biomedical Research: MedCalc® Statistical Software Version 19.5.3 (MedCalc Software Ltd, Ostend, Belgium; https://www.medcalc.org).

The thermal profile was collected from each animal. This comprised the summed composite of an individual animal's maximum eye temperature + the SD or measure of dispersion for that temperature + a landmarked skin mean temperature on the cheek + difference (delta T) between the max eye temperature and the skin site + a residual as well as a facial aspect difference in the ear position. These measures identified (from a single image) biometric values for a region of interest known to show early disease detection (the orbital area), plus a demonstration of thermal symmetry change (the difference between the orbital area and the cheek), as well as an ethological disposition change (ear angle), again known to show a predisposition to a disease or stress state. This profile contrasts with a single thermal pixel temperature point from a hand-held thermal radiometer. This data is illustrated in Table 2 and shows the thermal profile had an *r*-value or correlation coefficient with known TP or TN health status of $r = 0.93$ at $P < 0.01$. This shows the advantage of using a thermal profile plus a 3D facial aspect value (ears at different degrees vertical) in contrast to a single temperature measurement.

The days animals were identified by a veterinarian as symptomatic, ill or treated were referred to as a 'pull day' (PD), symptom or treatment day. The group of animals contained 14 animals with a comparatively high prevalence of BRD. There were nine calves demonstrating a TP condition and five with a TN condition. The correlation *r*-values for the thermal profile were comparatively high for 1 day (PD-1), 2 days before (PD-2) and 3 days before (PD-3) before the animals were pulled (PD-1 $r = 0.7$, PD-2 $r = 0.9$ and PD-3 $r = 0.8$) compared to the other single trait thermal parameters. This thermal image profile approach was both a more accurate representation of health status than a single measure and allowed earlier identification of symptomatic animals. As this study shows, IRT clearly has significant utility as a PLF tool in the identification of animals at risk. An automated, non-invasive, internet-enabled 3D/4D scan station is being developed which would incorporate a thermal profile plus other biometrics such as a micro-behaviour and or facial aspect measurement. This system could serve as a rapid screening technology based on parallel testing. This approach could augment subsequent testing using ELISA or other pathogen testing techniques.

8 Conclusions

The potential application of IRT in disease detection is well known. However, its application in farms is yet to become an established practice. Advances in camera design, software development, automated animal ID, cloud computing, machine learning and artificial intelligence are presently converging to promote the incorporation of thermal cameras into PLF.

In the systems discussed previously, IR imaging is either a continuous process in which images are recorded at pre-defined intervals, or the images are recorded of individual animals each time the animal uses a facility in the pen such as a feeder or drinking station. In either case, multiple images can be collected for individual animals. The temperature of any one IR image can be compared to the history of temperature recordings for that animal or group. A typical example might be continuous minute-by-minute imaging of animals, from which is derived the running average temperature over a defined period, such as the last hour or day. This can be compared to the previous average temperature for that time of day or daily average. Deviations from past temperature would indicate a problem for that animal or group. In theory, a non-normal reduction in temperature might indicate early signs of illness prior to the onset of a febrile response and observation of clinical symptoms. The drop in surface temperature as an infected animal develops a febrile response may be the first diagnostic temperature-based sign of infection. Detecting a drop in surface temperature against a backdrop of environmental effects or circadian rhythm is challenging. However, as the core body temperature rises, so does the animal's surface temperature, and this is probably detectable against the background of normal radiated temperature. Critical factors in this system are multiple measurements, the ability to make rapid comparisons among temperature variables across time and the correction for environmental effects.

9 Future trends in research

Considering humans have only been aware of the infrared component of the electromagnetic spectrum for about 200 years, there has already been significant development and utility found in infrared technology. The detection equipment is currently much more robust, user-friendly and affordable than ever. Perhaps equally important is the fact that computer platforms and internet systems for data capture and management are also equal to the task of managing large data files and images in real-time, which is a fundamental requirement for precision agriculture. The combination of reliable camera systems with the computer and internet-assisted management of data is now emerging in automated systems rather than simply handheld systems. These developments have thus converged to a point where the automated use of infrared systems in animal agriculture is now becoming a reality. The next point of evolution of this technology will thus be to continue to empirically validate the applications with large-scale commercial herds and flocks. Such information generated with these technological systems will become a valuable part of any blockchain technology.

10 Where to look for further information

In addition to the reference list provided with the chapter, the following are useful information sites:

1 For an historical perspective regarding infrared thermography. Thermography. 1975. History of infrared use in Animals. Ed. by Aarts, N. J. M., Gautherie, M. and Ring, E. F. J. S. Karger. N. Y. Proceedings of the 1st European Congress on Thermography.
2 Teaching Text Reference. Thermography. Current status and advances in Livestock animals and in veterinary medicine. 2013. Ed. by Luzi, F., Mitchell, M., Costa, L. N. and Redaelli, V. Brescia Foundation. Brescia Via Istria. Italy.
3 Training Modules. FLIR Comp. Infrared Training Center: Itc-canada@flir.com. See training modules plus level 1, 2 and 3 training courses.
4 International Proceedings. ECPA. International Society of Precision Agriculture website: www.ispag.org. Links to many articles on the use of infrared thermography in precision farming.

5 Classic Engineering Text. Infrared Imaging Systems Analysis. 1988. Shumaker, D. L., Wood, J. and Thacker, C. R. DCS Corporation. Alexandria, Virginia. Environmental Research Institute of Michigan Infrared Information Analysis Centre.

11 References

Aerts, J.-M., Wathes, C. M. and Berckmans, D. 2003. Dynamic data-based modelling of heat production and growth of broiler chickens: development of an integrated management system. *Biosystems Engineering* 84(3): 257–266.

Aerts, N. J. M., Gautherie, M. and Ring, E. F. J. 1975. Thermography. Proceedings of the First European Congress on Thermography. Amsterdam. 1974. S. Karger. Basel.

Alsaaod, M. and Buescher, W. 2012. Detection of hoof lesions using digital infrared thermography in dairy cows. *Journal of Dairy Science* 95(2): 735–742.

Alsaaod, M., Schaefer, A. L., Büscher, W. and Steiner, A. 2015. The role of infrared thermography as a non-invasive tool for the detection of lameness in cattle. *Sensors* 15(6): 14513–14525.

Astill, J., Dara, R. A., Fraser, E. D. G., Roberts, B. and Sharif, S. 2020. Smart poultry management: smart sensors, big data, and the internet of things. *Computers and Electronics in Agriculture* 170, 105291:1–8.

Baes, C. and Schenkel, F. 2020. The future of phenomics. *Animal Frontiers* 10(2 April): 1–5.

Barkema, H. W., von Keyserlingk, M. A. G., Kastelic, J. P., Lam, T. J. G. M., Luby, C., Roy, J. P., LeBlanc, S. J., Keefe, G. P. and Kelton, D. F. 2015. The importance of economies of scale and automation effects in the dairy industry. *Journal of Dairy Science* 98(11): 7426–7445.

Bench, C. J. and Schaefer, A. L. 2012. The use of a behavioural biometric in combination with infrared thermography for early detection of bovine respiratory disease in a spontaneous model. International Society of Animal Ethology, Proceedings. Banff, AB, Canada.

Bench, C. J. and Schaefer, A. 2017. Apparatus and method for using infrared thermography and behaviour information for identification of biologically important states in animals. USA Patent. US9565837B2.

Berckmans, D. 2006. Automatic on-line monitoring of animals by precision livestock farming. P 287–294. In: *Livestock Production and Society* Geers, R. and Madec, F. (Eds) Wageningen Academic Publishers. 220, 6700 A. E. Wageningen. The Netherlands.

Besteiro, R., Arango, T., Rodríguez, M. R., Fernández, M. D. and Velo, R. 2018. Estimation of patterns in weaned piglets' activity using spectral analysis. *Biosystems Engineering* 173: 85–92.

Bloch, V., Barchilon, N., Halachmi, I. and Druyan, S. 2019. Automatic broiler temperature measuring by thermal camera. *Biosystems Engineering* 199: 127–134. DOI: 10.1016/j.biosystemseng.2019.08.011.

Byrne, D. T., Berry, D. P., Esmonde, H., McGovern, F., Creighton, P. and McHugh, N. 2019. Infrared thermography as a tool to detect hoof lesions in sheep. *Translational Animal Science* 3(1): 577–588.

Casas-Alvaradoa, A., Mota-Rojasb, D., Hernández-Ávalosc, I., Mora-Medinad, P., Olmos-Hernándeze, A., Verduzco-Mendozae, A. and Martínez Burnes, J. 2019. Advances in infrared thermography: surgical aspects, vascular changes and pain 2 monitoring in veterinary medicine thermal biology. *International Journal of Life Sciences and Applied Sciences* 1(1): 58–69.

Church, J. S., Hegadoren, P. R., Paetkau, M. J., Miller, C. C., Regev-Shoshani, G., Schaefer, A. L. and Schwartzkopf-Genswein, K. S. 2013. Influence of environmental factors on infrared eye temperature measurements in cattle. *Research in Veterinary Science* 96(1): 220–226.

Collier, R. J., Coppola, C. and Wolfgram, A. 2003. Novel approaches for the alleviation of climatic stress in farm animals. P 61–71. In: *Interactions between Climate and Animal Production* Lacetta, N., Bermabucci, U., Khalifa, H. H., Ronchi, B. and Nardone, A. (Eds) Book Series EAAP European Association for Animal Production Technical Series. Waganingen Academic Publisher. 220. 6700 E. E. Wageningen. The Netherlands.

Cook, N. J., Bench, C. J., Liu, T., Chabot, B. and Schaefer, A. L. 2018. The automated analysis of clustering behaviour of piglets from thermal images in response to immune challenge by vaccination. *Animal: An International Journal of Animal Bioscience* 12(1): 122–133.

Cook, N. J., Chabot, B., Liu, T., Bench, C. J. and Schaefer, A. L. 2014. Infrared thermography detects febrile and behavioural responses to vaccination of weaned piglets. *Animal: An International Journal of Animal Bioscience* 9(2): 339–346.

Cook, N. J., Chabot, B., Liu, T., Froehlich, D. and Juàrez, M. 2021. Radiated temperature of weaned piglets and the response to in-feed antibiotics with correction for ambient conditions. *Canadian Journal of Animal Science* 101(2): 318–328.

Cook, N. J. and Schaefer, A. L. 2013. Infrared thermography and disease surveillance. P 79–92. In: *Thermography: Current Status and Advances in Livestock Animals and in Veterinary Medicine* Luzi, F., Mitchell, M., Costa, L. N. and Redaelli, V. (Eds). Brescia Foundation. 3/b -25125, Brescia. Italy.

Cook, N. J., Smykot, A. B., Holm, D. E., Fasenko, G. and Church, J. S. 2006. Assessing feather cover of laying hens by infrared thermography. *Journal of Applied Poultry Research* 15(2): 274–279.

Cook, N. J., Timsit, E., Colyn, J., Chabot, B., Liu, T., Wolfger, B. and Schaefer, A. 2016. Assessing consistency in radiated thermal output of beef steers by infrared thermography journal of imaging. *Journal of Imaging* 2(21): 1–17.

Cuthbertson, H., Tarrb, G., Loudonc, K., Lomaxa, S., Whited, P., McGreevyd, P., Polkinghornee, R. and Gonzáleza, L. A. 2020. Using infrared thermography on farm of origin to predict meat quality and physiological response in cattle (Bos taurus) exposed to transport and marketing. *Meat Science* 169: 108173.

de Olivier, D. and Keeling, L. J. 2018. Routine activities and emotion in the life of dairy cows: integrating body language into an affective state framework. *PLoS ONE* 13(5): 1–16.

European Food Safety Authority, EFSA 2012. Statement on the use of animal-based measures to assess the welfare of animals, EFSA Panel on Animal Health and Welfare (AHAW). *EFSA Journal* 10(6). DOI: 10.2903/j.efsa.2012.2767.

Ferreira, C. A., Jeune, S. S., Rayburn, M. C. and Chigerwe, M. 2019. Thermographic evaluation of primary closure and second intention healing in dairy calves. *Veterinary Surgery: VS* 48(5): 878–884.

Galen, R. S. and Gambino, S. R. 1975. *Beyond Normality: The Predictive Value and Efficiency of Medical Diagnoses*. J. Wiley: New York.

Gates, M. C., Holmstrom, L. K., Biggers, K. E. and Beckham, T. R. 2015. Integrating novel data streams to support bio-surveillance in commercial livestock production systems in developed countries: challenges and opportunities. *Frontiers in Public Health* 3: 1-13.

Gloster, J., Ebert, K., Gubbins, S., Bashiruddin, J. and Paton, D. J. 2011. Normal variation in thermal radiated temperature in cattle: implications for foot-and-mouth disease detection. BMC (Biomedcentral). *Veterinary Research* 7: 73.

Golzarian, M. R., Soltanali, H., Doosti Irani, O. and Ebrahimi, S. H. 2017. Possibility of early detection of bovine mastitis in dairy cows using thermal images processing. *Iranian Journal of Applied Animal Science* 4(7): 549-557.

Gorczyca, M. T., Milan, H. F. M., Maia, A. S. C. and Gebremedhin, K. G. 2018. Machine learning algorithms to predict core, skin, and hair-coat temperatures of piglets. *Computers and Electronics in Agriculture* 151: 286-294.

Guarino, P. A., Sartori, M. L. and Francesco, M. 2018. A feasibility study on the use of a structured light depth-camera for three-dimensional body measurements of dairy cows in free-stall barns. *Sensors* 18(Issue. 3): 673.

Guesgen, M. J. and Bench, C. J. 2017. What can kinematics tell us about the affective states of animals? *Animal Welfare* 26(4): 383-397.

Halachmi, I. and Guarino, M. 2016. Precision livestock farming: a per animal approach using advanced monitoring technologies. *Animal: An International Journal of Animal Bioscience* 10(9): 1482-1483.

Hansen, M. F., Smith, M. L., Smith, L. N., Salter, M. G., Baxter, E. M., Farish, M. and Grieve, B. 2018. Towards on-farm pig face recognition using convolutional neural networks. *Computers in Industry* 98: 145-152.

Hodge, S. Jr. 1987. *Thermography and Personal Injury Litigation* (vol. P71). Hoboken, NJ: John Wiley and Sons.

Hoffmann, G., Schmidt, M. and Ammon, C. 2016. First investigations to refine video-based infrared thermography as a non-invasive tool to monitor the body temperature of calves. *Animal: An International Journal of Animal Bioscience* 10(9): 1542-1546.

Hovinen, M., Siivonen, J., Taponen, S., Hänninen, L., Pastell, M., Aisla, A. M. and Pyörälä, S. 2008. Detection of clinical mastitis with the help of a thermal camera. *Journal of Dairy Science* 91(12): 4592-4598.

Hu, S. P., Zhang, Z., Liu, Y. G., Tian, Z. J., Wu, D. L., Cai, X. H. and He, X. J. 2013. Pathogenicity and distribution of highly pathogenic porcine reproductive and respiratory syndrome virus in pigs. *Transboundary and Emerging Diseases* 60(4): 351-359.

Jacob, F. G., Baracho, M. dos S., Nääs, I. de A., Souza, R. and Salgado, D. D. 2016. The use of infrared thermography in the identification of pododermatitis in broilers. *Engenharia Agrícola* 36(2): 253-259.

Jaddoa, M. A., Al-Jumaily, A. A., Gonzalez, L. A. and Cuthbertson, H. 2019. Automatic temperature measurement for hot spots in face region of cattle using infrared thermography. *ICINCO* 2019: 23.

Jahns, G. 2006. Automatic identification of farm animals utterances–animal call–recognition. *Landbauforschung Volkenrode* 56(1-2): 31-37.

Jerem, P., Jenni-Eiermann, S., McKeegan, D., McCafferty, D. J. and Nager, R. G. 2019. Eye region surface temperature dynamics during 2 acute stress relate to baseline glucocorticoids independently of environmental conditions. *Physiology and Behavior* 210: 112627.

Johnson, S. R. and Dunbar, M. R. 2011. FLIR technical series application note for research & science infrared thermography: its use and application for detecting infectious diseases in wildlife and domestic animals. FLIR Technical Bulletin.

Kashiha, M., Pluk, A., Bahr, C., Vranken, E. and Berckmans, D. 2013. Development of an early warning system for a broiler house using computer vision. *Biosystems Engineering* 116(1): 36-45.

Kastberger, G. and Stachl, R. 2003. Infrared imaging technology and biological applications. *Behavior Research Methods, Instruments, and Computers: A Journal of the Psychonomic Society, Inc* 35(3): 429–439.

Kunc, P. and Knizkova, I. 2012. The use of infrared thermography in livestock production and veterinary field. P 85–101. In: *Infrared Thermography Recent Advances and Future Trends* Meola, C. (Ed). Europe Soest, The Netherlands: Bentham Sci Pubs.

Lambert, H. and Carder, G. 2019. Positive and negative emotions in dairy cows: can ear postures be used as a measure. *Behavioural Processes* 158: 172–180.

LokeshBabu, D. S., Jeyakumar, S., Vasant, P. J., Sathiyabarathi, M., Manimaran, A., Kumaresan, A., Pushpadass, H. A., Sivaram, M., Ramesha, K. P., Kataktalware, M. A. and Siddaramanna 2018. Monitoring foot surface temperature using infrared thermal imaging for assessment of hoof health status in cattle: a review. *Journal of Thermal Biology* 78: 10–21.

Loughmiller, J. A., Spire, M. F., Dritz, S. S., Fenwick, B. W., Hosni, M. H. and Hogge, S. B. 2001. Relationship between mean body surface temperature measured by use of infrared thermography and ambient temperature in clinically normal pigs and pigs inoculated with *Actinobacillus pleuropneumoniae*. *American Journal of Veterinary Research* 62(5): 676-681.

Loughmiller, J. A., Spire, M. F., Tokach, M. D., Dritz, S. S., Nelssen, J. L., Goodband, R. D. and Hogge, S. B. 2005. An evaluation of differences in mean body surface temperature with infrared thermography in growing pigs fed different dietary energy intake and concentration. *Journal of Applied Animal Research* 28(2): 73–80.

Lowe, G., Mc Cane, B., Sutherland, M., Waas, J., Schaefer ,A., Cox, N. and Stewart, M. 2020. Automated collection and analysis of infrared thermograms for measuring eye and cheek temperatures in calves. *Animals: An Open Access Journal from MDPI* 10(2): 2–15.

Lowe, G. L., Sutherland, M. A., Waas, J. R., Schaefer, A. L., Cox, N. R. and Stewart, M. 2018. Physiological and behavioral responses as indicators for early disease detection in dairy calves. *Journal of Dairy Science* 102(6): 5389–5402.

MacMillan, K., Cook, N. J. and Colazo, M. G. 2019. Evaluation of infrared thermography to identify illness in transition cows. *WCDS Advances in Dairy Technology* 31: 332.

Maharjan, D., Rodas-González, A., Tanner, A. R., Kennedy, V. C., Kirsch, J. D., Gaspers, J., Negrin-Pereira, N., Fontoura, A. B. P., Bauer, M. L., Swanson, K. C., Reynolds, L. P., Stokka, G., Ward, A. K., Dahlen, C. R., Neville, B., Wittenberg, K., McGeough, E., Vonnahme, K. A., Schaefer, A. L., López-Campos, O., Aalhus, J. L. and Ominski, K. H. 2018. Impact of needle-free injection devoice on injection site tissue damage in beef sub–primals. ASAS. Proceedings 2018 Vancouver.

Marsot, M., Mei, J., Shan, X., Ye, L., Feng, P., Yan, X., Li, C. and Zhao, Y. 2020. An adaptive pig face recognition approach using convolutional neural networks. *Computers and Electronics in Agriculture* 173 1–10.

McLennan, K. M., Miller, A. L., Dalla Costa, E., Stucke, D., Corke, M. J., Broom, D. M. and Leach, M. C. 2019. Conceptual and methodological issues relating to pain assessment in mammals: the development and utilisation of pain facial expression scales. *Applied Animal Behaviour Science* 217: 1–15.

Metzner, M., Sauter-Louis, C., Seemueller, A., Petzl, W. and Klee, W. 2014. Infrared thermography of the udder surface of dairy cattle: characteristics, methods and correlation with rectal temperature. *Veterinary Journal* 199(1): 57–62.

Metzner, M., Sauter-Louis, C., Seemuellor, A., Petzl, W. and Zerbe, H. 2015. Infrared thermptgraphy of the udder after experimentally induced Escherichia coli mastitis in cows. *Veterinary Journal* 204(3): 360–362.

Moe, R. O., Bohlin, J., Flø, A., Vasdal, G., Erlandsen, H., Guneriussen, E., Sjökvist, E. C. and Stubsjøen, S. M. 2018a. Effects of subclinical footpad dermatitis and emotional arousal on surface foot temperature recorded with infrared thermography in turkey toms (*Meleagris gallopavo*). *Poultry Science* 97(7): 2249–2257.

Moe, R. O., Flø, A., Bohlin, J., Vasdal, G., Erlandsen, H., Guneriussen, E., Sjökvist, E. C. and Stubsjøen, S. M. 2018b. Experimental factors affecting the within- and between-individual variation of plantar foot surface temperatures in turkeys (Meleagris gallopovo) recorded with infrared thermography. *Infrared Physics and Technology* 92: 381–386.

Montanholi, Y. R., Palme, R., Haas, L. S., Swanson, K. C., Vander Voort, G. and Miller, S. P. 2013. On the relationships between glucocorticoids and feed efficiency in beef cattle. *Livestock Science* 155(1): 130–136.

Nasirahmadi, A., Edwards, S. A. and Sturm, B. 2017. Implementation of machine vision for detecting behaviour of cattle and pigs. *Livestock Science* 202: 25–38.

Nikkhah, A. 2015. Infrared thermography as a prognostic livestock agrotechnology: a critique. *Agrotechnology* 04(1): 1.

Novotna, I., Langova, L. and Havlicek, Z. 2019. Risk factors and detection of lameness using infrared thermography in dairy cows–a review. *Annals of Animal Science* 19(3): 563–578.

O'Brian, B., Hennessy, D. and Shalloo, L. 2019. Editorial. Precision Livestock Farming. 9th European Conference on Precision Farming. P7. Cork, Ireland. 26–29 August 2019. ISBN 978-1-84170-654-2.

Okada, K., Takemura, K. and Sato, S. 2013. Investigation of various essential factors for optimum infrared thermography. *Journal of Veterinary Medical Science* 75(10): 1349–1353.

Orman, A. and Endres, M. I. 2016. Use of thermal imaging for identification of foot lesions in dairy cattle. *Acta Agriculturae Scandinavica, Section A–Animal Science* 66(1): 1–7.

Pampariene, I., Veikutis, V., Oberauskas, V., Zymantiene, J., Zelvyte, R., Stankevicius, A., Marciulionyte, D. and Palevicius, P. 2016. Thermography based inflammation monitoring of udder state in dairy cows: sensitivity and diagnostic priorities comparing with routine California mastitis test. *Journal of Vibroengineering* 18(1): 511–521.

Parker, J. R. 1997. *Algorithms for Image Processing and Computer Vision*. New York: Wiley Computer Pub.

Poikalainen, V., Praks, J., Veermäe, I. and Estonian, E. K. 2012. Infrared temperature patterns of cow's body as an indicator for health control at precision cattle farming. *Agronomy Research Biosystem Engineering Special Issue* 1: 187–194.

Polat, B., Colak, A., Cengiz, M., Yanmaz, L. E., Oral, H., Bastan, A., Kaya, S. and Hayirli, A. 2010. Sensitivity and specificity of infrared thermography in detection of subclinical mastitis in dairy cows. *Journal of Dairy Science* 93(8): 3525–3532.

Proctor, H. S. and Carder, G. 2015. Measuring positive emotions in cows. Do visible eye whites tell us anything. *Physiology and Behavior* 147: 1-6.

Racewicz, P., Sobek, J., Majewski, M. and Różańska-Zawieja, J. 2018. The use of thermal imaging measurements in dairy cow herds. *Scientific Annals of Polish Society of Animal Production* 14(1): 55-69.

Rainwater-Lovett, K., Pacheco, J. M., Packer, C. and Rodriguez, L. L. 2009. Detection of foot-and-mouth disease virus infected cattle using infrared thermography. *Veterinary Journal* 180(3): 317-324.

Rekant, S. I., Lyons, M. A., Pacheco, J. M., Arzt, J. and Rodriguez, L. L. 2017. Veterinary applications of infrared thermography. *American Journal of Veterinary Research* 77(1): 98-107.

Sankur, B., Barendregt, W., Ince, C. and Bruining, H. A. 1992. Analysis of respiratory sounds induced ARDS in pigs. P 2537-2538. In: *Proceedings of the Annual International Conference of the IEEE Engineering in Medicine and Biology Society* (vol. 14). PTS 1-7 Morucci, J. P., Plonsey, R., Coatrieux, J. L. and Laxminarayan, S. (Eds).

Sathiyabarathi, M., Jeyakumar, S., Manimaran, A., Jayaprakash, G., Pushpadass, H. A., Sivaram, M., Ramesha, K. P., Das, D. N., Kataktalware, M. A., Prakash, M. A. and Kumar, R. D. 2016. Infrared thermography: a potential noninvasive tool to monitor udder health status in dairy cows. *Veterinary World* 9(10): 1075-1081.

Schaefer, A. L. and Cook, N. J. 2013. Heat generation and the role of infrared thermography in pathological conditions. P 69-78. In: *Thermography: Current Status and Advances in Livestock Animals and in Veterinary Medicine* Luzi, F., Mitchell, M., Costa, L. N. and Redaelli, V. (Eds.). Brescia: Brescia Foundation.

Schaefer, A. L., Cook, N. J., Bench, C., Chabot, J. B., Colyn, J., Liu, T., Okine, E. K., Stewart, M. and Webster, J. R. 2012. The non-invasive and automated detection of bovine respiratory disease onset in receiver calves using infrared thermography. *Research in Veterinary Science* 93(2): 928-935.

Schaefer, A. L., Cook, N. J., Church, J. S., Basarab, J., Perry, B., Miller, C. and Tong, A. K. W. 2007. The use of infrared thermography as an early indicator of bovine respiratory disease complex in calves. *Research in Veterinary Science* 83(3): 376-384.

Schaefer, A. L., Cook, N., Tessaro, S. V., Deregt, D., Desroches, G., Dubeski, P. L., Tong, A. K. W. and Godson, D. L. 2004. Early detection and prediction of infection using infrared thermography. *Canadian Journal of Animal Science* 84(1): 73-80.

Schaefer, A. L., Genho, D., Clisdell, R., von Gaza, H., Desroches, G., Heimer, L., Pelech, G., Grumpelt, B. and Paterson, R. 2018a. The automated and real time use of infrared thermography in the detection and correction of DFD and fevers in cattle. American Society of Animal Science Proceeding. Vancouver.

Schaefer, A. L., Ominski, K., Thompson, S., Crow, G., Bench, C., Colyn, J., Rodas-Gonzalez, A., Maharjan, D., Bollum, R., Cook, N. J., Basarab, J. and von Gaza, H. 2018b. Energy utilization in cattle with steady state and non-steady state methods: the importance of thermal neutrality. *Heliyon* 4(10): e00843: 1-22.

Schaefer, A. L., Jones, S. D. M., Tong, A. K. W. and Vincent, B. C. 1988. The effects of fasting and transportation on beef cattle. 1. Acid-base-electrolyte balance and infrared heat loss of beef cattle. *Livestock Production Science* 20: 15-24.

Schofield, C. P., Whittemore, C. T., Green, D. M. and Pascual, M. D. 2002. The determination of beginning and end-of-period live weights of growing pigs. *Journal of the Science of Food and Agriculture* 82(14): 1672-1675.

Scoley, G. E., Gordon, A. W. and Morrison, S. J. 2019. Use of thermal imaging in dairy calves: exploring the repeatability and accuracy of measures taken from different anatomical regions. *Translational Animal Science* 3(1): 564-576.

Sellier, N., Guettier, E. and Staub, C. 2014. A review of methods to measure animal body temperature in precision farming. *American Journal of Agricultural Science and Technology* 2(2): 74-99.

Soerensen, D. D. and Pedersen, L. J. 2015. Infrared skin temperature measurements for monitoring health in pigs: a review. *Acta Veterinaria Scandinavica* 57(1): 5.

Song, X. Y., Leroy, T., Vranken, E., Maertens, W., Sonck, B. and Berckmans, D. 2008. Automatic detection of lameness in dairy cattle–vision based trackway analysis in cow's locomotion. *Computers and Electronics in Agriculture* 64(1): 39-44.

Stewart, M., Schaefer, A. L., Sutherland, M. A. and Wilson, M. T. 2015. Automated applications of infrared thermography for disease detection in cattle. Proceedings P 711-717. In: 7th European Conference on Precision Farming Sept. 2015 M. Guarino and D. Berckmans (Eds). Milan, Italy.

Stewart, M., Wilson, M. T., Schaefer, A. L., Huddart, F. and Sutherland, M. A. 2017. The use of infrared thermography and accelerometers for remote monitoring of dairy cow health and welfare. *Journal of Dairy Science* 100(5): 3893-3901.

Unruh, E. M., Theurer, M. E., White, B. J., Larson, R. L., Drouillard, J. S. and Schrag, N. 2017. Evaluation of infrared thermography as a diagnostic tool to predict heat stress events in feedlot cattle. *American Journal of Veterinary Research* 78(7): 771-777.

Wolfert, S., Ge, L., Verdouw, C. and Bogaardt, M.-J. 2017. Big data in smart farming–a review. *Agricultural Systems* 153: 69-80.

Wouters, P. R., Geers, R., Parduyns, G., Goossens, K., Truyen, B., Goedseels, V. and Van der Stuyft, E. 1990. Image-analysis parameters as inputs for automatic environmental temperature control in piglet houses. *Computers and Electronics in Agriculture* 5(3): 233-246.

Zaninelli, M., Redaelli, V., Luzi, F., Bontempo, V., Dell'Orto, V. and Savoini, G. 2017. A monitoring system for laying hens that uses a detection sensor based on infrared technology and image pattern recognition. *Sensors* 17(6): 1195.

Zhao, Y., Xin, H. and Dong, B. 2013. Use of infrared thermography to assess laying-hen feather coverage. *Poultry Science* 92(2): 295-302.

Chapter 3

Developments in acoustic techniques to assess livestock health

Erik Vranken, SoundTalks NV, Belgium and KU Leuven M3-BIORES – Measure, Model & Manage Bioresponses, Belgium; Daniel Berckmans, KU Leuven M3-BIORES - Measure, Model & Manage Bioresponses and BioRICS NV, Belgium; and Wim Buyens and Dries Berckmans, SoundTalks NV, Belgium

1 Introduction

In the past two decades, an increasing amount of research has been elaborated in the field of precision livestock farming (PLF). PLF systems aim to offer a real-time monitoring and management system that focuses on improving the lives of the animals by warning when problems arise so that the farmer may take immediate action. Continuous, fully automatic monitoring and improvement of animal health and welfare, product yields and environmental impacts become possible. Precision Livestock Farming assumes that continuous direct monitoring or observation of animals will enable farmers to detect and control the health and welfare status of their animals at any given time.

Nowadays, farmers can use modern technologies to measure different parameters on the farm, such as indoor climate, feed and water consumption, ventilation and heating/cooling inputs, but few of the tools available to date

http://dx.doi.org/10.19103/AS.2021.0090.03

have focussed on the most important participant in the production process, the animal (Berckmans, 2014).

Sound analysis is a convenient method for automatic continuous recording of animal responses: it is non-invasive to the animal, as microphones measuring the sound can be mounted within the animal house without causing additional stress to the animal. Compared to image analysis, a single sound monitor covers a bigger area in the house, monitors more animals and is not influenced by light conditions. In the past decades, a lot of attention has been given to measuring and analyzing animal vocalizations and other environmental sounds, but so far most of the applications are mainly based on experiments at laboratory scale or research farms and the ones that use field measurements often lack scientific validation. Only a few systems have been commercialized yet.

In this chapter, we give an overview of sound monitoring and its applications for different livestock species (pigs, poultry and cattle) while performing a critical review on the status of the current technological developments.

2 Animal vocalizations and sounds produced

The majority of livestock sounds were initially recorded in common environmental conditions (e.g. in a livestock building) and were focussed mainly on evaluating the behaviour or health status of the animals (e.g. feeding, social conflict, respiratory diseases, etc.). On the other hand, laboratory experiments under well-controlled conditions were used to uncover the meaning of the observed sounds. For example, it has been revealed that an increased vocalization rate of pigs and calves is indicative of the excitement and their degree of fearfulness to novelty and social separation (Fraser, 1974; Von Borell and Ladewig, 1992).

There is a continuous search for new methodical approaches allowing more exact description and interpretation of animal vocalizations. These approaches range from periodic vocal-fold vibration to completely atonal turbulent noise. Between these states, non-linear dynamics appear in the sounds. It has been hypothesized that an increased appearance of non-linear phenomena is correlated with a worsened condition of an animal or with the effort that was put on the vocalization production (Manteuffel et al., 2004).

It can be stated that the applications to localize and monitor the health and well-being of livestock animals are feasible with sound analysis. For each application, the main effort is to develop an algorithm for automatic detection of sound events and then associate these specific sound events to health, distress and abnormal behaviour. However, the analysis of animal-produced sounds is still in an initial phase and a very limited number of technologies have been commercialized yet with installations in the field.

3 Sound recording, features and digital signal processing

Sound analysis is a valid method for automatic continuous recording of animal variables that is non-invasive to the animal, as the microphones measuring the sound can be mounted within the animal house without causing additional stress to the animal. Within the PLF research field and sound analysis, quite a lot of attention has been given for measuring the animal vocalizations.

The first step in the sound analysis is the acquisition of the raw sound data (i.e. the input data). Sound is recorded and saved in blocks for a certain time, which can easily be adapted. For recording the sounds, the sampling frequency can also be adapted and the most optimal position for mounting the microphone should be explored. This includes the height of the microphone and the relative position from the walls and disruptive sound sources (e.g. ventilation). The sound data also needs pre-processing, which is a sound analysis procedure to separate the clear sound sources, also called 'foreground sounds' (e.g. animal vocalizations and knocking sounds), from specific interfering noises, also called 'background noises' (e.g. noise from fan and heaters).

After completing data collection and removal of the background noises, the next step is to extract meaningful information from the signal. This process is commonly termed 'audio feature extraction'. This can be carried out in the time domain, but most algorithms focus on the time-frequency domain. To transform a signal from the time domain to the frequency domain, a discrete Fourier transform is often used. This Fourier transform breaks down a signal into several different sinusoidal functions, each with its own frequency, phase and amplitude values. When a signal is converted to the frequency domain, it is possible to extract several acoustic features, such as Mel frequency cepstrum coefficients (MFCCs), which gained considerable attention because of their success in human speech recognition algorithms.

To distinguish various types or individual characteristics of animal vocalizations, the following procedures have been applied:

1 Standard statistical methods (e.g. ANOVA, *t*-test, Wilcoxon test) if few selected features (e.g. fundamental frequency, duration and rate) are sufficient for discrimination.
2 Complex statistical methods to compare groups of features (e.g. correlation analysis, discriminant analysis and cluster analysis) if the complex quality of a call, described by a set of features, is important (Schrader and Hammerschmidt, 1997).
3 Neural Networks (NN) allow applications in noisy environments because they tolerate variations of a trained feature set (Schön et al., 2001., Chedad et al., 2001).

4 Hidden Markov Models (HMM) allow an arbitrary number of different vocalizations to be classified, but this demands a specific HMM for each type of vocalization. A given utterance is then attributed to the best-fitting HMM (Jahns et al., 1997a,b).

Recent trends in sound classification have implemented a variety of deep learning methods. A simple definition of deep learning refers to supervised and unsupervised machine learning algorithms that carry out a variety of tasks (such as classification, data generation, translation and prediction) using very large datasets (big data) and large neutral networks.

4 Sound applications for monitoring of pigs

Pig vocalizations are an important sound with valuable information for the farmer. For instance, high-frequency calls of pigs have already been linked to stressful situations (Manteuffel et al., 2004). Several studies on piglet vocalizations and sounds in stressful situations have focussed on analysing high-frequency calls caused by castration practices (White et al., 1995; Weary et al., 1998; Puppe et al., 2003), cold stress (Hillmann et al., 2004) or heat stress (Ferrari et al., 2013a). Other examples are screams due to the crushing of piglets (Illmann et al., 2013) or manipulation of animals with an electric shock (Düpjan et al., 2008). Several other studies conducted analyses on three pig sound events: screaming, squealing and grunting (Marx et al., 2003; Von Borell et al., 2009).

Moreover, animal sounds such as coughs could be linked to respiratory diseases and thus to the animal health status (Van Hirtum and Berckmans, 2003; Silva et al., 2008). A study of Berckmans Jr. et al. (2015) showed that a detection tool could give warnings 2 weeks earlier, compared to a situation where the pigs were observed by the farmer and the veterinarian. In addition to the cough detection, Van Hirtum and Berckmans (2002), Exadaktylos et al. (2008a,b) and Ferrari et al. (2008a,b) showed that it is possible to decipher cough between healthy and sick pigs.

Characterization of temporal and spatial patterns of clinical respiratory episodes in growing pigs using continuous pig cough monitoring was used to calculate an algorithm-based respiratory distress index. This technology was commercialized by SoundTalks NV, Belgium (Polson et al., 2018).

To conclude, continuous automated analysis of pig sounds can be useful for assessing the animal's condition in terms of welfare and health status.

5 Sound applications for monitoring poultry

The first research on chicken was focusing on the effect of sounds on chicken or chicken embryos (Stadelman, 1958). Only 20 years later research started on sound produced by chicken (Brackenbury, 1978). It lasted till 1994 before the research was done on the meaning of the sound produced by birds (Marler and Evans, 1996).

Chicken vocalizations have been shown to have specific patterns when the chicks are exposed to cold stress (Bugden and Evans, 1999). The thermal comfort of chicks has been shown to have links with the amplitude and frequency of calls (Moura et al., 2008a).

Also, chicken vocalizations are directly linked to other stressors. In an experimental design, it was demonstrated that distress calls of chickens were mediated by isolation (Feltenstein et al., 2002). In another study, a method has been developed that could be used for stress monitoring of a flock. The method was based on multiparametric sound analysis (Mair et al., 2001).

It also has been presented in the literature that different types of poultry calls, such as distress or happy calls, could be identified based on their frequency characteristics (Marx et al., 2001; Ghesquire et al., 2002). Also, attempts were made to automatically classify these calls and relate them to animal welfare (Pereira et al., 2011).

Furthermore, in relation to respiratory diseases in chickens, which are commonly observed in poultry flocks, several studies have used sound analysis as a detection method. One clinical sign often observed when respiratory diseases occur is sneezing, for example, Newcastle disease (Austic and Nesheim, 1990), respiratory disease complex (McMullin, 2004), infectious bronchitis (McMullin, 2004; Ignjatović and Sapats, 2000) and avian influenza (Spickler, 2015). In a recent study at KU Leuven, an algorithm enabled the monitoring of the number of sneezes in an experimental group of broilers (Carpentier et al., 2019). This work represents the first step towards the development of an automated sound-based monitoring system for poultry health.

However, the potential of the use of microphones in poultry is not limited to automatic recording of vocalizations but can, for example, also be used to measure growth rate (Fontana et al., 2015) and very accurate real feed uptake by broilers using the pecking sound (Aydin et al., 2014). The unique feature in the latter case is that a sound detection system was defined based on a sound sensor attached to the feeder pan. In this way, it was possible to perform and analyse sound measurements in real time in a fully non-invasive, but still automated, manner during the full growth process of this group of animals.

6 Sound applications for monitoring cattle

Several authors have tried to describe and classify the occurrence of different forms of vocalization in cattle using spectrographic methods (Kiley, 1972; Hall et al., 1988). Kiley (1972) identified six different calls occurring as a part of a continuum rather than as discrete calls. Also, the communicative functions of the vocalization in cattle were investigated, for example, the mother calls by calves (Barfield et al., 1994; Marchant-Forde et al., 2002). Jahns et al. (1997a) developed a method to identify cattle by sound analysis. With increasing interest in measures of welfare, the study of vocalizations of cattle is in progress. Watts and Stookey (2000) have studied the vocal behaviour of cattle as a potentially useful indicator of their physiological and psychological state. But communicative function and the causes of the vocal behaviour are rather poorly understood (Watts and Stookey, 2000), and a considerable individual variability in vocal response may complicate a clear evaluation concerning welfare (Watts and Stookey, 2001).

It has also been suggested that the scoring of vocalization can be used as a very simple indicator for handling equipment problems in slaughter plants (Grandin, 1998, 2001). The hot-iron branding of calves obviously causes severe pain and results in high-pitch vocalizations. It has also been demonstrated that both cows and calves vocalize in response to separation (Weary and Chua, 2000; Flower and Weary, 2001; Watts and Stookey, 2001; Marchant-Forde et al., 2002). Brief periods of social isolation in unfamiliar surroundings seem to be stressful to cows as indicated by an increased incidence of high-frequency vocalization and paralleled by activation of the hypothalamic-pituitary-adrenocortical axis (Rushen et al., 1999). Cow vocalizations after separation were able to release cardiac responses in calves (Marchant-Forde et al., 2002) indicating that also the perception of conspecific vocalizations could affect their welfare. Moreover, Thomas et al. (2001) demonstrated that some variations in the vocal response of calves to weaning and separation (e.g. an increase of incidence and fundamental frequency) were also correlated to milk deprivation. This indicates that vocal utterances of calves can include information on general needs like feed and distressing housing.

Green et al. (2017) provided an excellent review of the evolution of cattle vocal communication, as well as an overview of how these vocalizations relate to various welfare contexts. They separated cattle vocalization functions according to individuality of vocalizations, vocal recognition, calf separation, social isolation, oestrus, feeding and painful husbandry procedures. As known for all living organisms, individual differences also play a role in cattle. Cattle calls contain information on individuality due to high levels of inter-cow variability in the acoustic characteristic of their vocalizations. This allows for each animal to be identified by the 'uniqueness' of their call (Padilla de la Torre et al., 2016; Watts and Stookey, 2001). Cattle are herd animals, and isolation from their conspecifics results in physiological changes in the animal such as increased heart rate, salivary cortisol, urination and defaecation rates and an increase in vocal responses (Muller and Schrader, 2005). Green et al. (2017) could create a database of audio recordings of different cattle vocalizations and their related contexts. It was suggested that different machine learning algorithms could potentially be trained using labelled datasets in order to identify the vocalization, and thus the context in which it occurred. Cattle cough sounds have been classified using labelled data from a variety of recordings, which were identified by a human labeller using a combination of audio and visual scoring (Vandermeulen et al., 2016). They labelled a total of 205 min of sounds, resulting in 285 labelled calf coughs. They extracted features by calculating the Fast Fourier Transformations (FFT) of the incoming audio, removing the background noise and reducing the resolution in the spectrograms by summing the frequencies into 12 separate bands. They also calculated the duration of the cough. An example-based classifier was used to compare the rough reduced spectrogram of incoming audio with the reduced spectrogram of the labelled data. This was achieved by calculating the Euclidean distance between the two rough spectrograms. The lower the distance, the more it resembled its corresponding spectrogram. This research achieved a 98% specificity rate (true negatives) and 52% sensitivity rate (true positive). Despite the low sensitivity, the algorithm was still able to detect increased periods of coughing, allowing farmers to administer treatment for the respiratory disorder. Cattle grazing sounds have also been analysed in order to determine the relationship between behaviour and acoustics measurements with herbage dry matter intake (Galli et al., 2018). This was achieved by attaching microphones and cameras to a cow's forehead and exposing the cattle to different treatments which include varied plant species, two different plant heights, increasing the herbage mass and the number of bites it takes to finish (10–30). The sounds were analysed by extracting the energy flux density from the sounds. It was found that energy flux density related linearly to dry matter intake.

It can be concluded that sound processing techniques and complex statistical methods can help to analyse specific calls or call features of cattle. Thus, the state of an animal can be evaluated, or individuals may be discriminated (Jahns et al., 1997a; Ikeda and Ishii, 2001).

Definite solutions or systems, however, are apparently not yet commercially available.

7 From science to practical applications

Although interesting results have been obtained in the above-mentioned publications and citations, most of the research was mainly based on experiments in controlled environments, research farms or lab conditions. Still today, too many researchers are repeating work that has been published already some time ago, all in lab conditions. The extension towards the monitoring of vocalizations and other sounds in commercial farms is not straightforward and needs substantial investigational activities, such as validation against gold standards and accurate labelling in different conditions (house dimensions and types, building materials, type of animals, environmental noises etc.).

The European Commission has funded several research projects in PLF (e.g. BRIGHTANIMAL, BIOBUSINESS, ALL-SMART-PIGS, EU-PLF and SMARTAGRIHUBS). These projects have demonstrated that PLF (1) can indeed stay operational 24/7 in commercial farms and (2) has a significant potential to automate the continuous measurement of key indicators on farm using modern technologies such as image analysis, sound analysis and novel sensors. From the outcome of these projects, it became clear that sound analysis is a very interesting technology for groups of animals with the advantage that the animals are not touched or influenced by the technology while measurements can be done continuously, also during dark periods and from a relatively great distance with an only limited number of sensors for a high number of animals.

A commercial product needs a combination of hardware, working in a reliable way within the harsh environment of a livestock house, complying with certain technical and safety standards in combination with software, a good user interface, a backup solution to store data, an auto-restart function in case of power failure, manual and help functions, installers who can install and service the product, and so on.

SoundTalks NV (Belgium), a spinoff of KU Leuven and the University of Milan was one of the first companies that developed a commercial product ('Pig Cough Monitor') that was successfully introduced in 2012.

From literature and expertise in running and past research projects, we know that the analysis and interpretation of sounds collected in pig and poultry farms is a powerful tool when studying the behaviour, welfare, health and production performance of the animals. Deviations from optimal sound patterns can be detected by continuous automated monitoring and used by animal caretakers to decide on actions to be taken.

Moreover, to design robust algorithms for monitoring the health and welfare of piglets, more (labelled) data need to be collected with more recent (reliable) hardware, such as 'SOMO Gen 2' (Sound Monitor Generation 2, SoundTalks NV).

With the first version of the 'Pig Cough Monitor', SoundTalks was active in several EU and regional research projects, where the potential for practical implementation at farm level was evaluated:

- In the EU-projects 'ALL-SMART-PIGS' and 'EU-PLF', the technical and economic viability of PLF technologies in European pig farming was tested and demonstrated in commercial livestock houses. Living lab infrastructure was established for bringing innovative PLF technologies to two Hungarian, two Spanish, two Dutch, two French, one Italian and one North Irish pig farmers. In the EU-PLF project, SoundTalks also installed Sound Monitors in five broiler farms, ten pig farms and one cattle research farm, distributed over in total 12 European countries.

- Demonstration project 'PLF' in Belgium, in which health monitoring of fattening pigs through sound monitoring with established sound algorithms was done, with the purpose of training farmers and demonstrating PLF technology in ten Belgium farms.
- EU-project 'LESSAntibiotics': Its aim was to work towards the reduction of the use of antimicrobial compounds and avoid antimicrobial resistance in the food chain by combining novel developed feed additives and sound technology for health monitoring of piglets. First attempts were made to track coughs, sneezes and activities automatically with newly developed sound algorithms, but the performance was not yet optimal.

The expertise who developed these projects clearly showed and demonstrated (1) that sound monitoring as a PLF technology can work 24/7 in commercial farms and (2) that keeping such systems running is challenging due to the complexity of animal responses in developing early warning tools and the need to extend the application to other animal sounds and other species such as poultry. But all the projects showed and convinced the researchers and the companies about the high potential of the technology.

8 Status of the current sound applications for pigs, poultry and cattle

In Tables 1, 2 and 3, an overview of the most relevant sound applications for pigs, poultry and cattle based on a screening of relevant scientific literature on sound monitoring is given. The screening was done in Web of Science with key words 'sound' or 'sound analysis' in combination with 'pig or swine', 'poultry or chicken' and 'cattle or beef'. For each of the applications, the status of the development is indicated at different levels (listed below) in the tables:

1. Animal species;
2. Sound application;
3. Scale;
4. Validated;
5. Commercially available;
6. Summary.

Table 1 Relevant publications on pigs' sounds

Citation	Animal species	Sound application	Scale	Validated	Commercially available	Summary
General noise levels						
Da Silva et al. (2012)	Five piglets	General dB level	Lab experiment	No	No	Assess the influence of environmental variables (e.g. temp and RH) on sound level emitted by piglets and quantify the sound bands using dB compared to thermal comfort (= comfort condition – alert condition – thermal stress condition).
Wegner et al. (2019)	Pigs	General dB level	214 conventional commercial farms (Germany)	Unknown	Unknown	Reveal the noise status in the pig's environment. Overall noise level and evaluate the suitability of noise as an animal welfare indicator (e.g. noise analysis with age, feed, floors, and air systems). No statistical association between animal health (aggression) factors and noise levels.
Damasceno et al. (2019)	Piglets	General dB level distribution	Experimental farm (Brazil)	No	No	Evaluate welfare and spatial distribution of noise levels in swine nursery: first 3 weeks of life. Climate conditions were evaluated. The ambience condition inside the bay of the nursery does NOT provide comfort to piglets.
Stress-related vocalizations						
Schrader and Todt (1998)	Pigs	Vocalizations & stress	Laboratory conditions (Germany)	No	No	Vocalizations = reliable cues for particular endocrine stress responses? (By analysing blood samples & analysing recordings of vocalizations and non-vocal activities.) Four types of vocalizations were detected within two treatments. Possible effects of central stress reactions on both control of vocalization and activation of endocrine stress responses.

Schön et al. (2001)	Piglets, pigs, sows	Stress screams, grunts, nursing grunts	Unknown	Unknown	No	Development of a procedure that allows characterization, classification, and visualization of cluster structures of stress calls. Used to discriminate individual piglets by their scream characteristics and to classify stress calls versus other calls/background noise.
Schön et al. (2004)	Pigs	Levels of stress calls	Research institute and practical farming	Unknown	No 'as a stand-alone device' 'STREMODO' 'present development stage'	Development of a system to monitor and record levels of stress calls (indicates duration and intensity of stress screaming).
Moura et al. (2008b)	Piglets	Vocalizations (stressful and non-stressful situations)	Commercial swine farm	Yes (commercial swine farm)	No	Development of software (using Delphi 5) to monitor and analyse distinct sounds correlating the noise response with levels of stress for assessing welfare.
Borell et al. (2009)	Piglets	Grunting, squealing and screaming	Unknown	Yes (open-field study)	Unknown	Classification of three different classes of piglet vocalizations: grunting, squealing and screaming. Group-farrowing versus single-farrowing system differ for duration, energy & relative maximum levels.
Ferrari et al. (2013a)	Piglets	Vocalizations	Experimental farm	No	No	Determine whether heat stress induces specific vocalizations in piglets and how these vocalizations acoustically differ from normal vocalizations. Frequency, duration and amplitude showed clear differences.
Cough/respiratory disease/air quality						
Moshou et al. (2001)	Pigs	Cough	Laboratory conditions	No	No	Build an intelligent alarm system used for the early detection of coughs. New approach to distinguish coughs from grunts, metal clanging, etc. (neural network).

(Continued)

Table 1 (*Continued*)

Citation	Animal species	Sound application	Scale	Validated	Commercially available	Summary
Van Hirtum and Berckmans (2003)	Six piglets	Cough	Laboratory conditions	No	No	Acoustic indicator for appearance/absence of coughing based on time and frequency characteristics of sounds. Positive cough recognition of 92% and misclassification of 21%.
Aerts et al. (2005)	Ten pigs	Coughs (labelling)	Experimental farm (Belgium)	No	No	Give an overview of the attained results in cough recognition and define a methodology to label cough data. Underestimations of up to 94% in the number of coughs when labelling in an audio-visual way. Differences were found for different observers. Additional research is needed.
Silva et al. (2008)	Pigs	Cough (localization)	Unknown ('a pig stable in Milan') 'Stable in field condition'	Yes	No	Proposing an algorithm for localizing cough sounds by using the difference in arrival time of a sound signal between different microphones. -> Visualizing spread of respiratory diseases.
Ferrariet al. (2008a)	Pigs	Cough (distinguish between symptomatic or non-symptomatic)	Field and laboratory conditions	No	No	To describe the acoustic features of cough sounds from a lung infection and compare these with healthy cough sounds. Sick animals -> Lower root mean square error, lower peak frequency, longer duration and shorter intervals between coughing.
Exadaktylos et al. (2008b)	Pigs	Cough (sick & localization)	Laboratory conditions and field conditions ('a pig stable in Milan')	No	No	Automated online recognition and localization of sick coughs. Autoregression analysis for the sound signal. Correct identification ratio between 73% and 93%.

Exadaktylos et al. (2008a)	Pigs	Coughs	Laboratory conditions	No	No	Extend existing cough identification methods and propose a real-time algorithm for identifying sick coughs. (Based on frequency domain characteristics of the signal, improved procedure to extract the reference -> speed up diagnosis and treatment.) Eighty-five per cent overall correct classification ratio with 82% sick cough correctly identified.
Ferrari et al. (2008b)	Pigs	Cough (frequencies & duration)	Two intensive pig farms	No	No	Compare acoustic features (peak frequency and duration) of cough sounds (infectious versus healthy). Infected animals -> Considerably lower peak frequency and longer mean duration -> can be used in an algorithm-based alarm system.
Silva et al. (2009)	Pigs	Coughs	Unknown	No	No	Assess if the dynamics in the energy envelope of coughs are related to pathological conditions of the respiratory system. Time constant (duration) is significantly higher for the decay of coughs from sick pigs (*Pasteurella multocida*) compared to non-infected pigs.
Ferrari et al. (2009)	Pigs	Coughs (peak frequency and duration)	Laboratory and field conditions	No	No	To improve labelling of coughs giving physic features to specific sounds. In the next step, these characteristics are the input for an automated alarm (algorithm for cough detection). A distinction (sound) between pathological and healthy cough can be made (sick cough = lower peak frequency, longer duration of cough).
Gutierrez et al. (2010)	36 pigs	Cough	Commercial farm	No	Unknown	Classifying the different porcine wasting diseases through sound analysis (differences in acoustic footprints of coughs in PCV2, PRRS, MH infected pigs versus normal cough). Normal cough = highest pitch level, MH = highest intensity, all coughs statistically different from each other. No statistical difference regarding the duration of cough.

(*Continued*)

Table 1 *(Continued)*

Citation	Animal species	Sound application	Scale	Validated	Commercially available	Summary
Ferrari et al. (2013b)	Pigs	Cough	Unknown	Unknown	Unknown	A summary on how sound analysis may be used for intensive piggeries health monitoring: comparing the acoustic features (peak frequency, duration of sound, energy envelope and time constant) of different types of cough sounds, analysing their acoustic features and how they may be used in an algorithm-based alarm system to automatically identify cough sounds and provide farmers with an early warning about the health status of their herds.
Berckmans et al. (2015)	Pigs	Cough	Ten fattening pig farms across Europe	Yes	Yes	To present an example of a successful sound-based PLF product (the respiratory distress monitor) and to encourage the reader to consider performing sound-based PLF research in the future. The effectiveness of the monitor as an early warning is shown by discussion of five different use cases.
Polson et al. (2018)	Pigs	Cough	Commercial wean-to-finish farm (2400 pigs)	Yes	Yes	To characterize the temporal and spatial patterns of clinical episodes of swine respiratory diseases of growing pigs under large-scale commercial production conditions using a continuous audio monitoring system.
Polson et al. (n.d.)	Pigs	Cough	Three commercial wean-to-finish farms (1200 to 2400 pigs)	Yes	Yes	To evaluate the optimal placement and configuration of a continuous sound monitoring system in large airspace buildings in the United States containing growing pigs to enable both a high sensitivity for detection and establishing directionality of clinical respiratory episodes.

Zhuang et al. (2019)	Guinea pig	Coughs	Lab experiment	Unknown	Unknown	To validate a novel recording/analysis system that automatically counts coughs and differentiates cough patterns.
Wang et al. (2019)	84 pigs (weaners)	Coughs	Commercial farm (China)	No	No/unknown	Identify the relationship between sounds and air quality. Power spectral density significantly differed in sounds under different air qualities. The developed algorithm has an average recognition rate of 95%.
Pain-related vocalizations						
Marx et al. (2003)	70 piglets	Grunt, squeal, scream	Commercial farm (Germany)	No	No	Classify vocalizations during castration and assess alterations in vocalizations under local anaesthesia (multiparametric call analysis). Three types = grunt, squeal and scream. Without anaesthesia = double the number of screams (screams from castration= lower peak and main frequency than squeals).
Schön et al. (2006)	19 male piglets	Screams	Unknown	Unknown	Unknown	To examine various parameters of the vocalization of piglets during the three phases of the castration process. High frequency calls detected by STREMODO.
Vandermeulen et al. (2015)	27 grower pigs	Screams	Experimental farm (BE Merelbeke)	No	No	Understanding which sound features (formant structure, power, frequency content, variability and duration) define a scream -> a method to detect screams was developed. Seventy-two per cent sensitivity, 91% specificity and 83% precision.
Emotional state/human manipulation/nursing sounds						
Blackshaw et al. (1996)	Piglet/sow	Nursing sounds	Veterinary Science Farm (Australia)	No	No	Characterize individual nursing sounds using Canary 1.2 Software (spectrogram, auditory features) (min, max, main frequency, length, gap and volume). -> sows can be distinguished by their vocalizations.

(Continued)

Table 1 (*Continued*)

Citation	Animal species	Sound application	Scale	Validated	Commercially available	Summary
Marchant et al. (2001)	67 gilts	Single grunts, single squeals and rapidly repeated grunts	Unknown	No	No	Categorize and ascribe the function of vocalizations during a standard human approach test. Behaviour and sound were recorded -> 3 types of calls: single grunts (investigatory behavior/contact call), single squeals (higher level of arousal) and rapidly repeated grunts (greeting/threat function) (all three in short and long types).
Maigrot et al. (2018)	19 wild boars	Observations + recordings of grunts, screams and squeal	Wildlife park (Switzerland)	Unknown	Unknown	Investigating the vocal expression (energy quartiles, duration and formants) of emotional valence in wild boars. Wild boars can vocally express their emotional states. Could be used to tell something about their welfare. (Grunts = positive situation, squeal and scream = negative situation).

Table 2 Relevant publications on poultry sounds

Reference	Animal species	Sound application	Scale	Validated	Commercially available	Summary
Health monitoring						
Sadeghi et al. (2015)	Ross chickens	Vocalization	University	Yes (validation of network and classifier structure)	no	*Clostridium perfringens* type A (intestinal) detection through vocalization. Classification accuracy was 66.6% and 100% for days 16 and 22; i.e. two and eight days after the disease, respectively. The vocalization of infected chickens presented lower sound intensity than healthy ones but presented higher frequency within the main low-frequency range. Vocalization of the infected birds was less uniform and more dispersed than that of the healthy ones. the neural network was able to differentiate vocalization samples of healthy chickens from those of unhealthy ones with 100% accuracy
Banakar et al. (2016)	Ross 308 chickens	Sound signals in frequency and time-frequency domains,	Agricultural School (Tehran)	No	no	An intelligent device for diagnosing avian respiratory diseases. Newcastle disease, Bronchitis Virus & Avian Influenza. Sound signals = frequency and time-frequency domains. 41.35% and 83.33% or 91.15% accuracy. Main aim of this study = develop AI methodology based on signal processing. -> Developed an electronic device
Huang et al. (2019)	Leghorn chicken	Sound for avian influenza virus	Experimental setups (Belgium)	? (Four-fold cross validation)	no	Audio analysis-based detection methods for avian influenza as early as possible. Accuracy rate between 84% and 90%. (Endpoint detection used) 80% is detected in 22 h after inoculation.

(Continued)

Table 2 (*Continued*)

Reference	Animal species	Sound application	Scale	Validated	Commercially available	Summary
Carpentier et al. (2019)	Broilers	Sneezing	Experimental setups (China)	No	no	Developing an algorithm to monitor chicken sneezing sounds (sneezing is a key clinical sign for respiratory diseases) (where lots of birds are active and multiple noise sources). In an experimental group, dataset contained only 0.24% sneezes and they were classified with sensitivity 66.7% and 88.4% precision.
Mahdavian et al. (2020)	Ross and Cobb	Sick <> non sick calls	Standard (commercial poultry farming protocol)	No	no	A report on the successful application of some of the features involved in extracting healthy and non-healthy birds' calls from their sound signals. Vocal phrases were extracted using the presented algorithm. Error increased when age and onset of illness increased. Detection accuracy was calculated at 95% for healthy young birds and 72% for nonhealthy birds. (focus points were genotype, gender and health condition) (increasing age -> decrease voice activity and call energy).
Stress vocalizations						
Marx et al. (2001)	White Leghorn chicks	Distress calls, short peeps, warblers, pleasure notes	Laboratory conditions	No	No	91.20% of the recorded vocalizations = Distress calls (44%), short peeps (22%), warblers (19%), pleasure notes (6%) and 9% undefined. Single isolated animals showed the highest total vocalization. Vocalization is strongly dependent on social contacts. Different degrees of social deficits find their expression in discrete changes in pattern and elements of vocalization.

Lee et al. (2015)	Laying hens	Vocalization	Commercial farm (South Korea)	Yes (classification model was validated)	No	The detection and classification module detects the stress from changes in the sound and classifies it into subsidiary sound types, such as physical stress from changes in temperature and mental stress from fear. The accuracy in detecting stress approached 96.2%, and the classification model was validated, confirming that the average classification accuracy was 96.7% and that its recall and precision measures were satisfactory.
Abnormal vocalizations						
Du et al. (2018)	Laying hens	Abnormal vocalizations	Laboratory tests and small group tests (China)	No (but they did do small group tests to evaluate the algorithm)	No	The system (sound source localization) had an accuracy of 74.7% in laboratory tests and 73.6% in small poultry group tests for different area sound recognition. Flocks produced an average of 40 sounds per bird during feeding time in small group tests. Each normal chicken produced more than 53 sounds during the daytime (noon to 6:00 pm) (also more random vocalizations during the day than during night) and less than one sound at night (11:00 pm–3:00 am). Particularly, the flock made significantly more sounds while undergoing stress or suffering from starvation.
Liu et al. (2020)	Broilers	Abnormal sounds (cough and snore) and interfering sounds	Commercial farm (China)	? (HMM were trained and compared)	No	The developed recognition algorithm provided average accuracy so it can be used for respiratory assessment. Results are affected by noise interference (fan noise), quality of sound sample and algorithm. Also, age, weight and type of bird also have an effect.

(Continued)

Table 2 (*Continued*)

Reference	Animal species	Sound application	Scale	Validated	Commercially available	Summary
Growth						
Fontana et al. (2017)	Broilers	Vocalizations ~ weight	Intensive broiler farm	This is the validation study ('the aim of this study was to identify and validate a model that describes the growth rate of broiler chickens based on the peak frequency of their vocalizations and to explore the possibility to develop a tool capable of automatically detecting the growth of the chickens based on the frequency of their vocalizations during the production cycle.')		Aim of this study: identify and validate a model that describes the growth rate based on the peak frequency of their vocalizations and to explore the possibility to develop a tool automatically detecting growth of the chickens based on the frequency of vocalizations. No significant difference is shown between expected and observed weights along the entire production cycles -> weight can be predicted with a function of peak frequency (the more the animals grow, the lower the PFs and the higher the noise of fans and feeding lines).
Fontana et al. (2015)	Broilers	Vocalizations ~ weight	Indoor reared broiler farm (UK)	No	No (This is the basis for a further development of an automated growth monitoring tool)	Analysing sound recordings (peak frequency) -> significant correlation between frequencies of vocalizations and weight of broilers; the more they grow the lower the frequency of the sound emitted. The R2 indicates that the model accounts for 98% of the variation in peak frequency. This method may be used as an early warning system.

Feed consumption						
Aydin et al. (2014)	Broilers	Individual pecking sounds	Three laboratory experiments	Yes (feed intake of broilers was calculated and used for the validation of the proposed algorithm)	No	An algorithm was developed to detect individual pecking sounds of broiler chickens. Relation between pecking sounds and the amount of feed intake and uptake were investigated. Ninety-three per cent of the pecking sounds were correctly identified by algorithm, whereas 7% of the identification results were false positives. In addition to the high relation, 90% of feed intake was correctly monitored using sound analysis. This has the potential to be used as a tool to monitor the feed intake of chickens. The correlation among feed uptake, feed intake and number of peckings was calculated and a linear correlation was found among these three variables. Applying the method under field conditions will probably reveal problems about the accuracy of the algorithm and different types of feed can affect the sounds generated by birds.
Aydin and Berckmans (2016)	Broilers	Pecking sounds (feed uptake)	Laboratory testing	No	No	Short-term feeding behaviours (meal size, meal duration, meals per day and feeding rate) of broilers at group level. Microphone at feeder + camera at top of feeding pen + weighing system -> relationship between feeding behaviours by algorithm and by weighing scale and camera -> strong positive correlation. Accuracy between 89–95%. Only for 39-day-old Ross-308 broilers, not tested for other factors.

Table 3 Relevant publications on cattle sounds

Citation	Animal species	Sound application	Scale	Validated	Commercially available	Summary
Health monitoring						
Vandermeulen et al. (2016)	Calf	Cough	Research centre (AGRIC, Grange) (62 calves)	No	No	Develop an automated calf cough monitor as an early warning for BRD. A sound analysis algorithm was developed: 50.3% sensitivity, 99.2% specificity and 87.5% precision.
Carpentier et al. (2018)	Calf	Cough	Experimental farm (Ireland) (35 calves)	No	No	Algorithm development for detection of coughing sounds in calf house. Algorithm precision of more than 80% for five out of seven cases and 53.8–66.6% for two of the seven cases (false positives). Sensitivity = 41.4% and general precision of 94.2%?? The robustness of algorithms seems to be confirmed.
Volkmann et al. (2019)	Holsteins	Footfall sound (lame <> non lame)	Free-stall barn	No	No	Develop a system for automated diagnosis of claw lesions by analysing footfall sounds. (Standard deviation of volume = force of the cow's footsteps/gait pattern).
Grazing/feed consumption						
Clapham et al. (2011)	Fteers or heifers	Grazing	University farm (USA)	Yes (calibration and validation of automated bite detection)	Unknown	Acoustic recording and analysis system to automatically detect, classify and quantify injective events in free-grazing cattle (to estimate forage intake). (By using microphone).

Deniz et al. (2017)	Lactating cows	Ruminating and grazing	Laboratory environment + field experiment (Argentina)	Unknown	Unknown (protected under the international patent application)	An electronic system for real-time monitoring of feeding patterns in dairy cows. (Based on an embedded circuit with the algorithm) sound analysis + GPS tracker + long operation time. Classification: 78% (not tested for robustness yet).
Vanrell et al. (2018)	Grazing cattle	Foraging behaviour	Dairy facility in the Kellogg Biological Station (Michigan State University)	No	No	Algorithm proposed for long-term analysis of foraging behaviour (grazing and rumination bouts and resting bouts) 0.962, 0.891, 0.935 for segmentation, rumination classification, grazing classification and a narrow error distribution -> can be used in practical applications but needs lots more testing.
Behaviour-related vocalizations						
Meen et al. (2015)	Holstein	linking vocalization and behaviour	high production dairy farm (the Netherlands)	No	No	Determine if there is a correlation between cattle vocalizations and cattle behaviour. (Done by using camera and sound.) Significant difference in mean max frequencies (Hz) of different behaviours.

8.1 Overview

Reviewed papers that mentioned sound monitoring for pigs, poultry and cattle:

Species	# Papers	# Commercial scale	# Validated	# Commercially available
Pigs	32	16	5	4
Poultry	13	6	5	1
Cattle	7	4	0	0

9 Conclusion

This chapter was an effort to show the status of developments in acoustic techniques to assess livestock welfare and health. We can conclude that automated and continuous sound analysis is a very interesting technique to be used for livestock monitoring. Some of the advantages are as follows: relatively low cost and reliability of the sound monitoring devices in the harsh environment of livestock houses, no light is required, a limited amount of sensors can monitor a high number of animals and several functionalities can be selected from the sound produced by animals as shown in the chapter. At the same time, we have noted that most developments, testing and validations are so far only done in laboratory or research environments. The big challenge is to bring the technology in an accurate, reliable and affordable way to the farm, and not many companies are successful in realizing this product and service. This will take the required time, but the shown examples in this chapter are only the tip of the iceberg of the huge potential that animals sound analysis has in near future.

10 Where to look for further information

10.1 Review articles

- Review: Precision livestock farming: building 'digital representations' to bring the animals closer to the farmer. Anima l(2019),13:12, pp 3009-3017 © The Animal Consortium 2019 https://doi.org/10.1017/S1751731119 00199X3009.
- Editorial: Precision livestock farming: a 'per animal' approach using advancedmonitoringtechnologies.https://doi.org/10.1017/S1751731116 001142.
- Digital Livestock Farming. https://doi.org/10.1016/j.sbsr.2021.100408.

10.2 Websites

- M3-BIORES KULeuven: https://www.biw.kuleuven.be/biosyst/a2h/m3-biores.
- SoundTalks NV: https://www.soundtalks.com/.
- EAAP: European Federation of Animal Science: https://www.eaap.org/.
- VetmedUni PLF Hub: https://www.vetmeduni.ac.at/plf-hub.
- Wageningen Livestock Research: https://www.wur.nl/en/Research-Results/Research-Institutes/livestock-research.htm.
- ILVO: https://ilvo.vlaanderen.be/en/dossiers/precision-livestock-farming.

10.3 Journals

- Animal: The international journal of animal Bioscience. https://www.journals.elsevier.com/animal.
- COMPAG: Computers and electronics in agriculture. https://www.journals.elsevier.com/computers-and-electronics-in-agriculture.
- Frontiers in Animal Science: https://www.frontiersin.org/journals/animal-science/sections/precision-livestock-farming.

10.4 Conferences

- ECPLF: European Conference on Precision Livestock Farming.
- PDC: International Conference on Precision Dairy Farming.

10.5 International research projects

- Bright Farm by Precision Livestock Farming: https://cordis.europa.eu/project/id/311825/reporting.
- SmartAgriHubs: https://www.smartagrihubs.eu/.
- ICT-Agri-Food network (ERA-NET Cofund): https://ictagrifood.eu/.
- Healthy Livestock: https://healthylivestock.net/.
- Livestock sense: https://livestocksense.eu/livestocksense-the-applicability-of-digital-technologies-in-livestock-farming/.

11 References

Aerts, J.-M., Jans, P., Halloy, D., Gustin, P. and Berckmans, D. (2005). Labeling of cough data from pigs for on-line disease monitoring by sound analysis. *Transactions of the ASAE* 48(1), 351–354. https://doi.org/10.13031/2013.17948.

Austic, R. E. and Nesheim, M. C. (1990). *Poultry Production* (13th edn.). Philadelphia: Lea & Febiger.

Aydin, A., Bahr, C., Viazzi, S., Exadaktylos, V., Buyse, J. and Berckmans, D. (2014). A novel method to automatically measure the feed intake of broiler chickens by sound technology. *Computers and Electronics in Agriculture* 101, 17-23. https://doi.org/10.1016/j.compag.2013.11.012.

Aydin, A. and Berckmans, D. (2016). Using sound technology to automatically detect the short-term feeding behaviours of broiler chickens. *Computers and Electronics in Agriculture* 121, 25-31. https://doi.org/10.1016/j.compag.2015.11.010.

Banakar, A., Sadeghi, M. and Shushtari, A. (2016). An intelligent device for diagnosing avian diseases: Newcastle, infectious bronchitis, avian influenza. *Computers and Electronics in Agriculture* 127, 744-753. https://doi.org/10.1016/j.compag.2016.08.006.

Barfield, C. H., Tang-Martinez, Z., and Trainer, J. M. (1994). Domestic calves (*Bos taurus*) recognize their own mothers by auditory cues. *Ethology* 97(4), 257-264.

Berckmans, D. (2014). Precision livestock farming technologies for welfare management in intensive livestock systems. *Revue Scientifique et Technique* 33(1), 189-196.

Berckmans, D., Hemeryck, M., Berckmans, D., Vranken, E. and van Waterschoot, T. (2015). Animal sound… talks! Real-time sound analysis for health monitoring in livestock - SoundTalks. In: *Proceedings of the Animal Environment and Welfare* (pp. 215-222). Available at: https://www.soundtalks.com/paper/animal-sound-talks-real-time-sound-analysis-for-health-monitoring-in-livestock?fbclid=IwAR0BFm3RVVP9brQaq79G8iCPemZT7DOmbBrcD8kPNsa3-o2T_tarn5dO550.

Blackshaw, J. K., Jones, D. N. and Thomas, F. J. (1996). Vocal individuality during suckling in the intensively housed domestic pig. *Applied Animal Behaviour Science* 50(1), 33-41. https://doi.org/10.1016/0168-1591(96)01074-X.

Borell, E., Bünger, B., Schmidt, T. and Horn, T. (2009). Vocal-type classification as a tool to identify stress in piglets under on-farm conditions. *Animal Welfare* 18(4). Available at: https://www.researchgate.net/publication/263484246_Vocal-type_classification_as_a_tool_to_identify_stress_in_piglets_under_on-farm_conditions.

Brackenbury, J. H. (1978). Respiratory mechanics of sound production in chickens and Geese. *Journal of Experimental Biology* 72(1), 229-250.

Bugden, S. C. and Evans, R. M. (1999). The development of vocal thermoregulatory response to temperature in embryos of the domestic chicken. *Wilson Bulletin* 111(2), 188-194.

Carpentier, L., Berckmans, D., Youssef, A., Berckmans, D., van Waterschoot, T., Johnston, D., Ferguson, N., Earley, B., Fontana, I., Tullo, E., Guarino, M., Vranken, E. and Norton, T. (2018). Automatic cough detection for bovine respiratory disease in a calf house. *Biosystems Engineering* 173, 45-56. https://doi.org/10.1016/j.biosystemseng.2018.06.018.

Carpentier, L., Vranken, E., Berckmans, D., Paeshuyse, J. and Norton, T. (2019). Development of sound-based poultry health monitoring tool for automated sneeze detection. *Computers and Electronics in Agriculture* 162, 573-581. https://doi.org/10.1016/j.compag.2019.05.013.

Chedad, A., Moshou, D., Aerts, J. M., Van Hirtum, A., Ramon, H. and Berckmans, D. (2001). Recognition system for pig cough based on probabilistic neural networks. *Agricultural Engineering Research* 79(4), 449-457.

Clapham, W. M., Fedders, J. M., Beeman, K. and Neel, J. P. S. (2011). Acoustic monitoring system to quantify ingestive behavior of free-grazing cattle. *Computers and*

Electronics in Agriculture 76(1), 96–104. https://doi.org/10.1016/j.compag.2011.01.009.

Da Silva, M. K. O., Borges, G., Menegale, V. L. D. C. and Da Silva, I. J. O. (2012). Effect of environmental conditions on sound levels emitted by piglets. *Engenharia Agrícola* 32(3), 435–445.

Damasceno, F. A., Oliveira, C. E. A., Ferraz, G. AeS., Nascimento, J. A. Cd, Saraz, J. A. O. and Brandão, L. F. (2019). Welfare and spatial distribution of noise levels in swine nursery. *Revista Brasileira de Engenharia Agrícola e Ambiental* 23(3), 196–202.

Deniz, N. N., Chelotti, J. O., Galli, J. R., Planisich, A. M., Larripa, M. J., Leonardo Rufiner, H. and Giovanini, L. L. (2017). Embedded system for real-time monitoring of foraging behavior of grazing cattle using acoustic signals. *Computers and Electronics in Agriculture* 138, 167–174. https://doi.org/10.1016/j.compag.2017.04.024.

Du, X., Lao, F. and Teng, G. (2018). A sound source localisation analytical method for monitoring the abnormal night vocalisations of poultry. *Sensors* 18(9), 14. https://doi.org/10.3390/s18092906.

Düpjan, S., Schön, P.-C., Puppe, B., Tuchscherer, A. and Manteuffel, G. (2008). Differential vocal responses to physical and mental stressors in domestic pigs (*Sus scrofa*). *Applied Animal Behaviour Science* 114(1-2), 105–115.

Exadaktylos, V., Silva, M., Aerts, J.-M., Taylor, C. J. and Berckmans, D. (2008a). Real-time recognition of sick pig cough sounds. *Computers and Electronics in Agriculture* 63(2), 207–214. https://doi.org/10.1016/j.compag.2008.02.010.

Exadaktylos, V., Silva, M., Ferrari, S., Guarino, M., Taylor, C. J., Aerts, J. M. and Berckmans, D. (2008b). Time-series analysis for online recognition and localization of sick pig (Sus scrofa) cough sounds. *The Journal of the Acoustical Society of America* 124(6), 3803–3809. https://doi.org/10.1121/1.2998780.

Feltenstein, M. W., Ford, N. G., Freeman, K. B. and Sufka, K. J. (2002). Dissociation of stress behaviors in the chick social-separation-stress procedure. *Physiology and Behavior* 75(5), 675–679.

Ferrari, S., Costa, A. and Guarino, M. (2013a). Heat stress assessment by swine related vocalizations. *Livestock Science* 151, 29–34. https://doi.org/10.1016/j.livsci.2012.10.013.

Ferrari, S., Silva, M., Exadaktylos, V., Berckmans, D. and Guarino, M. (2013b). The sound makes the difference: The utility of real time sound analysis for health monitoring in pigs. In: Aland, A. and Banhazi, T. (Eds) *Livestock Housing: Modern Management to Ensure Optimal Health and Welfare of Farm Animals* (pp. 407–417). https://doi.org/10.3920/978-90-8686-771-4 (eISBN: 978-90-8686-771-4 | ISBN: 978-90-8686-217-7).

Ferrari, S., Silva, M., Guarino, M., Aerts, J. M. and Berckmans, D. (2008a). Cough sound analysis to identify respiratory infection in pigs. *Computers and Electronics in Agriculture* 64(2), 318–325. https://doi.org/10.1016/j.compag.2008.07.003.

Ferrari, S., Silva, M., Guarino, M. and Berckmans, D. (2008b). Analysis of cough sounds for diagnosis of respiratory infections in intensive pig farming. *Transactions of the ASABE* 51(3), 1051–1055. https://doi.org/10.13031/2013.24524.

Ferrari, S., Silva, M., Sala, V., Berckmans, D. and Guarino, M. (2009). Bioacoustics: A tool for diagnosis of respiratory pathologies in pig farms. *Journal of Agricultural Engineering* 40(1), 7. https://doi.org/10.4081/jae.2009.1.7.

Flower, F. C. and Weary, D. M. (2001). Effects of early separation on the dairy cow and calf: 2. Separation at 1 day and 2 weeks after birth. *Applied Animal Behaviour Science* 70(4), 275–284.

Fontana, I., Tullo, E., Butterworth, A. and Guarino, M. (2015). An innovative approach to predict the growth in intensive poultry farming. *Computers and Electronics in Agriculture* 119, 178–183. https://doi.org/10.1016/j.compag.2015.10.001.

Fontana, I., Tullo, E., Carpentier, L., Berckmans, D., Butterworth, A., Vranken, E., Norton, T., Berckmans, D. and Guarino, M. (2017). Sound analysis to model weight of broiler chickens. *Poultry Science* 96(11), 3938–3943. https://doi.org/10.3382/ps/pex215.

Fraser, D. (1974). The vocalizations and other behaviour of growing pigs in an "open field" test. *Applied Animal Ethology* 1(1), 3–16.

Galli, J. R., Cangiano,C. A., Pece, M. A., Larripa, M. J., Milone, D. H., Utsumi, S. A. and Laca, E. A., Utsumi, S. A. and Laca, E. A. (2018). Monitoring and assessment of ingestive chewing sounds for prediction of herbage intake rate in grazing cattle. *Animal* 12(5), 973-982. https://doi.org/10.1017/S1751731117002415.

Ghesquire, K., Van Hirtum, A., Buyse, J. and Berckmans, D. (2002). Non-invasive quantification of stress in the laying hen: A vocalization analysis approach. *Mededelingen Faculteit Landbouwkundige en Toegepaste Biologische Wetenschappen Universiteit Gent* 67(4), 55–58.

Grandin, T. (1998). The feasibility of using vocalization scoring as an indicator of poor welfare during cattle slaughter. *Applied Animal Behaviour Science* 56(2–4), 121–128.

Grandin, T. (2001). Cattle vocalizations are associated with handling and equipment problems at beef slaughter plants. *Applied Animal Behaviour Science* 71(3), 191-201.

Green, A. C., Johnston, I. N. and Clark, C. E. F. (2017). Invited review: The evolution of cattle bioacoustics and application for advanced dairy systems. *Animal* 12(6), 1250-1259. https://doi.org/10.1017/S1751731117002646.

Gutierrez, W. M., Kim, S., Kim, D. H., Yeon, S. C. and Chang, H. H. (2010). Classification of porcine wasting diseases using sound analysis. *Asian-Australasian Journal of Animal Sciences* 23(8), 1096–1104. https://doi.org/10.5713/ajas.2010.90483.

Hall, S. J. G., Vince, M. A., Shillito Walser, E. and Garson, P. J. (1988). Vocalisations of the chillingham cattle. *Behaviour* 104(1–2), 78–104.

Hillmann, E., Mayer, C., Schön, P. C., Puppe, B. and Schrader,L. (2004). Vocalisation of domestic pigs (*Sus scrofa domestica*) as an indicator for their adaptation toward ambient temperatures. *Applied Animal Behaviour Science* 89(3–4), 195–206.

Huang, J., Wang, W. and Zhang, T. (2019). Method for detecting avian influenza disease of chickens based on sound analysis. *Biosystems Engineering* 180, 16–24.

Ignjatović, J. and Sapats, S. (2000). Avian infectious bronchitis virus. *Revue Scientifique et Technique* 19(2), 493–508. http://doi.org/10.20506/rst.19.2.1228.

Ikeda, Y. and Ishii, Y. (2001). Characteristics of cow's voices in time and frequency domains for recognition. *Agriculture Biosystem Engineering* 2, 15–23.

Illmann, G., Hammerschmidt, K., Špinka, M. and Tallet, C. (2013). Calling by domestic piglets during simulated crushing and isolation: A signal of need? *PLoS ONE* 8(12), e83529. https://doi.org/10.1371/journal.pone.0083529.

Jahns, G., Kowalczyk, W. and Walter, K. (1997a). An application of sound processing techniques for determining condition of cows. In: Domañski, M. and Stasiñski, R. (Eds) *Proceedings of the Fourth International Workshop on Systems, Signals and Image Processing* (pp. 105–108), Poznan, Poland.

Jahns, G., Kowalczyk, W. and Walter, K. (1997b). Sound analysis to recognize different animals. *IFAC Proceedings Volumes*. In: Munack, A. and Tantau, H. J. (Eds) *Proceedings of the 3rd IFAC Workshop on Mathematical and Control Applications in Agriculture and Horticulture: Mathematical and Control Applications in Agriculture and Horticulture* 30(26), 169–173.

Kiley, M. (1972). The vocalizations of ungulates, their causation and function. *Zeitschriftfür Tierpsychologie* 31(2), 171–222.

Lee, J., Noh, B., Jang, S., Park, D., Chung, Y. and Chang, H. H. (2015). Stress detection and classification of laying hens by sound analysis. *Asian-Australasian Journal of Animal Sciences* 28(4), 592–598. https://doi.org/10.5713/ajas.14.0654.

Liu, L., Li, B., Zhao, R., Yao, W., Shen, M. and Yang, J. (2020). A novel method for broiler abnormal sound detection using WMFCC and HMM. *Journal of Sensors* 2020, 1-7. https://doi.org/10.1155/2020/2985478.

Mahdavian, A., Minaei, S., Yang, C., Almasganj, F., Rahimi, S. and Marchetto, P. M. (2020). Ability evaluation of a voice activity detection algorithm in bioacoustics: A case study on poultry calls. *Computers and Electronics in Agriculture* 168,, 105100. https://doi.org/10.1016/j.compag.2019.105100.

Maigrot, A. L., Hillmann, E. and Briefer, E. F. (2018). Encoding of emotional valence in wild boar (Sus scrofa) calls. *Animals: An Open Access Journal from MDPI* 8(6), 1-15. https://doi.org/10.3390/ani8060085.

Mair, J., Marx, G., Petersen, J. and Mennicken, L. (2001). Development of multi parametric sound analysis parameters as welfare indicator in chicken flocks. *Proceedings of the International Seminar on Modal Analysis* 3, 1523–1528.

Manteuffel, G.,Puppe, B. and Schon, P. C. (2004). Vocalization of farm animals as a measure of welfare. *Applied Animal Behaviour Science* 88(1–2), 163–182.

Marchant, J. N., Whittaker, X. and Broom, D. M. (2001). Vocalisations of the adult female domestic pig during a standard human approach test and their relationships with behavioural and heart rate measures. *Applied Animal Behaviour Science* 72(1), 23-39. https://doi.org/10.1016/S0168-1591(00)00190-8.

Marchant-Forde, J. N., Marchant-Forde, R. M. and Weary, D. M. (2002). Responses of dairy cows and calves to each other's vocalisations after early separation. *Applied Animal Behaviour Science* 78(1), 19–28.

Marler, P. and Evans, C. (1996). Bird calls: Just emotional displays or something more? *Ibis* 138(1), 26–33.

Marx, G., Horn, T., Thielebein, J., Knubel, B. and Von Borell, E. (2003). Analysis of pain-related vocalization in young pigs. *Journal of Sound and Vibration* 266(3), 687–698. https://doi.org/10.1016/S0022-460X(03)00594-7.

Marx, G., Leppelt, J. and Ellendorff, F. (2001). Vocalisation in chicks (Gallus gallus dom.) during stepwise social isolation. *Applied Animal Behaviour Science* 75(1), 61–74. https://doi.org/10.1016/S0168-1591(01)00180-0.

McMullin, P. (2004). *A Pocket Guide to Poultry Health and Disease*. Sheffield.

Meen, G. H., Schellekens, M. A., Slegers, M. H. M., Leenders, N. L. G., van Erp-van der Kooij, E. and Noldus, L. P. J. J. (2015). Sound analysis in dairy cattle vocalisation as a potential welfare monitor. *Computers and Electronics in Agriculture* 118, 111–115. https://doi.org/10.1016/j.compag.2015.08.028.

Moshou, D., Chedad, A., Van Hirtum, A., De Baerdemaeker, J., Berckmans, D. and Ramon, H. (2001). Neural recognition system for swine cough. *Mathematics and Computers in Simulation* 56(4-5), 475-487. https://doi.org/10.1016/S0378-4754(01)00316-0.

Moura, D. J., Nääs, IdA., Alves, E. CdS., Carvalho, T. M. Rd, Vale, M. Md. and Lima, K. A. Od. (2008a). Noise analysis to evaluate chick thermal comfort. *Scientia Agricola* 65(4), 438-443.

Moura, D. J., Silva, W. T., Naas, I. A., Tolón, Y. A., Lima, K. A. O. and Vale, M. M. (2008b). Real time computer stress monitoring of piglets using vocalization analysis. *Computers and Electronics in Agriculture* 64(1), 11-18. https://doi.org/10.1016/j.compag.2008.05.008.

Muller, R. and Schrader, L. (2005). Behavioural consistency during social separation and personality in dairy cows. *Behaviour* 142(9-10), 1289-1306. https://doi.org/10.1163/156853905774539346.

Padilla de la Torre, M., Briefer, E. F., Ochocki, B. M., McElligott, A. G. and Reader, T. (2016). Mother-offspring recognition via contact calls in cattle (Bos taurus). *Animal Behaviour* 114, 147-154. https://doi.org/10.1016/j.anbehav.2016.02.004.

Pereira, E. M., Naas, I. A. and Jacob, F. C. (2011). Using vocalization pattern to assess broiler's well-being. In: Lokhorst, K. and Berckmans, D. (Eds) *Precision Livestock Farming*, July 2011, Prague, Czech Republic (Vol. 11, pp. 11-14).

Polson, D. D., Playter, S., Berckmans, D., Cui, Z. Y., Quinn, B., Genzow, M. and Duran, O. (2018). Characterization of temporal and spatial patterns of clinical respiratory episodes in growing pigs using continuous sound monitoring and an algorithm-based respiratory distress index - SoundTalks. In: *The 15th International Symposium of Veterinary Epidemiology and Economics*. Available at: https://www.soundtalks.com/paper/characterization-of-temporal-and-spatial-patterns-of-clinical-respiratory-episodes-in-growing-pigs-using-continuous-sound-monitoring-and-an-algorithm-based-respiratory-distress-index/?fbclid=IwAR0IN0Y1IEUn7eNfhx5mykzI6SKvf.

Puppe, B., Schön, P. C., Tuchscherer, A. and Manteuffel, G. (2003). The influence of domestic piglets' (Sus scrofa) age and test experience on the preference for the replayed maternal nursing vocalisation in a modified open-field test. *Acta Ethologica* 5(2), 123-129.

Rushen, J., Boissy, A., Terlouw, E. M. C. and De Passillé, A. M. B. (1999). Opioid peptides and behavioural and physiological responses of dairy cows to social isolation in unfamiliar surroundings. *Journal of Animal Science* 77(11), 2918-2924.

Sadeghi, M., Banakar, A., Khazaee, M. and Soleimani, M. (2015). An intelligent procedure for the detection and classification of chickens infected by *Clostridium perfringens* based on their vocalization. *Revista Brasileira de Ciência Avícola* 17(4), 537-544.

Schön, P. C., Puppe, B. and Manteuffel, G. (2001). Linear prediction coding analysis and self-organizing feature map as tools to classify stress calls of domestic pigs (Sus scrofa). *The Journal of the Acoustical Society of America* 110(3 Pt 1), 1425-1431. https://doi.org/10.1121/1.1388003.

Schön, P. C., Puppe, B. and Manteuffel, G. (2004). Automated recording of stress vocalisations as a tool to document impaired welfare in pigs. *Animal Welfare* 13, 105-110.

Schön, P. C., Puppe, B., Tuchscherer, A. and Manteuffel, G. (2006). Changes of the vocalization during the castration of the domestic pig are indicators of pain. *Zuchtungskunde* 78, 44-54. Available at: https://www.researchgate.net/publication

/285936176_Changes_of_the_vocalization_during_the_castration_of_the_domestic_pig_are_indicators_of_pain.
Schrader, L. and Hammerschmidt, K. (1997). Computer-aided analysis of acoustic parameters in animal vocalisation: A multi-parametric approach. *Bioacoustics* 7(4), 247-265.
Schrader, L. and Todt, D. (1998). Vocal quality is correlated with levels of stress hormones in domestic pigs lars. *Ethology* 105, 859-876.
Silva, M., Exadaktylos, V., Ferrari, S., Guarino, M., Aerts, J. M. and Berckmans, D. (2009). The influence of respiratory disease on the energy envelope dynamics of pig cough sounds. *Computers and Electronics in Agriculture* 69(1), 80-85. https://doi.org/10.1016/j.compag.2009.07.002.
Silva, M., Ferrari, S., Costa, A., Aerts, J.-M., Guarino, M. and Berckmans, D. (2008). Cough localization for the detection of respiratory diseases in pig houses. *Computers and Electronics in Agriculture* 64(2), 286-292. https://doi.org/10.1016/j.compag.2008.05.024.
Spickler, A. R. (2015). Avian Influenza: Technical factsheet. Available at: https://www.cfsph.iastate.edu/Factsheets/pdfs/highly_pathogenic_avian_influenza.pdf.
Stadelman, W. J. (1958). The effect of sounds of varying intensity on hatchability of chicken egg. *Poultry Science* 37(1), 166-169.
Thomas, T. J., Weary, D. M. and Appleby, M. C. (2001). Newborn and 5-week-old calves vocalize in response to milk deprivation. *Applied Animal Behaviour Science* 74(3), 165-173.
Van Hirtum, A. and Berckmans, D. (2002). Assessing the sound of cough towards vocality. *Medical Engineering and Physics* 24(7-8), 535-540.
Van Hirtum, A. and Berckmans, D. (2003). Fuzzy approach for improved recognition of citric acid induced piglet coughing from continuous registration. *Journal of Sound and Vibration* 266(3), 677-686. https://doi.org/10.1016/S0022-460X(03)00593-5.
Vandermeulen, J., Bahr, C., Johnston, D., Earley, B., Tullo, E., Fontana, I., Guarino, M., Exadaktylos, V. and Berckmans, D. (2016). Early recognition of bovine respiratory disease in calves using automated continuous monitoring of cough sounds. *Computers and Electronics in Agriculture* 129, 15-26. https://doi.org/10.1016/j.compag.2016.07.014.
Vandermeulen, J., Bahr, C., Tullo, E., Fontana, I., Ott, S., Kashiha, M., Guarino, M., Moons, C. P., Tuyttens, F. A., Niewold, T. A. and Berckmans, D. (2015). Discerning pig screams in production environments. *PLoS ONE* 10(4), e0123111. https://doi.org/10.1371/journal.pone.0123111.
Vanrell, S. R., Chelotti, J. O., Galli, J. R., Utsumi, S. A., Giovanini, L. L., Rufiner, H. L. and Milone, D. H. (2018). A regularity-based algorithm for identifying grazing and rumination bouts from acoustic signals in grazing cattle. *Computers and Electronics in Agriculture* 151, 392-402. https://doi.org/10.1016/j.compag.2018.06.021.
Volkmann, N., Kulig, B. and Kemper, N. (2019). Using the footfall sound of dairy cows for detecting claw lesions. *Animals: An Open Access Journal From MDPI* 9(3), 11. https://doi.org/10.3390/ani9030078.
Von Borell, E., Bünger, B., Schmidt, T. and Horn, T. (2009). Vocal-type classification as a tool to identify stress in piglets under on-farm conditions. *Animal Welfare* 18, 407-416.

Von Borell, E. and Ladewig, J. (1992). Relationship between behaviour and adrenocortical response pattern in domestic pigs. *Applied Animal Behaviour Science* 34(3), 195–206.

Wang, X., Zhao, X., He, Y. and Wang, K. (2019). Cough sound analysis to assess air quality in commercial weaner barns. *Computers and Electronics in Agriculture* 160, 8–13. https://doi.org/10.1016/j.compag.2019.03.001.

Watts, J. M. and Stookey, J. M. (2000). Vocal behaviour in cattle: The animal's commentary on its biological processes and welfare. *Applied Animal Behaviour Science* 67(1–2), 15–33.

Watts, J. M. and Stookey, J. M. (2001). The propensity of cattle to vocalise during handling and isolation is affected by phenotype. *Applied Animal Behaviour Science* 74(2), 81–95.

Weary, D. M., Braithwaith, L. A. and Fraser, D. (1998). Vocal response to pain in piglets. *Applied Animal Behaviour Science* 56(2–4), 161–172.

Weary, D. M. and Chua, B. (2000). Effects of early separation on the dairy cow and calf. 1. Separation at 6 h, 1 day and 4 days after birth. *Applied Animal Behaviour Science* 69(3), 177–188.

Wegner, B., Spiekermeier, I., Nienhoff, H., Große-Kleimann, J., Rohn, K., Meyer, H., Plate, H., Gerhardy, H., Kreienbrock, L., grosse Beilage, E., Kemper, N. and Fels, M. (2019). Status quo analysis of noise levels in pig fattening units in Germany. *Livestock Science* 230, 103847. https://doi.org/10.1016/j.livsci.2019.103847.

White, R. G., DeShazer, J. A., Tressler, C. J., Borcher, G. M., Davey, S., Waninge, A., Parkhurst, A. M., Milanuk, M. J. and Clemens, E. T. (1995). Vocalization and physiological response of pigs during castration with or without a local anesthetic. *Journal of Animal Science* 73(2), 381–386.

Zhuang, J., Zhao, L., Gao, X. and Xu, F. (2019). An advanced recording and analysis system for the differentiation of guinea pig cough responses to citric acid and prostaglandin E2 in real time. *PLoS ONE* 14(5), e0217366. https://doi.org/10.1371/journal.pone.0217366.

Chapter 4

Machine vision techniques to monitor behaviour and health in precision livestock farming

C. Arcidiacono and S. M. C. Porto, University of Catania, Italy

1 Introduction

Automated visual analysis of livestock in the farm environment has developed rapidly to overcome the limitations of traditional human-based procedures [1-4]. Direct observation of the herd - even by skilled operators - suffers from a number of problems, including:

http://dx.doi.org/10.19103/AS.2021.0090.04

- those related to the presence of the observer, which may affect and thus modify animal behaviour; and
- challenges in the visual recognition of animal behaviour, which is time consuming and subject to variability of interpretation and human error.

To avoid these problems, automated monitoring of livestock animals has been used for various tasks such as:

- monitoring an animal's health status;
- early detection of diseases;
- welfare assessment;
- observation of behaviour; and
- animal identification.

All of these tasks aim at reducing the need for human intervention and improving the efficiency of livestock management. Monitoring systems have evolved with developments in technology, such as the development of increasingly precise and miniaturised internet of things (IoT) sensors and devices. The emergence of artificial intelligence and big data techniques has made it possible to analyse large amounts of data from IoT devices and cameras.

The advantages of computer vision (CV)-based systems over other sensor-based systems relate to their non-invasive nature. They do not require devices to be fitted on animals, do not obstruct animal movement and are not vulnerable to damage by animals. They are also independent of any constraints that usually result from the way physical devices are mounted on an animal. The typical vision tasks include [5]:

- recognition of environments and objects;
- motion and pose assessment; and
- behaviour analysis.

A variety of techniques have been developed to achieve these objectives. This chapter reviews advances in CV-based technologies for precision livestock farming. It also reviews how automation in image analysis can promote smart management of livestock housing. The chapter discusses the main devices for data acquisition in CV-based systems and the range of tasks that CV techniques can perform. It reviews key steps such as initialisation, tracking, pose detection and recognition. The chapter includes illustrative case studies of precision livestock farming (PLF) applications based on existing CV techniques. The chapter concludes by reviewing the recent advances in CV techniques, including cases based on artificial neural network (ANN) techniques, and future challenges.

2 Devices for data acquisition in computer vision-based systems

A fundamental activity in all CV techniques is data acquisition. The choice of hardware depends on a number of conditions, including:

- the environment to be filmed;
- the animal species to be monitored;
- the task to be performed;
- the location of the cameras; and
- the micro-environmental variables (e.g. light conditions and dust levels).

In the case of pigs, for example, a view from above is preferred, while for dairy cattle, a side view is quite often used. However, in side-view images, the background in the image is not stable, which is considered a disadvantage. As a result, even an easy-to-calculate feature from a side view (e.g. the back arch of a cow) is better undertaken using an additional top-view image due to the stable background [6].

The hardware used in CV systems is based on 2D and 3D cameras. Two-dimensional cameras are cheaper and can provide high-resolution images. However, 2D cameras are vulnerable to poor image quality in conditions of low light or excessive brightness. This is a significant problem in livestock buildings, which are subject to wide variations in light conditions. They are also vulnerable to occlusions caused, for example, by dirt, insects or deterioration of the camera's protective glass. This is the case in livestock environments where dust from feed, corrosive gases from livestock manure emissions, mud, flies etc. are likely to damage the hardware components. Although much more expensive than 2D cameras, and thus less used in commercial barns, 3D cameras have significant advantages as they are less affected by changes in brightness and occlusions and can work in darker environments.

A key feature in measuring depth is separation of the animal's shape from the background for image segmentation. The kind of camera capable of handling this is currently used in research studies for estimating the weight of large animals and for pose detection. Recent studies [7, 8] have investigated the use of Kinect sensors using input devices for motion detection, developed in 2010 by Microsoft, for gaming. The Kinect sensor is composed of RGB cameras, infrared projectors and detectors for map depth through either structured light or time-of-flight calculations, and a microphone array. Specific software and artificial intelligence perform real-time gesture, speech and body skeletal recognition. Kinect has found unexpected application in scientific and commercial applications (e.g. in robotics, medicine and health care), as the Kinect sensor is cheaper and has proved more robust than other depth-sensing technology applications. It has been shown that a 3D camera capturing top-view images in a livestock setting has a high level of accuracy compared to a 2D camera capturing side-view images [9]. Single-frame photographic images and video images are profoundly different types of data. While the former are static, video images are dynamic, particularly in analysing the action of a subject. Tasks such as pose estimation, object tracking or movement require the use of video equipment. Working with video equipment requires higher computational power and costs more compared to working with single-frame images.

3 Animal species and tasks analysed in computer vision systems for precision livestock farming

Most research applying CV techniques has been focused on three animal species: pigs, poultry and dairy cattle. Different animals show different behavioural characteristics, and above all, very different physical characteristics. This affects the choice of hardware and software solutions in CV systems. The positioning of cameras also depends on the target species, and different tasks are associated with different species of animals. The tasks generally addressed by CV techniques are:

1. **Reproduction and mounting detection:** Tasks related to reproduction include the identification of optimal time for artificial insemination, oestrus detection [10], and identifying difficulties in calving [11, 12] that might require human assistance. Mounting detection is important because such intimate, continuous and intense interactions between animals can generate serious health problems such as bruises, fractures, lameness or psychological distress. This makes it important to detect and intervene in such behaviour if it becomes dangerous and causes health issues and economic losses [13].
2. **Lameness and movement tracking:** Lameness is usually a manifestation of pain in animals. A lame animal moving with an irregular gait could be affected by a serious disease, making early detection of lameness necessary for the detection of conditions that could seriously compromise the health of the animal. Computer vision techniques can be critical, as the human eye may not detect gait characteristics indicating lameness until a disease is at a more advanced stage and becomes more difficult to treat [7, 9, 12, 14–18].
3. **Daily management and health monitoring**: These tasks include identification [19-22], water consumption monitoring [23], nutrition monitoring [22] and other activities related to animal welfare monitoring [24].

4 **Detection of aggressive behaviour**: The timely identification of aggressive behaviours is an important goal, considering both the economic aspects and the animal's health. One of the most analysed aspects of aggressive behaviour concerns pigs. Pigs are hierarchical animals, and the practice of mixing animals, carried out in intensive farming, favours aggressive interactions within the groups - that last until hierarchical dominance is established in the groups. Tail biting, for instance, has been extensively studied since this behaviour can result in infection, welfare and reproduction problems [25, 26].
5 **Weight estimation:** Estimating the weight of animals is a key aspect of monitoring their health status. This activity is challenging if managed manually, especially for large animals such as pigs. Computer vision techniques can speed up the whole process and allow regular measurement [8, 18, 27–29].

These tasks affect the way a CV-based system is constructed.

4 Key elements of computer vision-based systems: initialisation

The fundamental steps (Fig. 1) for the construction of a system based on artificial vision include [30]:

- initialisation;
- tracking;
- pose estimation; and
- recognition.

These steps are based on developing a CV system to capture human movement but can form the basis for a CV-based system for precision livestock farming

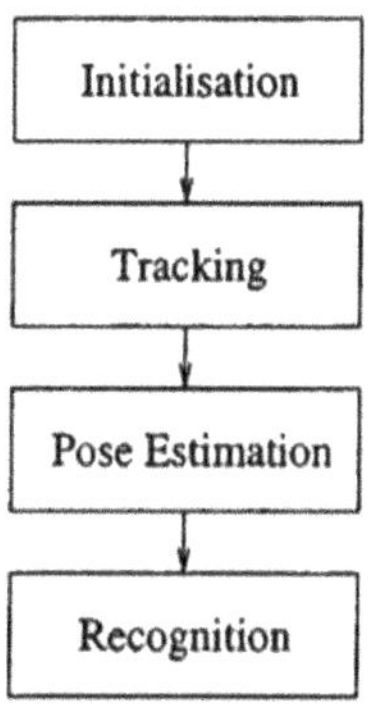

Figure 1 Computer vision system, 'A Survey of Computer Vision-Based Human Motion Capture' [30].

tasks. These steps are discussed in this and the following sections, starting with initialisation.

The initialisation phase includes all the preliminary actions needed for the system to be able to capture livestock activity, including hardware and software modifications. In terms of hardware, initialisation includes the calibration of devices used to acquire data. Calibration involves parameters such as contrast, exposure, the position of the camera, distance to the subject and the type of lens [31] needed to optimise image quality. This type of activity is generally carried out offline. Software initialisation involves calibration of software parameters to correctly perform tasks such as tracking (segmentation), pose estimation and recognition. There are two fundamental aspects of software calibration:

- the initial pose of a subject; and
- the model representing the subject.

If necessary, stitching is carried out, that is, the images acquired by several cameras or mirror images from one camera are combined, essentially creating a composite set of images of the subject in three-dimensional form.

5 Key elements of computer vision-based systems: tracking – image segmentation

The tracking phase is fundamental for the development of a CV system. It involves defining the relationships between the subject and the frames (individual images). The tracking phase is usually divided into three phases:

1 Segmentation;
2 Feature extraction; and
3 Object tracking.

Tracking the movements of animals cannot be performed if the input data (which can be an image or a video) are not first segmented. Segmentation makes it possible to divide the initial data into parts, making it possible to distinguish the various objects in the scene. After segmentation, object detection is required to locate and distinguish individual objects in the scene or the environment under investigation. After object localisation, feature extraction is carried out with the aim of identifying the fundamental characteristics of the objects which will be used in the next phases. Finally, it may be necessary to track the identified object between consecutive frames (this is essential if, for instance, the task requires tracing movement). Unlike the segmentation and extraction phases which are essential, tracking is only required for movement-related tasks.

Segmentation is a fundamental element in any pattern recognition and CV. It aims to separate the object or subject of interest from the background. In this case, we want to separate the animal from all surrounding objects or subjects (e.g. humans). Researchers have recently invested a lot of time and effort in the problem of image/video data segmentation. The development of increasingly efficient segmentation techniques has had a significant impact on new algorithms for the recognition of patterns and applications.

There are two types of segmentation:

- semantic segmentation; and
- instance segmentation.

Semantic segmentation is a technique that detects, for each pixel, the category of objects to which it belongs. Besides performing the same function as semantic segmentation, instance segmentation identifies, for each pixel, the instance of the object to which it belongs. Instance segmentation basically differentiates two objects belonging to the same category. Brief descriptions of the different segmentation approaches for images and videos are discussed in this and the following sections, as well as related segmentation techniques specific to videos and camera images.

Image segmentation is used to segment parts sharing similar visual and semantic features or attributes to isolate objects of interest from the background. Multiple image segmentation techniques have been proposed, dividing images into multiple parts based on certain image characteristics such as pixel intensity value, colour, texture etc. Image segmentation can be categorized into:

- local segmentation, which affects a specific portion or region of the image; and
- global segmentation, in which segmentation is applied to the entire image.

The type of image to be segmented and the task to be performed are basic elements in choosing the segmentation method to be used. The level of detail at which the segmentation is performed depends on the objectives of the system. In some situations, it may be sufficient to isolate objects belonging to the same category from the background. In other cases, it may be necessary to distinguish between objects belonging to the same class. The accuracy of the segmentation determines the success or failure of subsequent computer analysis.

In general, image segmentation approaches can be based on two properties: discontinuity and similarity. It is possible to distinguish two categories [32, 33] (Fig. 2):

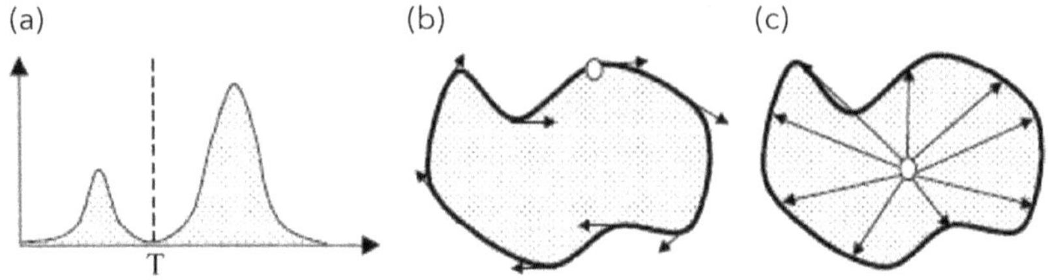

Figure 2 Typical segmentation techniques (a) thresholding, (b) edge-based segmentation and (c) region-based segmentation [67].

- The first category is characterized by segmentation based on the 'discontinuities' in images; that is, the 'segments' or regions are distinguished based on discontinuities. The image is generally segmented based on 'abrupt changes' in intensity. This category includes methods performing **edge-based detection.** These methods assume that the boundaries and edges of regions are closely related, as there is often a sharp adjustment in intensity at regional boundaries.
- The second category is characterized by segmentation based on the 'similarity' between the regions. This category includes **thresholding** techniques, **region-based** techniques (e.g. region growing, region splitting and merging) and **clustering** techniques.

5.1 Edge-based segmentation

Discontinuity-based segmentation methods measure abrupt local changes in the intensity of the images represented in greyscale, for example, changes in terms of brightness, colour and structure. These methods are called edge- or boundary-based methods. Image segmentation methods for detecting discontinuities belong to the category of boundary-based methods. The types of image features of interest are isolated edges, lines and points. Edge pixels are the pixels where the intensity function of the image changes abruptly, while the edge segments are sets of connected edge pixels. Edge detectors are used to identify edge pixels. As an example, the grey level in the edge between two adjacent regions is not continuous, which means this discontinuity can be identified by using differential operators. The best-known first-order differential operators are the Prewitt operator, Roberts operator and Sobel operator, while second-order (non-linear) operators are the Laplacian, Kirsch, and Wallis operators.

5.2 Thresholding techniques

The threshold-based segmentation method is the simplest method of image segmentation and one of the most common segmentation methods. This method relies on a clip layer (or threshold value) to transform a greyscale image into a binary image. The key to this method is to identify the most suitable threshold value (or values when multiple levels are selected). Thresholds can be local or global, depending on the specific context. The application of a global threshold involves a division of the image into two macro-regions:

- the region that includes the target; and
- the background.

In contrast, the use of local thresholds divides the image into multiple target regions and backgrounds. An example of threshold identification is the Otsu method, which selects an optimal threshold at a global level by maximising the interclass variance. The underlying assumption of this method is that only two classes are known and can be discriminated according to the intensity values of their pixels. This method is based on image histograms. The intraclass variance is defined as the weighted sum of the variances of the two classes. The weight represents the probability that the two classes are separated by the threshold t and by the σ^2 variance; hence, minimising intraclass variance maximises interclass variance. The other threshold identification methods are:

- entropy-based threshold segmentation;
- the minimum error method;
- the co-occurrence matrix method;
- the moment preserving method;
- the simple statistical method;
- the probability relaxation method; and
- the fuzzy set method.

Threshold-based segmentation methods are widely used since they are based on simple calculations, reducing computational cost and increasing speed. Threshold-based methods are particularly efficient when the target and the background have high contrast. However, in many real applications, there is seldom a substantial difference between target and background, making this approach less feasible. Examples of PLF applications using thresholding techniques have been reported by several authors [13, 20, 23].

5.3 Region-based techniques

While threshold-based techniques distinguish regions based on discontinuities or colour properties, region-based techniques determine regions directly. The most important region-based techniques are region growing, region merging and spitting:

- *Region growing*: This technique groups pixels or sub-regions into increasingly larger regions (hence the term 'growing'). Sub-regions or pixels can be aggregated only if they meet established criteria. The whole process starts from a set of points defined as 'seeds.' The surrounding pixels are then added – if they have characteristics similar to the seed – to form large regions. If the seed is not available a priori for each pixel, a set of known properties is calculated and used to assign the pixels belonging to the neighbourhood of the seeds to the regions. The criteria used to verify the similarity are based on the particular task and image. Drawbacks include the decision about when to stop the growing process; or the lack of connectivity, in some cases, which could generate meaningless segmentations.
- *Region splitting and merging:* These techniques are also part of the region-based macro category and represent an alternative to region-growing techniques. The basic principle involves the subdivision of an image into a set of arbitrary but disjoint regions and then merging or separating them into regions that satisfy specific segmentation conditions.

5.4 Clustering techniques

These segmentation methods segment images into partitions (or clusters), that is, pixels with similar characteristics are considered to belong to the same cluster. Clustering methods can be divided into two types:

- hierarchical methods; and
- partition-based methods.

While the former are based on the concept of tree hierarchies, the latter generally start with random partitions that are iteratively refined.

The main clustering-based image segmentation algorithms are k-means, fuzzy c-means clustering and expectation maximization (EM). An example of a PLF application using clustering techniques is described in Zhuang et al. [34].

6 Key elements of computer vision-based systems: tracking – video object segmentation

Segmentation and tracking of video objects/subjects are two fundamental activities in the field of CV and their issues are closely related. The task of object segmentation involves subdivision of the pixels in a video frame into two subsets:

- the foreground target; and
- the background region.

The object-segmentation mask, which is necessary for the recognition of the behaviour of the target objects, is then generated. Object tracking aims at determining the exact position of the target object in the video and generates a bounding box, which identifies the object in the scene by spatially delimiting the object in the scene. Generally, video object segmentation is represented by the acronym VOS, while the video object has the acronym VOT. A taxonomy of VOS and tracking has been proposed by Yao et al. [35]. They divided the VOS and tracking methods into five categories:

- unsupervised tracking;
- semi-supervised tracking;
- interactive tracking;
- weakly supervised tracking; and
- segmentation-based tracking.

In precision livestock farming, the most used methods are the unsupervised tracking techniques, particularly the background subtraction and point trajectory methods. Unsupervised VOS methods do not require user input. It is assumed that the objects to be segmented and traced appear frequently in the images or have different appearance times on the scene.

6.1 Background subtraction [35]

One of the most common approaches to segmenting objects is through background modelling/subtraction techniques. Background subtraction techniques are unsupervised VOS techniques and, as previously mentioned, do not need any input (annotation) from the user. The goal of background subtraction techniques is to extract the foreground objects for each video frame by estimating a scene model without target objects and a background model (Fig. 3). Background subtraction thus consists of two fundamental steps:

- construction of the background model (i.e. representation of the scene); and
- a search for deviations intended as moving objects for each frame with respect to the background.

These techniques are based on the key assumption that the background is static. However, this is not always applicable in real scenarios when, as in livestock housing, there are often changes in light intensity, lighted and shadow regions and brightness which the background may change. There are several approaches to background subtraction:

- **Frame difference**: To extract a moving object, this method uses pixel differences between consecutive frames. The advantage of the method

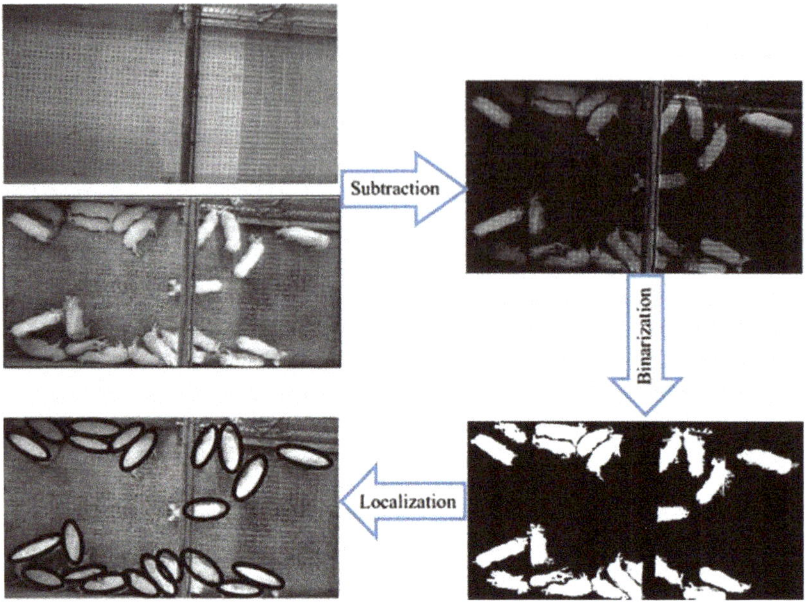

Figure 3 Background subtraction in 'Automatic detection of mounting behaviours among pigs using image analysis' [13].

is that it is adaptable to a dynamic environment, typical of videos where there are moving objects. The disadvantage is that episodes showing holes may occur inside moving objects.

- **Reference image**: This method only works well in some particular speed and frame rate conditions. Background estimation is done only in the first frame. The frames observed are then compared with the background. This comparison involves subtraction between pixels. For each pixel of the frames obtained, it takes the value of the pixel and subtracts the corresponding pixel (in the same position) of the background image. Subsequently, the value obtained from this subtraction is compared with a threshold (Th). Choosing the wrong threshold may not give reliable results; if the scene movement speed is high, a high threshold is needed.
- **Gaussian Mixture Model (GMM)**: This is a method for dynamic backgrounds. It is widely used because it can manage periodic and gradual lighting changes. The main disadvantages of this method are that it is not tolerant to sudden changes in brightness and irregular background movements. Being a parametric model, the performance also depends on the parameters selected. In GMM, the values of a particular pixel over time are modelled with a mixture of Gaussian distributions. The model assumes that pixel values belong to the background if there is no Gaussian distribution that can include them. In this case, they are considered as the

foreground [36]. Over the years, several variations of this method have been proposed.

Examples of PLF studies using background subtraction techniques are reported by several authors [13, 15, 37-39].

6.2 Point trajectory [35]

These methods perform segmentation by analysing movement over a fairly long period. The movement is analysed over a long period and is represented by precise trajectories. These methods generate punctual trajectories which are grouped in a matrix. The clustering trajectory is used to perform segmentation on the video.

Optical flow segmentation is based on movement. It uses the characteristics of flow to detect independently moving objects even in the presence of camera motion. Optical flow is a dense field displacement vector used to define the pixels of each region. To obtain good results with these methods, it is necessary that the brightness remains constant and that movements do not change drastically as time varies; that is, the variation between consecutive frames is very small. These methods are computationally more complex and costly and are sensitive to noise. Examples of PLF studies using optical flow have been provided by a number of authors [40, 41].

7 Key elements of computer vision-based systems: feature extraction

Feature extraction is one of the most important operations, as it consists of extracting the distinctive characteristics of an image. In CV, the term 'feature' is generally used to refer to visual characteristics or elements in images or videos which are distinguishable and discriminative of objects and their parts. For instance, a uniform white patch would not be described as a 'feature,' as it is not specific to any single object, while a circular shape may represent a useful feature in identifying an object such as a car since it has wheels. Features are classified into:

- general features; and
- domain-specific features.

While domain-specific features depend on the task required, general features are independent of the task. In videos and images, the most important general features are colour, shape, texture and motion. General and specific domain features can be evaluated at different levels:

- low level: that is, the pixel;
- intermediate level: that is, portions of images; and
- high level: that is, the image is considered in its entirety.

7.1 Colour features [42]

Colour is a very important general feature. In an image, it is defined based on colour spaces (e.g. RGB and LUV). The most frequently used method for the extraction of colour features is the colour histogram, which shows the distribution of colour. This method is not affected by the rotations, translations, viewing angle and dimensions of the image considered. Other methods are colour correlogram, colour coherence vector, Zernike chromaticity distribution moments and colour moments [43].

7.2 Texture features [42]

While pixels are used to extract colour features, pixel groups are used for texture features. Texture contains very important information for recognition as it describes the physical composition of a surface with homogeneous properties. Feature textures are classified into:

- spatial feature textures, and
- spectral feature textures.

While spatial feature textures are extracted by carrying out operations directly on the image pixels, the latter are calculated in the frequency domain. It is necessary to transform the image in the frequency domain and then perform the calculations for extraction. The most used extraction methods are:

- first-order statistics: third moment, smoothness, uniformity, mean, standard deviation and entropy;
- grey level co-occurrence matrix;
- local binary pattern and variants [44];
- Gabor filter (transform-based) [44];
- wavelet (transform-based); and
- Fourier approaches (transform-based).

7.3 Shape features [42, 45]

One of the most important features in the recognition and identification of objects is shape. Extraction of shape features can be carried out by considering the whole object or the boundary. Extraction techniques for shape features are divided into region-based and contour-based techniques. It is not always easy to describe the shape of an object due to noise, camera occlusion and distortions. According to Yang et al. [45], the most used methods are:

- *One-dimensional function for shape representation*: These approaches capture the perceptual feature of the shape; these methods are also called 'shape signature.' The shape is represented by the one-dimensional function acquired from information from the extracted boundary. Shape signatures are not affected by rotations and translations but are sensitive to even minor noise and variations in the boundaries of shapes. A shape signature includes features such as major axis, minor axis, diameter, centroid distance function, area function and contour curvature.
- *Polygonal approximation*: With this kind of approach, the shape boundary is 'decomposed' into line segments; that is, the shape is approximated by polygons. Polygonal approximation is a simple method to approximate the shape and boundaries, unaffected by translations, rotations and scale variations, and able to reduce noise. One of the methods in this category is the distance threshold method.
- *Region moments*: These methods, based on the concept of mechanical moments defined in physics, make it possible to analyse boundaries and regions. One of the methods in this category is the Zernike moments method [46].
- *Shape transforms domains*: This method involves the transformation of the image in the frequency domain, i.e. the outline of the shape or the region is projected into another domain and then the shape is described based on its frequency. Methods based on transforms are unaffected by translations, rotations and scale variations. However, they are affected by affine transformations and noise, and they require a high-to-medium computing power to run. Methods include Fourier descriptors and wavelet transforms.

Other techniques usually used for shape features extraction in PLF are the following:

- Hough transform [34, 47];
- Canny operator [47];
- Laplace operator;
- active shape algorithm [48]; and
- ellipse fitting [20, 23, 28, 34].

7.4 Motion features

The selection of motion features depends on the proposed task. For instance, when it comes to motion recognition, in addition to colour, texture and shape, it is also important to consider features capable of describing movement characteristics. Optical flow is commonly used in motion-based segmentation and tracking applications (as described in the discussion on segmentation). Optical flow involves a dense field of displacement vectors defining the translation of each pixel in a region. Computing optical flow assumes that the brightness of corresponding pixels remains constant in consecutive frames. In precision livestock applications, some methods have been used to extract features directly from optical flow vectors, such as modified angular histograms [41, 49] and Doane's formula [50]. Another method used for motion feature extraction is the motion history image [25]. This is a static image model that compresses temporal movement information. The intensity of the pixels is a function of the history of the movement in a certain position; that is, the brighter values indicate a more recent movement. Therefore, with a single image, it is possible to predict the moving parts in the video.

7.5 Object tracking

The goal of object tracking is to find the trajectory of movement and identify the position of an object between the frames of a video. Object tracking involves activities such as detecting moving objects, classifying objects and tracking objects frame by frame. Figure 3 shows steps in a video analysis flow. Object detection is the identification of objects from a video frame. This is very common in precision livestock farming where there is often a need to identify livestock between the video frames. Feature selection and extraction are required to classify objects. After an object has been detected and classified, it is possible to track its movement.

There are three main object-tracking methods [51] (Fig. 4):

- point tracking;
- kernel tracking; and
- silhouette tracking.

7.5.1 Point tracking

In point tracking, the detected objects are represented by points across frames. Generally, tracking is performed by evaluating the state of points in terms of position and motion, frame by frame. Tracking is complicated by occlusions, misdetections and other suboptimal conditions. Point tracking methods are divided into deterministic and statistical methods. In precision livestock farming, the most common methods for object tracking are:

- Kalman filters; and
- particle filters.

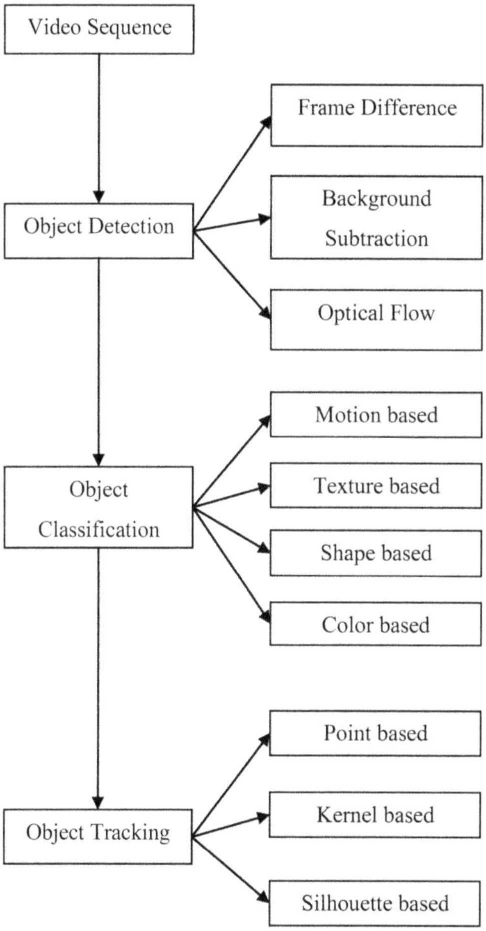

Figure 4 Steps for object detection and tracking from 'A Survey on Moving Object Tracking Using Image Processing' [51].

Both are statistical methods for point tracking. The Kalman filter is a single-object state estimation procedure. It involves two key steps: prediction and correction of the system state. The Kalman filter method is precise and reliable, and less affected by occlusions, but is limited by the assumption that the state variables must be normally distributed. It is only possible to apply the Kalman filter for tracking objects with linear movements or movements with small variations in terms of direction and speed. The particle filter method does not require normally distributed state variables. Both techniques assume a single measurement each time, meaning it is only possible to estimate the state of one object. There are several approaches to estimate the state of multiple objects with Kalman and particle filters [52].

Examples of PLF studies using point tracking approaches are reported by several authors [53, 54].

7.5.2 Kernel tracking

In kernel tracking, the detected objects can be represented across frames by geometric forms (e.g. rectangles and ellipses). Generally, object motion is tracked in parametric forms such as translation, conformal, affine or the dense flow fields computed in subsequent frames [52]. A common method is the mean-shift filter.

7.5.3 Silhouette tracking

An alternative technique is silhouette tracking. Since locating objects with simple geometric shapes may not always be practicable in the real world, this method focuses on an accurate description of the shape of the object. It locates the region of the object in each frame, with the help of an object model obtained from previous frames.

8 Key elements of computer vision-based systems: pose estimation and behaviour recognition

Pose estimation refers to the process of detecting key points from the target object and approximating their postures based on known geometric shapes. Pose estimation can be used after object tracking or it can be part of the tracking phase. There are many levels of accuracy in the estimation of the pose, depending on the task. While a rough estimate (information on the head or on the centre of mass of the body) may be sufficient in some applications, in others, it may be necessary to have more precise information (e.g. the orientation and the width) or to have this information for specific parts of the body (e.g. head or limbs). Pose estimation is an important stage to understand the behaviour of an object. For instance, it is possible to detect lameness or other health problems. Traditional human-based methods for pose estimation have been used for livestock [55]. The most famous of these approaches are the pictorial structure model and loopy structured model.

The behaviour recognition phase consists of the classification of behaviours based on movement data acquired in the previous steps. The main algorithms used are:

- Support vector machine (SVM): this is a supervised learning model. Input data must be annotated for training. The SVM uses binary classifications by assigning new examples to one of two classes. SVMs use a non-probabilistic

linear binary classifier. It is possible to perform non-linear classifications using the kernel method, in which the inputs are mapped into a multi-dimensional feature space [34, 44, 56].
- Viola–Jones [4, 57]
- Linear discriminant analysis [25]
- Transfer functions [23, 28]
- Principal component analysis [44]

9 Case studies of precision livestock farming applications based on traditional computer vision techniques

Most of the existing research studies in precision livestock farming have made use of traditional CV techniques. A selection of case studies in the literature are reported here with the aim of highlighting the CV methods used and the objectives achieved.

Kashiha et al. [23] monitored water consumption in growing pigs using CV techniques. The purpose of the study was to quantify water use and relate it to the drinking behaviour of the pigs. After collecting the data, image segmentation was undertaken by binarising the image to eliminate the background using threshold-based methods. The extraction of the animal's body was carried out using the ellipse method. After extraction, the body was analysed, with a specific focus on the head. The posture and position of the pigs were key features. The estimate of water use was achieved by using a transfer function.

Kashiha et al. [20] investigated the feasibility of CV techniques to identify marked pigs in a pen and to analyse their behaviour. The pigs were marked with paint patterns. The image was binarised to eliminate the background. Binarisation was performed by filtering the image through a 2D low-pass filter. A global threshold was calculated with the Otsu method, then a morphological closure operator was used to remove small objects from the image. Extraction of the pig bodies was carried out by ellipse fitting. The extraction of patterns was performed in the same way as localization. Fourier description was used to describe the characteristics of the patterns.

Nasirahmadi et al. [13] studied automatic mounting detection between pigs using image analysis. A background subtraction method was used to extract the objects of interest from the scene. The greyscale images were binarised using the Otsu method, thus establishing a global threshold. Small objects were eliminated by applying a morphological closing operator. Ellipse fitting was used to locate the body of each pig.

Viazzi et al. [25] assessed the monitoring of aggressive behaviour in a group of pigs. Spatial and temporal movements were described by the Motion History Image. The average intensity of movement and occupancy index features were extracted from the segmented region and classified by linear discriminant analysis.

Porto et al. [57] modelled detectors for analysing cow lying, feeding and standing behaviours from top-view panoramic images. Detectors were modelled and validated through a methodology based on the Viola–Jones algorithm. Detectors were able to calculate the Cow Lying Index, Cow Feeding Index and Cow Standing Index with real-time detection of behavioural changes. A multi-camera video recording system was used to obtain top-view panoramic images. No image-enhancement techniques were required in the training and execution phases.

Kashiha et al. [28] investigated a system to estimate individual pig weight to assess pig growth and weight variation. Two steps were used to extract the outline of the body area. Pigs were first localized and segmented using an ellipse fitting algorithm; then, to maximise correlation to body weight (BW), the head and neck were separated. Pigs were marked for identification to estimate individual pig weight. Weight estimation was made by using a transfer function.

Gronskyte et al. [41] analysed the movement of pigs through video recordings at a slaughterhouse by using statistical analysis of optical flow (OF) patterns. Optical flow vectors used modified angular histograms to detect and locate stationary pigs.

Huang et al. [44] investigated the identification of group-housed pigs. The images of individual pigs were extracted with an adaptive method from the top-view videos to avoid occlusions. Gabor features (obtained with Gabor filters) and local binary pattern features were combined. Principal component analysis (PCA) was used to reduce the features extracted, while a support vector machine (SVM) was used for pig identification (Fig. 5).

10 Advances in computer vision techniques: deep learning

Recent research in the field of precision livestock farming has increasingly focused on deep learning techniques (DL). Deep learning is a subfield of machine learning (ML) and is based on ANN. The number of studies using

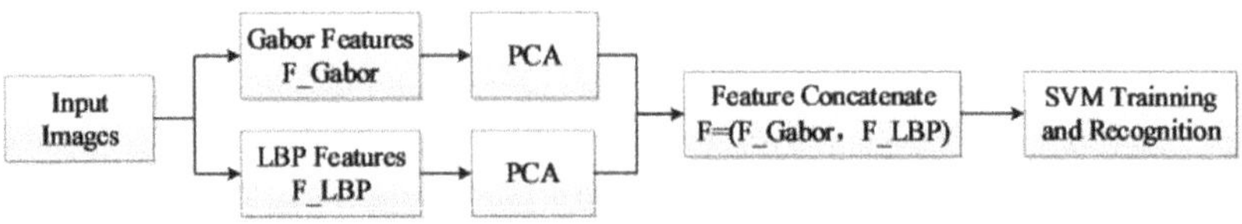

Figure 5 Method proposed for 'Identification of group-housed pigs based on Gabor and local binary pattern features' [44].

methods based on ANN has steadily increased. These techniques are gradually starting to replace the traditional CV techniques described earlier.

In the field of computer vision, DL techniques have shown a clear improvement in terms of performance and accuracy. Among the many advantages of DL techniques is that it is no longer necessary to define the features and perform operations for 'engineering of the features,' as neural networks (NN) are able to learn features from the analysis of a dataset. The phases described earlier - segmentation, object tracking, features extraction, pose estimation and behaviour recognition - can be carried out using models based on ANN (Fig. 6).

There are three typologies of learning:

- Supervised: this method uses labelled data (annotations) as model data input or a label that summarises the nature of the specific data. Classification tasks are addressed through supervised approaches, since the model can compare a predicted class with the true class that is known from the label of each data available. Although the supervised approach is very effective, its application is limited if labels are not available.
- Unsupervised: this method uses non-labelled data as model input. An unsupervised machine-learning algorithm reclassifies and organises data input based on common features to make statistical predictions on future outputs. Since data input does not contain any preliminary information, the algorithm is called upon to create 'knowledge discovery,' that is, new knowledge. Without labelling, the reliability of the unsupervised approach cannot be assessed because the model output cannot be compared with labelled data. The unsupervised approach is applied to extraction of not-yet-known information rather than in a classification task.
- Semi-supervised: this approach can be defined as a 'hybrid' between the previous two approaches, since, in the training phase, the dataset is

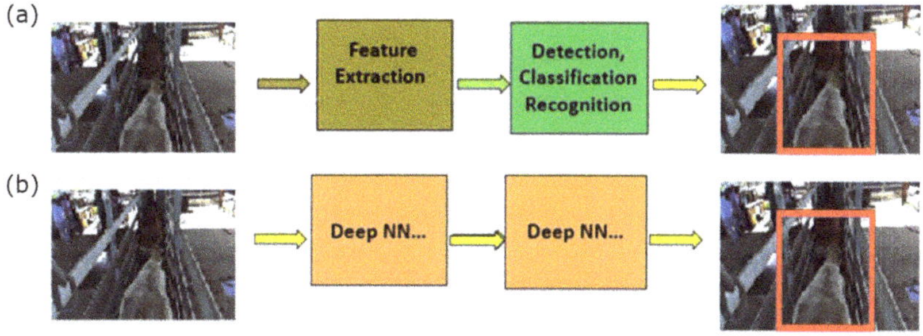

Figure 6 Typical flow chart with traditional techniques (a), and flow chart with the use of deep learning techniques (b). Cattle image from [62].

composed of a small set of labelled data and a bigger set of unlabelled data. This approach is useful when the available tagged data are not sufficient for a high-performing model.

As previously stated, DL techniques are based on the use of NNs. As demonstrated by George Cybenko's 'Universal Approximation' theorem (1989), they can be defined as universal approximators of continuous functions. Neural networks draw inspiration from the way the mammalian brain processes information and learns. Unlike mammalian neurons, however, information propagates in ANNs in a predetermined direction and the number of connections between neurons is defined. The basic architecture of a deep NN can be synthesized into: an input layer, a hidden layer and an output layer.

The main types of deep neural networks are:

- Feedforward neural networks: In this kind of NN, connections between nodes do not form loops, that is, nodes and connections form an acyclic graph. The most famous feedforward NN are convolutional neural networks (CNNs). These are currently used in CV for segmentation, object tracking, classification, pose estimation etc. CNNs assume that the inputs are images. Unlike other types of networks, the layers of a convolutional network have neurons arranged in the three dimensions, typical of an image: width, height and depth. Neurons in a convolutional network receive inputs and perform scalar products on them, using weights that are learned during training; then, a non-linearity function is applied to the result produced. The application of a non-linearity function is required because it adds non-linearity to the network and limits the output. This allows the network to learn complex patterns in the data. A particular feature is the convolution layer, where input is divided into many small parts and a filter called a kernel is superimposed on them. It is then possible to extract features or the fundamental characteristics of the image from each portion of the input. Each input portion is processed in the same way since connections between neurons are local and shared. Finally, the information is propagated to the following layers in the form of a feature map. There are a number of examples of PLF studies using CNNs [22, 58].
- Recurrent neural networks: These are a type of NN that, unlike others, contain feedback cycles within the network. The output provided in some layers becomes the input for the same layer or for lower layers. This interconnection between the various levels makes it possible to maintain a 'state memory.' Recurrent neural networks (RNNs), by giving a temporal sequence as input, make it possible to model dynamic temporal behaviour. Given information at time *t*, it is possible to 'predict' what will happen at

the instant of time *(t+1)*. While CNNs are designed to handle images, and therefore model space, RNNs are suitable for modelling time. RNNs allow working with sequences of arbitrary length, unlike CNNs that have input layers with a predefined size. Recurring networks are now used in natural language processing, sentiment analysis and automatic translation. There are various examples of PLF research using RNNs [59, 60].

The advantages and disadvantages of DL approaches have been compared to traditional CV techniques in the study by Mahony et al. [61].

11 Case studies of precision livestock farming applications based on deep learning techniques

This section includes a selection of learning-based case studies developed for livestock systems to highlight the specific features of the particular application.

Qiao et al. [62] developed a method for semantic segmentation of cattle images based on Mask R-CNN and a method for contour line extraction. Mask R-CNN is a framework that makes it possible to perform semantic segmentation on images. The approach involved the extraction of 'key' frames (containing images where the animal moves, changes posture or behaviour) and image improvements to reduce the influence of shadows and lighting. The result obtained with Mask R-CNN were compared with the results obtained by using other frameworks for learning-based segmentation, such as Deep Mask and Sharp Mask. Segmentation is not influenced by the colour of the animal's coat or posture.

Shen et al. [63] described a method for the identification of cattle based on convolutional networks. Whole images of cattle were used to train the convolutional network. To suppress background noise and 'interference' from different cattle, the images were inputted to the YOLO object detection model in order to correctly locate each cattle in real time. A fine-tuned AlexNet convolutional network was then used for identification.

Achour et al. [22] assessed real-time image analysis for automatic biometric identification and monitoring of dairy cow feeding behaviour on a commercial farm. Different models based on CNNs were developed. They were trained to identify cows, check the availability of food in the feeder and identify its type. The region of interest used was the top of the head. The acquired images were selected by a specially developed algorithm based on motion detection and background subtraction. A similarity index of consecutive images was also computed.

Qiao et al. [21] developed a DL-based framework to identify beef cattle by using CNNs and long-short-term memory networks. The CNN selected was Inception-V3, pretrained on ImageNet. It was used to extract features from a rear-view cattle video dataset. After being extracted, the features were used as input of a long short-term memory network to capture temporal information.

Liu et al. [26] developed a CV-based approach to automatically identify and locate tail biting in group-housed pigs. The method used a tracking-by-detection algorithm to extract the individual pig's motion patterns in a group of pigs to associate detection (bounding box) across frames. The CNNs and recurrent NNs were then jointly used to extract spatial and temporal features to detect tail-biting behaviour. A fine-tuned single-shot detector (SSD) was used for object detection, since it works in real time.

Zhang et al. [64] have assessed a detection algorithm for monitoring real-time sow behaviour, specifically drinking, urination and mounting. The algorithm included three stages: image pre-processing, a MobileNet classification and an SSD detection network. While VGG16 is used for classification, these authors chose MobileNet because it uses a deep separable convolution layer that is able to obtain similar results to VGG16 in a shorter time. The SSD detector was selected for its capability of simultaneously detecting and classifying to improve speed and accuracy.

Xu et al. [65] used a Mask R-CNN pipeline to process RGB images captured by a quadcopter to detect and count grazing cattle. The counting from Mask R-CNN was compared with other methods such as Faster R-CNN, Yolo-v3 and SSD.

Zheng et al. [66] have suggested a DL detector for automatic recognition of lactating sow postures using depth images. Faster R-CNN was used to detect and classify sow posture.

12 Conclusion

This chapter has described several CV techniques and their applications to precision livestock farming. Computer vision techniques allow continuous, automatic and non-invasive monitoring of livestock without the need for human operators. As suggested by the European Animal Welfare Quality Project, this reduces the risk of disease transmission and offers more objective animal-based measurement. Computer vision techniques make it possible to detect small changes in typical patterns of movements and behaviours which indicate potential health and welfare issues. Traditional computer vision techniques have been supplemented by recent advances in deep learning, improving output speed and quality. The use of graphics processing units (GPU), for instance, has allowed many operations to be performed in parallel, improving efficiency. Despite remarkable progress, there are still many challenges facing precision livestock farming. For instance, precise identification of livestock within the barn remains a problem when uneven lighting conditions or high variability during the day result in suboptimal or unstable levels of floor lighting, thus causing difficulties in background segmentation and feature extraction. Other improvements should focus on monitoring and tracking of large groups of animals since the high performance achieved for small groups is not always possible for larger groups. There is a need to improve both accuracy and precision as well as usability in commercial settings, that is, to design smart and low-cost systems usable in harsh environments such as livestock buildings.

13 References

1. Gregor, G. and Scholz, A. Evaluation of intramuscular fat-content in the longissimus-dorsi muscle by video image-analysis and discussion of objectives in pig breeding, *Archiv fur Tierzucht-Archives of Animal Breeding* 36(13-4), 383-395 1993.
2. Velasco, H., Aligué, F. J. L., Orellana, C. G., Macías, M. and Sotoca, M. I. A. Application of ANN techniques to automated identification of bovine livestock, *chez IWANN* 1 1999.
3. Muller, R. and Schrader, L. A new method to measure behavioural activity levels in dairy cows, *Applied Animal Behaviour Science* 83(4), 247-258 2003.
4. Porto, S. M. C., Arcidiacono, C., Anguzza, U. and Cascone, G. A computer vision-based system for the automatic detection of lying behaviour of dairy cows in free-stall barns, *Biosystems Engineering* 115(2), 184-194 2013.
5. Vanhoof, F., Sluyts, H., Paulussen, J., Berchmans, D. and Bloemen, H. Evaluation of a biomonitor based on the phototactic behavior of daphnia-magna using infrared detection and digital image-processing, *Water Science and Technology* 110, 79-86 1994.
6. Van Hertem, T., Schlageter Tello, A., Viazzi, S., Steensels, M., Bahr, C., Bites Romanini, C. E. B., Lokhorst, K., Maltz, E., Halachmi, I. and Berckmans, D. Implementation of an automatic 3D vision monitor for dairy cow locomotion in a commercial farm, *Biosystems Engineering* 173, 166-175 2018.
7. Stavrakakis, S., Li, W., Guy, J. H., Morgan, G., Ushaw, G., Johnson, G. R. and Edwards, S. A. Validity of the Microsoft Kinect sensor for assessment of normal walking patterns in pigs, *Computers and Electronics in Agriculture* 117, 1-7 2015.
8. Kongsro, J. Estimation of pig weight using a Microsoft Kinect prototype imaging system, *Computers and Electronics in Agriculture* 109, 32-35 2014.
9. Viazzi, S., Bahr, C., Van Hertem, T., Schlageter-Tello, A., Romanini, C. E. B., Halachmi, I., Lokhorst, C. and Berckmans, D. Comparison of a three-dimensional and two-dimensional camera system for automated measurement of back posture in dairy cows, *Computers and Electronics in Agriculture* 100, 139-147 2014.
10. Tsai, D.-M. and Huang, C.-Y. A motion and image analysis method for automatic detection of estrus and mating behavior in cattle, *Computers and Electronics in Agriculture* 104, 25-31 2014.
11. Cangar, Ö., Leroy, T., Guarino, M., Vranken, E., Fallon, R., Lenehan, J., Mee, J. and Berckmans, D. Model-based calving monitor using real time image analysis. *Precision Livestock Farming 2007-Papers Presented at the 3rd European Conference on Precision Livestock Farming*, pp. 291-298.
12. Cangar, Ö., Leroy, T., Guarino, M., Vranken, E., Fallon, R., Lenehan, J., Mee, J. and Berckmans, D. Automatic real-time monitoring of locomotion and posture behaviour

of pregnant cows prior to calving using online image analysis, *Computers and Electronics in Agriculture* 64(1), 53–60 2008.
13. Nasirahmadi, A., Hensel, O., Edwards, S. A. and Sturm, B. Automatic detection of mounting behaviours among pigs using image analysis, *Computers and Electronics in Agriculture* 124, 295–302 2016.
14. Poursaberi, A., Bahr, C., Pluk, A., Van Nuffel, A. and Berckmans, D. Real-time automatic lameness detection based on back posture extraction in dairy cattle: shape analysis of cow with image processing techniques, *Computers and Electronics in Agriculture* 74(1), 110–119 2010.
15. Ahrendt, P., Gregersen, T. and Karstoft, H. Development of a real-time computer vision system for tracking loose-housed pigs, *Computers and Electronics in Agriculture* 76(2), 169–174 2011.
16. Viazzi, S., Bahr, C., Schlageter-Tello, A., Van Hertem, T., Romanini, C. E. B., Pluk, A., Halachmi, I., Lokhorst, C. and Berckmans, D. Analysis of individual classification of lameness using automatic measurement of back posture in dairy cattle, *Journal of Dairy Science* 96(1), 257–266 2013.
17. Nasirahmadi, A., Richter, U., Hensel, O., Edwards, S. and Sturm, B. Using machine vision for investigation of changes in pig group lying patterns, *Computers and Electronics in Agriculture* 119, 184–190 2015.
18. Hansen, M. F., Smith, M. L., Smith, L. N., Abdul Jabbar, K. and Forbes, D. Automated monitoring of dairy cow body condition, mobility and weight using a single 3D video capture device, *Computers in Industry* 98, 14–22 2018.
19. Hyeon, T. K., Ikeda, Y. and Hong, L. C. The identification of Japanese black cattle by their faces, *Asian-Australasian Journal of Animal Sciences* 18(6), 868–872 2005.
20. Kashiha, M., Bahr, C., Ott, S., Moons, C. P. H., Niewold, T. A., Ödberg, F. O. and Berckmans, D. Automatic identification of marked pigs in a pen using image pattern recognition, *Computers and Electronics in Agriculture* 93, 111–120 2013.
21. Qiao, Y., Su, D., Kong, H., Sukkarieh, S., Lomax, S. and Clark, C. Individual cattle identification using a deep learning based framework, *IFAC-PapersOnLine* 52(30), 318–323 2019.
22. Achour, B., Belkadi, M., Filali, I., Laghrouche, M. and Lahdir, M. Image analysis for individual identification and feeding behaviour monitoring of dairy cows based on Convolutional Neural Networks (CNN), *Biosystems Engineering* 198, 31–49 2020.
23. Kashiha, M., Bahr, C., Haredasht, S. A., Ott, S., Moons, C. P. H., Niewold, T. A., Ödberg, F. O. and Berckmans, D. The automatic monitoring of pigs water use by cameras, *Computers and Electronics in Agriculture* 90, 164–169 2013.
24. Tullo, E., Fontana, I., Gottardo, D., Sloth, K. H. and Guarino, M. Technical note: validation of a commercial system for the continuous and automated monitoring of dairy cow activity, *Journal of Dairy Science* 99(9), 7489–7494 2016.
25. Viazzi, S., Ismayilova, G., Oczak, M., Sonoda, L. T., Fels, M., Guarino, M., Vranken, E., Hartung, J., Bahr, C. B. and Berckmans, D. Image feature extraction for classification of aggressive interactions among pigs, *Computers and Electronics in Agriculture* 104, 57–62 2014.
26. Liu, D., Oczak, M., Maschat, K., Baumgartner, J., Pletzer, B., He, D. and Norton, T. A computer vision-based method for spatial-temporal action recognition of tail-biting behaviour in group-housed pigs, *Biosystems Engineering* 195, 27–41 2020.
27. Schofield, C. P. Evaluation of image analysis as a means of estimating the weight of pigs, *Journal of Agricultural Engineering Research* 47, 287–296 1990.

28. Kashiha, M., Bahr, C., Ott, S., Moons, C. P. H., Niewold, T. A., Ödberg, F. O. and Berckmans, D. Automatic weight estimation of individual pigs using image analysis, *Computers and Electronics in Agriculture* 107, 38–44 2014.
29. Lu, M., Norton, T., Youssef, A., Radojkovic, N., Peña Fernández, A. and Berckmans, D. Extracting body surface dimensions from top-view images of pigs, *International Journal of Agricultural and Biological Engineering* 11(5), 182–191 2018.
30. Moeslund, B. and Granum, E. A survey of computer vision-based human motion capture, *Computer Vision and Image Understanding* 81(13), 231-268 2001.
31. Tillett, R., McFarlane, N., Wu, J., Schofield, C., Ju, X. and Siebert, J. Extracting morphological data from 3D images of pigs, chez Proceedings of the International Conference on Agricultural Engineering, AgEng, Leuven, Belgium 2004.
32. Gonzalez, R. C. and Woods, R. E. Digital, *Image Processing, Pearson* 2008.
33. Kaur, D. and Kaur, Y. Various image segmentation techniques: a review, *International Journal of Computer Science and Mobile Computing* 3(15), 809-814 2014.
34. Zhuang, X., Bi, M., Guo, J., Wu, S. and Zhang, T. Development of an early warning algorithm to detect sick broilers, *Computers and Electronics in Agriculture* 144, 102–113 2018.
35. Yao, H., Lian, D., Cao, Y., Wu, Y. and Zhou, T. Predicting academic performance for college students, *ACM Transactions on Intelligent Systems and Technology* 10(3), 1–21 2019.
36. Goyal, K. and Singhai, J. Review of background subtraction methods using Gaussian mixture model for video surveillance systems, *Artificial Intelligence Review* 50(2), 241–259 2018.
37. Van Hertem, T., Alchanatis, V., Antler, A., Maltz, E., Halachmi, I., Schlageter-Tello, A., Lokhorst, C., Viazzi, S., Romanini, C. E. B., Pluk, A., Bahr, C. and Berckmans, D. Comparison of segmentation algorithms for cow contour extraction from natural barn background in side view images, *Computers and Electronics in Agriculture* 91, 65–74 2013.
38. Chung, Y., Kim, H., Lee, H., Park, D., Jeon, T. and Chang, H. A cost-effective pigsty monitoring system based on a video sensor, *KSII Transactions on Internet and Information Systems (TIIS)* 4(14), 1481–1498 2014.
39. Pluk, A., Bahr, C., Poursaberi, A., Maertens, W., Van Nuffel, A. and Berckmans, D. Automatic measurement of touch and release angles of the fetlock joint for lameness detection in dairy cattle using vision techniques, *Journal of Dairy Science* 95(4), 1738–1748 2012.
40. Dawkins, M. S., Lee, H., Waitt, C. D. and Roberts, S. J. Optical flow patterns in broiler chicken flocks as automated measures of behaviour and gait, *Applied Animal Behaviour Science* 119(3–4), 203–209 2009.
41. Gronskyte, R., Clemmensen, L. H., Hviid, M. S. and Kulahci, M. Monitoring pig movement at the slaughterhouse using optical flow and modified angular histograms, *Biosystems Engineering* 141, 19-30 2016.
42. Salau, A. O. and Jain, S. Feature extraction: a survey of the types, techniques, applications *International Conference on Signal Processing and Communication (ICSC)*, pp. 158–164 2019.
43. Zhu, W., Guo, Y., Jiao, P., Ma, C. and Chen, C. Recognition and drinking behaviour analysis of individual pigs based on machine vision, *Livestock Science* 205, 129–136 2017.

44. Huang, W., Zhu, W., Ma, C., Guo, Y. and Chen, C. Identification of group-housed pigs based on Gabor and Local Binary Pattern features, *Biosystems Engineering* 166, 90–100 2018.
45. Yang, M., Kpalma, K. and Ronsin, J. Shape-based invariant feature extraction for object recognition, *Intelligent Systems Reference Library* 29, 255–314 2012.
46. Zhu, W., Zhu, Y., Li, X. and Yuan, D. The posture recognition of pigs based on Zernike moments and support vector machines, *chez 2015 10th International Conference on Intelligent Systems and Knowledge Engineering (ISKE)* 2015.
47. Baek, H., Chung, Y., Ju, M., Chung, Y. and Park, D. Segmentation of group-housed pigs using concave points and edge information, *2017 19th International Conference on Advanced Communication Technology (ICACT)*, pp. 563–565 2017.
48. Velasco, H. M. G., Orellana, C. J. G., Aliguè, F. J. L., Macías, M. M. and Caballero, R. G. Segmentation of bovine livestock images using GA and ASM in a two-step approach, *International Journal of Pattern Recognition and Artificial Intelligence* 17(4), 601–616 2003.
49. Fernández-Carrión, E., Martínez-Avilés, M., Ivorra, B., Martínez-López, B., Ramos, Á. M. and Sánchez-Vizcaíno, J. M. Motion-based video monitoring for early detection of livestock diseases: the case of African swine fever, *PLoS ONE* 12(9), e0183793 2017.
50. Chen, C., Zhu, W., Ma, C., Guo, Y., Huang, W. and Ruan, C. Image motion feature extraction for recognition of aggressive behaviors among group-housed pigs, *Computers and Electronics in Agriculture* 142, 380–387 2017.
51. Balaji, S. R. and Karthikeyan, S. A survey on moving object tracking using image processing, *11th International Conference on Intelligent Systems and Control (ISCO)*, pp. 469–474 2017.
52. Alper, Y., Omar, J. and Sha, M. Object tracking: a survey, *ACM Computing Surveys* 38 2006.
53. Fujii, T., Yokoi, H., Tada, T., Suzuki, K. and Tsukamoto, K. Poultry tracking system with camera using particle filters, chez 2008 IEEE International Conference on Robotics and Biomimetics, Bangkok, Thailand, pp. 1888–1893 2009.
54. Wang, K., Zhu, D., Guo, H., Ma, Q., Su, W. and Su, Y. Automated calculation of heart girth measurement in pigs using body surface point clouds, *Computers and Electronics in Agriculture* 156, 565–573 2019.
55. Brünger, J., Gentz, M., Traulsen, I. and Koch, R. Panoptic segmentation of individual pigs for posture recognition, *Sensors* 20(13) 2020.
56. Weixing, Z. and Jin, Z. Identification of abnormal gait of pigs based on video analysis, *chez 2010 Third International Symposium on Knowledge Acquisition and Modeling*, pp. 394–397 2010.
57. Porto, S. M. C., Arcidiacono, C., Anguzza, U. and Cascone, G. The automatic detection of dairy cow feeding and standing behaviours in free-stall barns by a computer vision-based system, *Biosystems Engineering* 133, 46–55 2015.
58. Hansen, M. E., Smith, M. L., Smith, L. N., Salter, M. G., Baxter, E. M., Farish, M. and Grieve, B. Towards on-farm pig face recognition using convolutional neural networks, *Computers in Industry, Web of Science, View ResearcherID* 98, 145–152 2018.
59. Maloy, H., Aamodt, A. and Misimi, E. A spatio-temporal recurrent network for salmon feeding action recognition from underwater videos in aquaculture, *Computers and Electronics in Agriculture* 167 2019.

60. Qiao, Y., Su, D., Kong, H., Lomax, S. S. and Clark, C. BiLSTM-based individual cattle identification for automated precision livestock farming, *chez 2020 IEEE 16th International Conference on Automation Science and Engineering (CASE)*, Hong Kong 2020.
61. Mahony, N. O., Campbell, S., Carvalho, A. and Harapanahalli, S. Deep learning vs. traditional computer vision. *Science and Information Conference*, pp. 128–144 2019.
62. Qiao, Y., Sukkarieh, S. and Truman, M. Cattle segmentation and contour extraction based on mask R-CNN for PLF, *Computers and Electronics in Agriculture* 165, 104958 2019.
63. Shen, W., Hu, H., Dai, B., Wei, X., Sun, J., Jiang, L. and Sun, Y. Individual identification of dairy cows based on convolutional neural networks, *Multimedia Tools and Applications* 79(21–22), 14711–14724 2020.
64. Zhang, Y., Cai, J., Xiao, D., Li, Z. and Xiong, B. Real-time sow behavior detection based on deep learning, *Computers and Electronics in Agriculture* 163, 104884 2019.
65. Xu, B., Wang, W., Falzon, G., Kwan, P., Guo, L., Chen, G., Tait, A. and Schneider, D. Automated cattle counting using Mask R-CNN in quadcopter vision system, *Computers and Electronics in Agriculture* 171, 105300 2020.
66. Zheng, C., Zhu, X., Yang, X., Wang, L., Tu, S. and Xue, Y. Automatic recognition of lactating sow postures from depth images by deep learning detector, *Computers and Electronics in Agriculture* 147, 51–63 2018.
67. Sun, D.-W. Inspecting pizza topping percentage and distribution by a computer vision method, *Journal of Food Engineering* 44(4), 245–249 2000.

Chapter 5

Developments in activity and location technologies for monitoring cattle movement and behaviour

N. A. Lyons, NSW Department of Primary Industries, Australia; and S. Lomax, The University of Sydney, Australia

1 Introduction

Livestock industries across most regions throughout the world have undergone a process of significant consolidation, with the number of farms decreasing and average farm and herd size increasing. For example, since 1990 the number of dairy farms in Australia and New Zealand has reduced by 67% and 24%, respectively, and the average herd size has increased by around 160% and 175%, respectively (Dairy Australia, 2020; Livestock Improvement Corporation Limited and DairyNZ Limited, 2020). The same trend with varying intensity has also been observed in Argentina and the United States (Lazzarini et al., 2019; USDA-NASS, 2013). A greater average herd size and a greater reliance on employed labour, together with the difficulty of attracting and retaining skilled labour, means that monitoring and managing individual animals has become more complex. When we consider this alongside the growing consumer demand for transparency in production systems, the significant challenge of optimising and tracking animal production, health and welfare into the future will require continuous innovation, which can be alleviated with technological advancements.

http://dx.doi.org/10.19103/AS.2021.0090.05

The behaviour and location of an animal can provide important information as to its health and well-being and enable more accurate decision-making around animal management. It can aid in the day-to-day running of farming operations by providing early alerts, as well as generating key indicators for either individual or herd performance regardless of the scale of operation (Eastwood et al., 2016). Traditionally, the collection of this information has relied on human observation, which limits the extent, objectivity and detail of data collected. The development and adoption of a range of technologies enabling automated measurement and reporting of individual animal behaviour and location have led to exponential changes in farmers' understanding of animal's status, and in turn in the way, animals are managed on-farm. In recent decades, advancements in the capability and functionality of technology (mainly driven by developments in other industries such as health, processing, mining or military that are then adapted for livestock production and agriculture), coupled with a relative decrease in cost, make the potential value proposition for greater adoption of technology on-farm and at an industry level much more appealing.

The incorporation of technology on-farm should not be driven by a desire to completely replace human management but rather aim to complement or enhance the underlying management skills, intuition and experience currently present within the operation, with objective, systematic, routinely collected metrics that can enable earlier, data-driven management decisions to drive profitability. The aim is, however, to reduce or even eliminate systems or processes that are either labour intensive, repetitive, time consuming or subjective. An example of this is the increased adoption of robotic milking systems on dairy farms that has taken place in most dairy-producing regions since they were first introduced in the early 1990s (de Koning, 2011). Robotic milking technology replaces the need for human labour dedicated to the repetitive task of milking cows, which therefore creates time for farm managers to better monitor, manage and evolve their business. This is particularly important when considering that the milk harvesting procedure itself can account for over 50% of labour time on a dairy farm (O'Brien et al., 2007). This can also aid in diversifying farm enterprises which in turn can have positive impacts on buffering environmental and industry challenges such as managing the impact of climate variability, the pressures around the social licence to operate and the rise of input costs.

Bewley and Russell (2010) defines precision dairy farming as 'the use of technologies to measure physiological, behavioral, and production indicators on individual animals to improve management strategies and farm performance'. It has also been described as 'the use of information and communication technologies for improved control of fine-scale animal and physical resource variability to optimise economic, social, and environmental dairy farm performance' (Eastwood et al., 2012), whereas some other authors include the concept of 'real-time monitoring and managing system for farmers' (Berckmans, 2017).

Across the wide range of technological innovations currently available, this chapter will focus particularly on the applications and advancements in technologies for monitoring dairy and beef cattle activity and location, though we recognise there is a growing field of application to other livestock species, including pigs, poultry and sheep. Here we review the range of options for monitoring cattle location and activity and highlight the application of these technologies rather than the technical aspects or algorithms behind them. Additionally, the focus will be placed on the research, development and adoption of these technologies from the authors' experience in the Australian context.

2 Activity and location technologies

2.1 Why measure activity and location?

Animal-based indicators provide specific information about an individual or group of animals that can be used to assess animal productivity, health and welfare and to optimise management. Sensors that support health management on dairy farms were grouped into four levels in a review by Rutten et al. (2013): '(I) techniques that measure something about the cow (e.g. activity); (II) interpretations that summarize changes in the sensor data (e.g. increase in activity) to produce information about the cow's status (e.g. estrus); (III) integration of information where sensor information is supplemented with other information (e.g. economic information) to produce advice (e.g. whether to inseminate a cow or not); and (IV) the farmer makes a decision or the sensor system makes the decision autonomously (e.g. the inseminator is called)'. Additionally, Rutten et al. (2013) mentioned that most fertility sensors (automated detection of oestrus, primarily using activity as the lead indicator) reached level II, whereas most of the locomotion ones (weight distribution or walking behaviour and activity) remained on level I. These same levels could apply to many other sensors currently available and to many other applications other than health management.

Both activity and location can have either a direct, level I, or an indirect, level II, application (Rutten et al., 2013). The direct measurement relates to either the location of an animal within an allocated area on the farm or the postural changes and physical movements of an animal over a given period of time. Traditionally, direct visual observation of animal behaviour, activity and location has been used in both research and production contexts. However, this is both time consuming and labour intensive, with the key limitation of omitting continuous data through time; it is simply not feasible or practical to do so, particularly as herd sizes continue to increase. The increasing availability of simple technologies to provide real-time, remote data on animal activity and location provides the opportunity to better manage individual animals within a herd and make data-driven decisions. The development and application of technologies to collect individual animal data are exponentially increasing. In dairy research alone, the number of journal publications (in English) has grown in the past 20 years from an average of 4 per year (2001–2010) to 12 per year (2011–2020) (search carried out in Scopus using keywords activity or location and dairy in the title of the article).

Herd-based measures are useful for larger-scale decision-making, such as grazing management. But decisions made on the individual level, accounting for variability between and within individuals, can optimise both farm profitability and animal well-being. Here lies one of the greatest potential value propositions between traditional and technology-driven approaches to animal monitoring. Many commercially available sensors work by 'learning' the normal or routine patterns in the behaviour of individual animals within certain contexts and then use this information to detect variations from this pattern to predict periods of reduced or compromised health or welfare. This accounts for individual differences in both the way animals respond to their environment and the effects of genotype, phenotype, age, parity, liveweight or body condition score, when determining the type and severity of the event. Indeed, there is a growing interest in the 'digital phenotype' of animals (Cerri et al., 2021) that can only be unlocked through the access to libraries of individual animal data and the internet of things (IoT).

Activity incorporates the movement, posture or behaviour of an animal. Within the literature to date, the most reported application of activity monitoring in dairy systems is to detect oestrus or reproductive events objectively and remotely (Aungier et al., 2012; Beauchemin, 2018; Chanvallon et al., 2014; Chapa et al., 2020; Kamphuis et al., 2012). Reproductive success, particularly in a dairy context, is limited by the ability to accurately and timely detect oestrus to enable artificial insemination or natural joining. Traditionally, oestrus detection has relied on visual observation of mounting behaviour in cows which is limited due to the variable duration and intensity of oestrus in modern dairy cows and the impact on labour of increasing herd size. This was replaced with simple technologies, including tail strips which change colour when a cow is mounted by a conspecific; however, this still requires visual monitoring. Cows in oestrous have shown to have an increase in activity and/or a decrease in daily lying time (Mayo et al., 2019). In many cases, software algorithms can also suggest an optimal breeding time, where conception is more likely to occur. Over the years, activity has branched out to focus on other reproductive, health and feeding applications. It has been used to detect or predict calving (Rutten et al., 2017; Clark et al., 2015) and to detect, predict or avoid health events

(Chapa et al., 2020; Smith et al., 2015), such as displaced abomasum (Talukder et al., 2015). Lying time can be a strong predictor of health events, where an increase has been linked to incidences of lameness and a decrease as an indicator of mastitis (Tucker et al., 2021). Some of these sensors also monitor and track rumination (the regurgitation of partially digested plant material from the rumen to enable the further breakdown of the cellulose components through chewing), feeding, drinking and social behaviour. Some studies have even highlighted the potential to use rumination and activity, through the development of a health score index, to identify cows with metabolic and digestive disorders, mastitis caused by *E. coli* and cows with severe cases of metritis (Stangaferro et al., 2016a,b,c).

Deviations from normal behaviour profiles are increasingly being used as a measure of negative welfare and to assess the impact of management and the environment on individual animals (Chapa et al., 2020). Such data can then be used objectively to ensure that layout and housing designs meet animal needs related to behaviour and welfare (Shepley et al., 2020). These sensors can also be used for early detection and intervention of production-related physiological challenges, including heat stress in dairy (Bar et al., 2019; Ramón-Moragues et al., 2021) and feedlot cattle (Islam et al., 2020).

Despite most applications being on mature milking cows, there have been some recent studies of its usefulness in calves (Costa et al., 2021). This includes its application, with some degree of effectiveness, for early detection of neonatal calf diarrhoea (Lowe et al., 2019) as well as to either indicate (Duthie et al., 2021), detect (Borghart et al., 2021) or evaluate recovery following treatment for bovine respiratory diseases (Cantor et al., 2022).

On the other hand, knowing an animal's location within both indoor and outdoor farming systems has been used to understand cows' use or preference of different areas, enrichment or bedding materials, feed or even social interactions (Frost et al., 1997). Location provides information on the time budget of an animal, which in turn can be used to inform health, welfare and management. For example, cows and heifers have different preferences for their environment at calving (indoors vs. on pasture) (Edwards et al., 2020), and when provided access to an outdoor open area, cows use it more, particularly during the night (Smid et al., 2020). By exploring the location, variables related to movement or activity can sometimes also be calculated. In robotic milking systems, location systems are also used to find a cow that must be fetched and manually herded to the robot, given her milking interval has exceeded a certain threshold. Location has also been used as a proxy for other things, for example, time spent in the feed alley as a proxy for feeding time. In more extensive production systems, knowing animal location can provide an understanding of the interaction between the animal and the environment and its grazing behaviour and enable more accurate herding for management purposes

(Swain et al., 2011). Tracking animal movement can provide information on distances travelled, which in turn enables decision-making around nutritional requirements (McGavin et al., 2018). A recent review by Aquilani et al. (2022) also mentioned the potential use of location technologies to monitor the theft of animals.

2.2 Type of devices?

There are a wide range of technologies available for measuring and monitoring activity and location in livestock, with the most common approaches being the use of triaxial accelerometers (activity) and GPS (location). Broadly, these technologies can be classified into 'on-animal' and 'off-animal' monitoring devices.

2.2.1 On-animal devices

A recent review by Halachmi et al. (2019) showed that wearable technologies still dominate the market of sensors for livestock production. In cattle, the most common behaviours that are monitored are steps and rumination. These repetitive, rhythmic actions are easy to classify in the raw data derived from wearable sensors. Further, rumination is an essential behaviour that can be used as an early indicator of reduced health or welfare (Marchesini et al., 2018). Other ways of quantifying behaviours associated with jaw and head movement to detect rumination with a good level of precision include a noseband halter with an in-built pressure system (Steinmetz et al., 2020).

Activity is mainly determined through sensors placed on the cows' ears, legs or neck that have embedded pedometers, activity meters or triaxial accelerometers. The majority of literature and commercially available technology is based on the use of accelerometers, predominantly triaxial accelerometers. These measure changes in acceleration and force that enable the determination of movement and orientation (Chapa et al., 2020). Simply, these can be used to measure the time an animal spends lying or standing, which can be indicative of pain or discomfort (Brown et al., 2015; Marti et al., 2018; Theurer et al., 2012), reduced health (Borchers et al., 2016; Costa et al., 2021) or good welfare (Chapa et al., 2020). In more complex logging intervals, accelerometers with on-board algorithms can provide information on an animal's time budget such as rumination, eating and activity (Molfino et al., 2017).

As discussed previously, in dairy systems, collar-based devices are routinely used to monitor oestrus and provide health alerts. This works well within these systems as animals are normally being monitored and handled twice daily (during milking) and therefore any issues with fit, battery or data logging can be easily identified and addressed then. When evaluating the extension into beef systems, the limitation of collar-based devices is the reduced handling of animals, the extensive nature of some of these systems (where animals might be several kilometres away from the farmer and go for weeks or months without direct contact with farm team) and the requirement for the fitting of the device to a growing animal. Here we see the value of ear tag devices, which are becoming increasingly available as technology allows for smaller-sized components at a cheaper price. The simpler the technology, the better in regard to battery life; therefore there is an emergence of studies evaluating the use of accelerometer-based ear tags for livestock monitoring (Bikker et al., 2014; Borchers et al., 2016; Costa et al., 2021; Krieger et al., 2019; Pereira et al., 2018; Reiter et al., 2018; Islam et al., 2020). In addition to monitoring health and welfare, the data from these systems can be used to identify individual animal variations, which can enable the selection of more resilient phenotypes in the future (Islam et al., 2020). This may become essential as we face the challenges of a changing climate and the impact on extensive and outdoor systems. The advantage of such technology allows for simple fitting to a juvenile animal and the continuous monitoring of behaviour and activity through time. Future developments in battery life, data transfer (with the growth of LoRaWAN® networks) and on-board algorithms will enhance applications of this technology to revolutionise individual animal monitoring. Although most of these sensors are non-invasive, there still are boluses and implants available that can monitor some of these behaviours.

The location of animals is normally determined using GPS, low-frequency or ultra-wideband RFID, radar–radio signals or through triangulation with fixed placed routers/antennas/receivers (Wang et al., 2018; Vázquez Diosdado et al., 2018; Halachmi et al., 2019). GPS is generally more viable for pasture-based and extensive systems, whereas triangulation is generally simpler and cheaper, and sometimes more accurate, and therefore better for indoor systems. Bluetooth Low Energy (BLE) provides a low-cost system for locating cows within a dairy barn (Bloch and Pastell, 2020).

Tracking of animal behaviour was historically used in wildlife research, with extension into grazing animal behaviour research (Swain et al., 2011). Swain et al. (2011) provide a detailed review of the history of GPS for animal tracking and its applications for livestock production and research, demonstrating an exponential rise in the past two decades in the use of GPS. However, this technology is limited by battery, data transmission capacity and cost. With a shift in focus to the IoT systems, integration of GPS with other simpler technologies such as Bluetooth can achieve similar outcomes in herd monitoring (Maroto-Molina et al., 2019).

There are other forms of technology that incorporate GPS such as virtual fencing. Virtual fencing aims to replace fixed or physical fencing with animal-mounted devices that control animal movement and location through the delivery of cues or stimuli which the animal learns through associative learning. Conceptually speaking, virtual fencing works by training animals to associate non-visual stimuli (sounds, vibrations) with aversive stimuli (usually an electric pulse) to remain within or excluded from specific areas. There have been many approaches to developing virtual fencing systems over the past two decades (Anderson, 2007; Anderson et al., 2014) including the use of buried wires or lasers which trigger an electric pulse or aversive sound signal when an animal crosses a boundary. Most recently, research has focused on the use of pre-commercial neckbands (eShepherd, Agersens Melbourne Australia) for the control of beef and dairy cattle in extensive and intensive grazing scenarios. These animal-mounted neckbands use GPS to set a perimeter around an exclusion or inclusion area and deliver an audio cue to warn an animal when it is approaching a boundary. If the animal continues forward into the exclusion area, an aversive electric pulse is delivered, with the aim of training the animal to respond to the audio cue alone. The technology has been successfully used to contain (Campbell et al., 2017b), move (Campbell et al., 2017a) or exclude grazing beef cattle from riparian zones (Campbell et al., 2018). Further, dairy cattle can be trained to the technology (Colusso et al., 2020), and grazing dairy cattle can be contained in pasture allocations using virtual fencing (Lomax et al., 2019; Langworthy et al., 2021; Verdon et al., 2021). Animal location is central to the success of this technology; however, currently, the drift and error with GPS have implications for groups of cattle on highly controlled pasture allocations which can have implications for adequate feed provision and animal welfare. Furthermore, future integration of activity sensors, such as a triaxial accelerometer, would improve the understanding of the impact on cattle behaviour and therefore animal welfare (Lee et al., 2018). This integration is yet to be commercially successful, and delivering products that provide a complete package for on-animal monitoring, while overcoming the limitations of battery life and on-board data storage, is a key area of research and development.

As shown, to date, in the writing of this chapter, the key limitation to wearable GPS devices in commercial settings remains to be the issue of battery life. GPS devices require significant energy to enable communication with a satellite and the relay of information back to the device to pinpoint the animal's relative location in space. This can take between seconds and minutes depending on the environment (e.g. buildings, tree cover) and the weather (cloud cover). Currently, there are no commercially available devices that operate on renewable energy (e.g. solar); therefore there is a requirement for the replacement of batteries at regular intervals that preclude the technology from being widely used at larger scales or in more extensive environments where animals are not handled often. As technology develops and solar batteries and cells become smaller, this challenge may be overcome. In the interim, there is a growing focus on novel approaches to monitoring animal location using off-animal devices.

2.2.2 Off-animal devices

Off-animal systems that use video or sensors (such as laser, LiDAR or RGB cameras) that are either fixed around the farm or mounted on the unmanned ground or aerial vehicles are an area of peak technological development in livestock agriculture (Tamas et al., 2019; Clark et al., 2014; Chen et al., 2018). Applications of these sensors include animal tracking, identification, measurement/weighing and health recording, with the capacity to monitor individual animals continuously. For example, object gait tracking using Time of Flight sensors that create two-dimensional near-infrared images can be used to automate lameness detection in dairy cows (Gardenier et al., 2018). Current limitations to deployment in extensive environments limit the use of indoor or fixed housing systems (dairy, pig production, poultry). Advanced machine learning using convoluted neural networking (CNN) means that the use of still images can be applied to cattle tracking and augmentation in more extensive environments (Qiao et al., 2020; Li et al., 2021). Advancement in the technology underpinning imagery collected from satellites, manned aircraft and unmanned aircraft systems (UASs) is enabling high-resolution animal detection, even with commercial satellites (Wang et al., 2019). With the integration of machine learning techniques, automated detection and recognition at the species-specific level is going to be possible in the near future (Wang et al., 2019). From an economic point of view, these technologies could have a greater value proposition than on-animal devices, with the potential to cover large numbers of individuals in short periods of time without the need to handle animals. Technological developments will mean that off-animal sensors will be available at scale, where for example one camera could 'monitor' a large group of animals, which will reduce the cost below that of individual on-animal sensors. This will depend on the use or location of these off-animal systems. For example, farmers would only need one camera at the exit of a dairy shed to monitor the herd for lameness detection (Gardenier et al., 2018), given all cows will walk past it typically twice a day. Whereas they might need several cameras if monitoring feeding behaviour across different feeding areas or housing barns on one property, unless the camera is mounted on some type of ground or aerial vehicle.

Although off-animal devices offer a potentially greater value proposition, particularly from a cost perspective, fewer technologies currently available fall in this category in comparison to on-animal sensors, both for cattle as well as other livestock species (Aquilani et al., 2022).

2.3 Data transfer

As shown in Fig. 1, most of these systems transmit data from the sensor, through an antenna or receiver, to a central controller and then to a computer software or dashboard on a handheld device. The raw measurement would be equivalent to what is observed in the level I sensors in Rutten et al. (2013). These measurements typically need to be analysed through models, equations, algorithms and settings to produce valuable information or insight. This is typically achieved by comparing the individual observations with a baseline for that particular animal as well as with animals in either the same group or within the same farm. When deviations are observed, either absolute or relative, and a certain threshold is exceeded (again this is a configuration setting that determines the accuracy of these sensors - that can and should be adjusted in each installation) an alarm is triggered. Information is usually shown in counts (such as steps, posture, time, proportion, bouts, visits or proportion of time doing a particular activity) or as an index (usually proprietary with little understanding of what lies behind it). Farmers can also typically create a series of reports and graphs that display this information. This insight would be equivalent to what is observed in the level II sensors in Rutten et al. (2013). The farmer then needs to attend to this animal or animals in order to correctly determine the type of intervention required.

2.4 Availability of technology

The marketplace of cattle monitoring technologies is very complex. Often products are the same but are provided to farmers under different brand names dependent on the distributor or global location (Stachowicz and Umstatter, 2020). Some products have a spectrum of optional functionalities, and each technology tends to have its own software and two-way transfer of data between different brands is not a given (may occur automatically, manually or not at all).

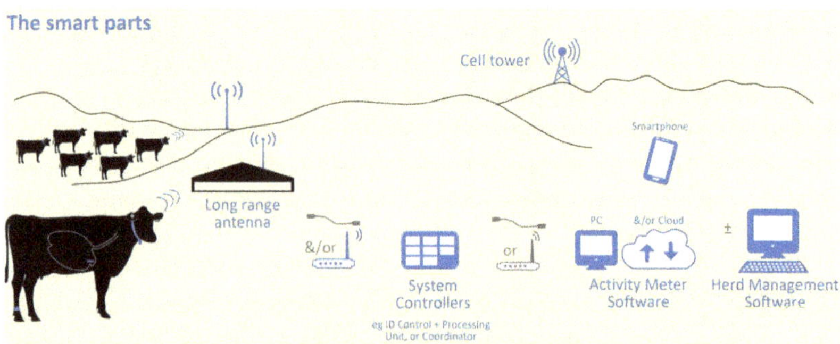

Figure 1 TechKISS Topic: Activity Meters, the smart parts (Harris Park Group, 2019).

The TechKISS project (Harris Park Group, 2019) was supported by the NSW Dairy Industry Fund (NSW Department of Primary Industries, NSW, Australia) and explored how dairy farmers were adopting technologies that assisted with day-to-day animal health and management tasks. The project aimed to identify and share the key things that led to success - how farms achieve the outcomes they want. The target technologies were electronic cow ID, in-line milk metering and analysis, automatic drafting, automatic feeding systems, activity meters and the relevant linking software. The first objective of the TechKISS project was to conduct a desktop audit of these target technologies in consultation with major equipment providers to create the TechMatrix. This is a list of products available in Australia with an outline of the main characteristics of each. It is presented in the same format as the EU 4D4F TechWarehouse (Innovation for Agriculture, 2019). The TechMatrix (Harris Park Group, 2019) included about 80 products from 20 manufacturers. Looking at activity meters alone, there were 13 companies that offered 17 products. This example related to technologies available in Australia is expected to be very similar to what could be seen in other markets or even within other technologies currently available. The 4D4F TechWarehouse (Innovation for Agriculture, 2019) had 17 companies offering 28 heat detection systems and 11 companies offering 11 location systems.

In Table 1 we have summarised some of the key commercially available systems for monitoring cattle activity and/or location and their applications for farmers. More detailed systematic reviews have been conducted (Stygar et al., 2021; Stachowicz and Umstatter, 2020, 2021) demonstrating the wide range of technologies available globally for use in livestock monitoring. Additionally, Table 1 highlights a lack of robust scientific validation of technologies and their applications. A review by Stygar et al. (2021) highlighted that a greater proportion of technologies that included an accelerometer (typical of on-animal activity-type sensors) had been validated, in comparison to cameras, load cells or other sensors (typical of off-animal type sensors). Indeed, Stygar et al. (2021) found that only 30% of accelerometer-based sensors available on the market had validation records, and only 18% of all currently retailed sensors had been externally validated. Validation should be conducted independently of the company developing the technology. In most cases, this occurs in research or commercial settings (Pereira et al., 2021), with sensitivity and specificity impacted by multiple factors, the main one being the definition of the golden standard. For example, when wanting to identify oestrus in dairy cattle, the measurements are usually compared against visual observations of cattle standing to be mounted, but they can also be compared against the outcome of ultrasound technology or from monitoring of hormonal profile in either blood or milk. All of these will have a different impact on the sensitivity and specificity of the technology. Other factors include the number of test animals, adequate placement of sensor (ensuring it is properly fit/installed but also assigned to

Table 1 Commercially available wearable activity and location technologies and their applications for farmers

Technology	Sensor capabilities	Commercial applications	Target animal	Citations
Moovement[1] GPS	GPS, LoRa communication, NFC (rather than RFID) tag for data exchange and accelerometer in an ear tag	Tracking cattle location	Dairy/beef	NA
Moovement[1] Bluetooth	Bluetooth, LoRa communication ear tag	Monitor cattle activity/ location by cattle check-in at water or feed points or other points of interest	Dairy/beef	NA
CowManager/Sensoor	Accelerometer ear tag mounted sensor	Monitor cattle movement including feeding, rumination, rest and activity. Oestrus detection, herd health and nutrition	Dairy	Bikker et al. (2014), Borchers et al. (2016), Pereira et al. (2018), Wolfger et al. (2015), Zambelis et al. (2019)
Allflex Sensehub™[3](formerly Heatime HR LD by SCR)	Accelerometer ear or neck tag	Monitor rumination, eating and activity. Commercial oestrus and health alerts for dairy and beef herds; non-commercial heat stress monitoring in feedlot cattle	Dairy, beef	Islam et al. (2020, 2021a), Merenda et al. (2019, 2020)
SmaXtec[4]	pH meter and/or accelerometer in a rumen bolus	Monitor feeding, drinking and rumination activity for detection of health and metabolic conditions, oestrus heat stress and calving; pH monitoring allows for early detection of ruminal acidosis and ration optimisation	Dairy	Klevenhusen et al. (2014)
Smartbow[5]	Accelerometer ear tag	Monitor rumination behaviour for detection of health and oestrus	Dairy	Borchers et al. (2016), Reiter et al. (2018)

MooMonitor[6]	Accelerometer collar	Monitor rumination, feeding, activity and rest behaviour; oestrus detection and health monitoring	Dairy	Werner et al. (2019), Moore et al. (2021)
Ceres Tag[7]	GPS and accelerometer in an ear tag with satellite connectivity	Activity and location monitoring; mortality and high activity alerts for remote cattle monitoring; cattle theft; data and software integration	Beef	NA

[1]https://www.moovement.com.au/products
[2]https://www.cowmanager.com
[3]https://www.allflex.global/na/product_cat/livestock-monitoring/
[4]https://smaxtec.com
[5]https://www.smartbow.com
[6]https://www.dairymaster.com/products/moomonitor/
[7]https://www.cerestag.com/

the right animal in the herd management software), appropriate recording/observation of behaviour of interest, settings and thresholds in the software, cow/farm or management-related factors and farmer interpretation of alerts (Rutten et al., 2013; Elischer et al., 2013; Hendriks et al., 2020). Therefore, there is a critical need for technologies to be validated by trusted external research partners, on different animals, herds and settings to ensure their potential application.

Furthermore, not all sensors are available in every country, but with well over 100 sensors available for livestock monitoring available commercially (Stygar et al., 2021), it is evident that purchasing decisions can be extremely challenging for farmers. This further raises the need for data integration and consultancy to enable efficient selection for on-farm purposes, which may address issues with adoption and uptake. In order for value to be generated from the adoption of technology, it is critical that raw measurements of variables are transformed into relevant information and insights for farmers. If this does not occur, farmers are usually left with a huge amount of data that becomes overwhelming and farmers stop utilising the system, causing a lower level of satisfaction with the technology. It is therefore not surprising that farmers express a higher degree of satisfaction with automation technologies, such as cup removers or drafting systems, in comparison to data-capturing technologies given the former are less operator-dependent, whereas the latter relies on the interpretation of data and decisions to be made by the farmer (Dela Rue et al., 2020).

3 Adoption of activity and location technologies

In general adoption has lagged behind the development of most technologies currently available. It is evident that this could be linked to farmer perception of value, with clear limitations to data processing, the integration between different technologies and the presentation of results effectively to enable decision-making (Shalloo et al., 2018; Borchers and Bewley, 2015). Factors such as uncertainty, lack of understanding of available options, observation of what peers are or are not doing, technology's capabilities and reliability, cost, unclear value proposition, ease of use, skills, timing, integration to other farm technologies and management practices, and local service and support explain to some degree the lower levels of adoption observed across many technologies (Steeneveld et al., 2017; Eastwood and Renwick, 2020; Borchers and Bewley, 2015; Eckelkamp, 2019; Jago et al., 2013; Lovarelli et al., 2020).

Adoption curves vary for different technologies and across dairy-producing regions, though generally adoption of sensors on-farm is low (Rutten et al., 2018). In a 2013 survey by Borchers and Bewley (2015) that included responses from 109 farmers across the US, 21% of them had technologies that detected standing oestrus, and 8.3% either animal position and location or lying and standing behaviour technologies. In that same survey standing oestrus and cow activity were rated within the most useful parameters, whereas animal position and location were rated among the least useful. In a 2014 survey of reproduction management on Canadian dairy farms, the adoption of automated oestrus detection systems represented 28% of herds, but this figure varied according to the region and housing system (Cerri et al., 2021; Denis-Robichaud et al., 2016). In an Australian survey conducted in 2014 (Beggs et al., 2015) assessing relationships between herd size and known or proposed risk factors for adverse animal welfare outcomes, larger herd sizes were more likely to use pedometers or activity meters. This contradicted the findings of a later survey conducted in 2015 evaluating current and intended technology investments on almost 200 Australian dairy farms (Gargiulo et al., 2018), which showed that although in general larger farms adopted more technology, farm size had no impact on the adoption of automatic oestrus detection systems (average 7% of surveyed farms). This last survey also demonstrated that there were no significant differences between either farm sizes or roles in the dairy industry (farmer or service provider) on perception towards future adoption of this same technology. That same survey indicated there was general optimism as to the future role of technology on-farm, with almost 80% of farmers expected an increasing adoption of automatic oestrus detection systems, which actually made it the technology with the highest expected adoption (followed by automatic sorting gates and automated mastitis detection tools with a 75% and 66% expected adoption, respectively). In 2018, the TechKISS project (Harris Park Group, 2019) aimed to understand on-farm adoption and adaptation of different target technologies. Twenty-six percent of farms surveyed (gathered through 102 responses to a survey and 39 individual interviews) used activity meters, particularly larger farms (13% of farms with less than 300 cows, 39% of farms with 301–700 cows and 83% of farms with more than 701 cows). These were mostly installed over the previous 2 years, were predominantly accelerometer-based collars, purchased with the aim of simplifying and improving oestrus detection as well as providing cow health alerts. Satisfaction with this technology was high, with most responders agreeing with 'accurately identifies cows', 'does everything we want' and 'made a significant difference to farm business'. Rutten et al. (2018) evaluated adoption by modelling investment decisions and net present value (NPV). Perceived value is impacted by the rate of technological progress; thus uncertainty in future sensor performance and the availability of support in decision-making were highlighted as important drivers in the decision to invest in sensor technology now or in the future (Rutten et al., 2018).

Another consideration is that the adoption of most of these technologies is scalable, which means farmers can decide to purchase less of them at the beginning, only for some animals, and go on to purchase more later on as they gain confidence or see the benefit in it. This means that they can choose to use them on certain animals during a period of time (activity meters, for example, to detect oestrus on cows, and once pregnancy is confirmed they can then move the sensor onto another animal; or use it on the same animal until it is culled or dies). Some of these technologies offer multiple functionalities that in time can either be turned on or off or added through hardware and software upgrades.

While it is generally accepted that precision technologies for monitoring livestock will improve farm efficiency and productivity, detailed studies on the economic sustainability of technology adoption are yet to be conducted. Lovarelli et al. (2020) highlight the need for Life Cycle Assessment (LCA), Life Cycle Cost (LCC) and Social Life Cycle Assessment (SLCA) methods which would contribute to policy approaches that may incentivise and improve the ease of adoption.

4 Integration of technology into the farm system and industry

There is a vast amount of data collected from monitoring technologies; however, utilisation of this data is sub-optimal, given the limitation of methods and systems for data integration and processing (Stygar et al., 2021). Most research to date has focused on individual technologies that in most cases measure one or only a few variables and not on integration with other technologies, the farming system itself or the value it might bring to the whole supply chain. However, the integration of activity and location sensors with other on-farm sensors would improve the accuracy of predictions of certain events. For example, activity has effectively been used in combination with the expected calving date to predict the start of calving (Rutten et al., 2017). In another study, lameness was detected with a high level of accuracy when the output of a neck-mounted accelerometer that gathered behavioural metrics was combined with parity, days in milk, milk production and liveweight (Borghart et al., 2021). A recent review by Cerri et al. (2021) also highlighted that combining measurements within one system is potentially a better alternative rather than combining measurements from different sensors.

Future progress on the usefulness of sensor technologies will be based on their integration in decision support software that brings together data produced or captured by different devices and sensors and is coupled with a powerful data analytics tool. This will allow farmers to make better use of data to improve on-farm decisions, leading to enhanced productivity and profitability. The integration of data on animal performance and behaviour with other on-farm data such as feed quality or environmental factors is possible through the use of the internet of thing (IoT) solutions using cloud-based data exchange (Michie et al., 2020). Michie et al. (2020) demonstrated the impact that decision support tools can have on operational efficiency in dairy systems by integrating data using cloud-based systems to significantly improve oestrus and disease detection and feed efficiency. By combining data from milk conductivity meters with collar-derived rumination and feeding behaviour, detection of mastitis was significantly improved, with 90% of cases detected as effectively as close clinical observations (Michie et al., 2020). Borghart et al. (2021) facilitated early lameness detection by scoring cows as being either sound (locomotion score 1) or unsound (locomotion scores 2 to 5) by combining data from an accelerometer with other production and animal data.

There are also a lot of potential opportunities to integrate data from different farms to gain a better understanding of variability within and across farms. This could enable not only benchmarking between farms but also the development of decision support systems that integrate different data from different farms and, through statistical analysis and models, enable a better understanding of drivers of system performance but also help farmers optimise certain parameters on-farm. This has recently been shown to be possible and effective and well-received by farmers, with data generated from multiple commercial farms operating robotic milking systems and the development of a novel decision support system (Gargiulo et al., 2022).

Anecdotally, farmers have highlighted the need for greater awareness across the whole industry and supply chain of the potential applications and challenges associated with these technologies. Engagement of service providers and farm consultants with the target technologies appears to be relatively low. Although most believe technology is the way to go, they find it difficult to access data from herd management software, as they are faced with multiple software packages and versions, each requiring a different approach. Furthermore, obtaining reliable, insightful and practical data that farmers can act upon with a low level of false negatives and false positives (high specificity and sensitivity) is key for farmers to continue using and relying on the technology as a tool to enable better decision-making on-farm. This limits their use of data to provide advice. To date, no one appears to be leading a commercial opportunity to provide independent advice about either selection, investment or management of technology or data on-farm.

5 Future trends in research and development

Further development of all these technologies continues and is still warranted, particularly in the areas of size, durability, longevity and data transfer (Costa et al., 2021; Rutten et al., 2013). Most importantly is the development of advanced data analytics and the continuous refinement of algorithms that lead to faster and more accurate prediction of biologically relevant events (Eckelkamp, 2019). Machine learning techniques can identify patterns in data that the human eye and brain cannot easily or timely recognise. This has led to the emergence of a vastly multi-disciplinary approach to agriculture in the past decade, increasingly integrating engineering, computer and data scientists.

Additionally, off-animal monitoring through commercial satellite or sensors mounted on unmanned ground or aerial vehicles (UAV) offer the potential to integrate the measurement of non-animal variables on-farm such as pasture condition (Tamas et al., 2019) to provide a whole-systems approach to livestock monitoring. Other on-farm management tasks such as herding animals to the dairy (Clark et al., 2014) or identifying and spraying weeds and fertilising paddocks (Islam et al., 2021b) can be incorporated. Other technologies such as wearable or implantable biosensors that provide detailed information on an animal's physiological state are also expected to be further developed and adopted on-farm (Neethirajan, 2017).

Farmers will continue to adopt sensors, robotics and automation that will replace or enhance many of the management skills, tasks and decisions on-farm (Britt et al., 2018). This will also impact the type and nature of work done by farm consultants and service providers that will have to adapt their service offering and business model to remain relevant. Additionally, we should expect to see a shift in the type and application of technology, some of which are currently being observed already. For example, from on-animal to off-animal sensors; from the use on milking cows to include calves, heifers and dry cows; from sensors that measure one single variable to those that measure - and integrate - multiple variables; from measuring variables to suggesting or implementing changes or interventions autonomously wherever possible (with boundaries defined by the farm manager); from stand-alone technology to integrated technology; from monitoring cows primarily during or around milking to monitoring cows 24/7; from detecting events to predicting events to preventing events; and from monitoring or optimising production traits to monitoring and tracking environmental and animal welfare traits. Many of these things are strongly aligned with the four levels of sensors as described by Rutten et al. (2013).

Most importantly, growing demand from consumers for transparency and provenance in their food sources continues to drive technological adoption across the livestock production industries to enable objective, quantifiable best practices. For this to be an effective tool we need to see improvements in data flowing from the farm to the consumer through the whole supply chain, incorporating social research to meet the needs of consumers. This in turn can lead to more profitable and sustainable production systems in the future.

6 Conclusions

Monitoring animal location and activity provides important information to both producers and consumers as to the productivity, health and welfare of livestock under various production environments. The vast range of technologies available is becoming more affordable and accurate as technology exponentially improves. This will drive adoption on-farm, which will enable systems for early detection and intervention to prevent disease, make timely management decisions and benchmark animal welfare on-farm. The use of all these technologies also has the potential to improve farmer well-being by reducing time spent on repetitive tasks and providing opportunities for innovation and diversification that will increase efficiency and profitability. There are risks with rapid technological developments, however, as with increasing amounts of data available, the requirement for more efficient data processing and packaging to enable targeted decision-making is essential. Yet there is still work to be done in the validation, integration and cost-benefit or value proposition of many of these technologies. The accelerated development and adoption of these technologies also present opportunities for the emergence of new employment opportunities in data translation consultancy. The rise of off-animal sensing technologies presents the greatest potential for animal monitoring, as this will preclude the need for individualised devices. In turn, this can improve the environmental impact of farming systems, while making technology more accessible to producers across all regions of the world.

7 References

Anderson, D. M. (2007). Virtual fencing - past, present and future. *Rangeland Journal* 29(1): 65-78.

Anderson, D. M., Estell, R. E., Holechek, J. L., Ivey, S. and Smith, G. B. (2014). Virtual herding for flexible livestock management - a review. *Rangeland Journal* 36(3): 205-221.

Aquilani, C., Confessore, A., Bozzi, R., Sirtori, F. and Pugliese, C. (2022). Review: precision livestock farming technologies in pasture-based livestock systems. *Animal* 16(1): 100429.

Aungier, S. P. M., Roche, J. F., Sheehy, M. and Crowe, M. A. (2012). Effects of management and health on the use of activity monitoring for estrus detection in dairy cows. *Journal of Dairy Science* 95(5): 2452-2466.

Bar, D., Kaim, M., Flamenbaum, I., Hanochi, B. and Toaff-Rosenstein, R. L. (2019). Technical note: accelerometer-based recording of heavy breathing in lactating and dry cows as an automated measure of heat load. *Journal of Dairy Science* 102(4): 3480-3486.

Beauchemin, K. A. (2018). Invited review: current perspectives on eating and rumination activity in dairy cows. *Journal of Dairy Science* 101(6): 4762-4784.

Beggs, D. S., Fisher, A. D., Jongman, E. C. and Hemsworth, P. E. (2015). A survey of Australian dairy farmers to investigate animal welfare risks associated with increasing scale of production. *Journal of Dairy Science* 98(8): 5330-5338.

Berckmans, D. (2017). General introduction to precision livestock farming. *Animal Frontiers* 7(1): 6-11.

Bewley, J. M. and Russell, A. (2010). Reasons for slow adoption rates of precision dairy farming technologies: evidence from a producer survey. In: *Proceedings of the first North American conference on precision dairy management, Toronto, Canada, 2-5 March 2010*, pp. 30–31. Available at http://precisiondairy.com/proceedings/s1bewley.pdf (accessed 24 May 2022).

Bikker, J. P., van Laar, H., Rump, P., Doorenbos, J., van Meurs, K., Griffioen, G. M. and Dijkstra, J. (2014). Technical note: evaluation of an ear-attached movement sensor to record cow feeding behavior and activity. *Journal of Dairy Science* 97(5): 2974–2979.

Bloch, V. and Pastell, M. (2020). Monitoring of cow location in a barn by an open-source, low-cost, low-energy bluetooth tag system. *Sensors (Basel, Switzerland)* 20(14): 3841.

Borchers, M. R. and Bewley, J. M. (2015). An assessment of producer precision dairy farming technology use, prepurchase considerations, and usefulness. *Journal of Dairy Science* 98(6): 4198–4205. DOI: 10.3168/jds.2014-8963.

Borchers, M. R., Chang, Y. M., Tsai, I. C., Wadsworth, B. A. and Bewley, J. M. (2016). A validation of technologies monitoring dairy cow feeding, ruminating, and lying behaviors. *Journal of Dairy Science* 99(9): 7458–7466.

Borghart, G. M., O'Grady, L. E. and Somers, J. R. (2021). Prediction of lameness using automatically recorded activity, behavior and production data in post-parturient Irish dairy cows. *Irish Veterinary Journal* 74(1): 4.

Britt, J. H., Cushman, R. A., Dechow, C. D., Dobson, H., Humblot, P., Hutjens, M. F., Jones, G. A., Ruegg, P. S., Sheldon, I. M. and Stevenson, J. S. (2018). Invited review: learning from the future–a vision for dairy farms and cows in 2067. *Journal of Dairy Science* 101(5): 3722–3741.

Brown, A. C., Powell, J. G., Kegley, E. B., Gadberry, M. S., Reynolds, J. L., Hughes, H. D., Carroll, J. A., Burdick Sanchez, N. C., Thaxton, Y. V., Backes, E. A. and Richeson, J. T. (2015). Effect of castration timing and oral meloxicam administration on growth performance, inflammation, behavior, and carcass quality of beef calves. *Journal of Animal Science* 93(5): 2460–2470.

Campbell, D. L. M., Haynes, S. J., Lea, J. M., Farrer, W. J. and Lee, C. (2018). Temporary exclusion of cattle from a riparian zone using virtual fencing technology. *Animals* 9(1): 5.

Campbell, D. L. M., Lea, J. M., Farrer, W. J., Haynes, S. J. and Lee, C. (2017a). Tech-savvy beef cattle? How heifers respond to moving virtual fence lines. *Animals* 7(9): 72.

Campbell, D. L. M., Lea, J. M., Haynes, S. J., Farrer, W. J., Leigh-Lancaster, C. J. and Lee, C. (2017b). Virtual fencing of cattle using an automated collar in a feed attractant trial. *Applied Animal Behaviour Science* 200: 71–77. DOI: 10.1016/j.applanim.2017.12.002.

Cantor, M. C., Renaud, D. L., Neave, H. W. and Costa, J. H. C. (2022). Feeding behavior and activity levels are associated with recovery status in dairy calves treated with antimicrobials for Bovine Respiratory Disease. *Scientific Reports (Nature Publisher Group)* 12(1): 4854.

Cerri, R. L. A., Burnett, T. A., Madureira, A. M. L., Silper, B. F., Denis-Robichaud, J., LeBlanc, S., Cooke, R. F. and Vasconcelos, J. L. M. (2021). Symposium review: linking activity-sensor data and physiology to improve dairy cow fertility. *Journal of Dairy Science* 104(1): 1220–1231.

Chanvallon, A., Coyral-Castel, S., Gatien, J., Lamy, J. M., Ribaud, D., Allain, C., Clément, P. and Salvetti, P. (2014). Comparison of three devices for the automated detection of estrus in dairy cows. *Theriogenology* 82(5): 734–741.

Chapa, J. M., Maschat, K., Iwersen, M., Baumgartner, J. and Drillich, M. (2020). Accelerometer systems as tools for health and welfare assessment in cattle and pigs - a review. *Behavioural Processes* 181: 104262.

Chen, Y., He, D. and Song, H. (2018). Automatic monitoring method of cow ruminant behavior based on spatio-temporal context learning. *International Journal of Agricultural and Biological Engineering* 11(4): 179-185.

Clark, C. E. F., Garcia, S. C., Kerrisk, K. L., Underwood, J. P., Nieto, J. I., Calleija, M. S., Sukkarieh, S., Cronin, G. M. (2014). Herding cows with a robot: the behavioral response of dairy cows to an unmanned ground vehicle. 2014 ADSA-ASAS-CSAS Joint Annual Meeting, USA.

Clark, C. E. F., Lyons, N. A., Millapan, L., Talukder, S., Cronin, G. M., Kerrisk, K. L. and Garcia, S. C. (2015). Rumination and activity levels as predictors of calving for dairy cows. *Animal* 9(4): 691-695.

Colusso, P. I., Clark, C. E. F. and Lomax, S. (2020). Should dairy cattle be trained to a virtual fence system as individuals or in groups? *Animals* 10(10): 1767.

Costa, J. H. C., Cantor, M. C. and Neave, H. W. (2021). Symposium review: precision technologies for dairy calves and management applications. *Journal of Dairy Science* 104(1): 1203-1219.

Dairy Australia (2020). Dairy Australia. In: *Focus 2020*. The Australian Dairy Industry. Dairy Australia Limited, 2020, Victoria.

de Koning, K. (2011). Automatic milking: common practice on over 10,000 dairy farms worldwide. Dairy Research Foundation Symposium 2011. The University of Sydney, Camden, Australia, 14-31.

Dela Rue, B. T., Eastwood, C. R., Edwards, J. P. and Cuthbert, S. (2020). New Zealand dairy farmers preference investments in automation technology over decision-support technology. *Animal Production Science* 60(1): 133-137.

Denis-Robichaud, J., Cerri, R. L. A., Jones-Bitton, A. and LeBlanc, S. J. (2016). Survey of reproduction management on Canadian dairy farms. *Journal of Dairy Science* 99(11): 9339-9351.

Duthie, C. A., Bowen, J. M., Bell, D. J., Miller, G. A., Mason, C. and Haskell, M. J. (2021). Feeding behaviour and activity as early indicators of disease in pre-weaned dairy calves. *Animal* 15(3): 100150.

Eastwood, C. R., Chapman, D. F. and Paine, M. S. (2012). Networks of practice for co-construction of agricultural decision support systems: case studies of precision dairy farms in Australia. *Agricultural Systems* 108: 10-18.

Eastwood, C. R., Jago, J. G., Edwards, J. P. and Burke, J. K. (2016). Getting the most out of advanced farm management technologies: roles of technology suppliers and dairy industry organisations in supporting precision dairy farmers. *Animal Production Science* 56(10): 1752-1760.

Eastwood, C. R. and Renwick, A. (2020). Innovation uncertainty impacts the adoption of smarter farming approaches. *Frontiers in Sustainable Food Systems* 4. Available at: https://www.frontiersin.org/articles/10.3389/fsufs.2020.00024/full.

Eckelkamp, E. A. (2019). Invited review: current state of wearable precision dairy technologies in disease detection. *Applied Animal Science* 35(2): 209-220.

Edwards, E. M., Krawczel, P. D., Dann, H. M., Schneider, L. G., Whitlock, B. and Proudfoot, K. L. (2020). Calving location preference and changes in lying and exploratory behavior of preparturient dairy cattle with access to pasture. *Journal of Dairy Science* 103(6): 5455-5465.

Elischer, M. F., Arceo, M. E., Karcher, E. L. and Siegford, J. M. (2013). Validating the accuracy of activity and rumination monitor data from dairy cows housed in a pasture-based automatic milking system. *Journal of Dairy Science* 96(10): 6412–6422.

Frost, A. R., Schofield, C. P., Beulah, S. A., Mottram, T. T., Lines, J. A. and Wathes, C. M. (1997). A review of livestock monitoring and the need for integrated systems. *Computers and Electronics in Agriculture* 17(2): 139–159.

Gardenier, J. S., Underwood, J. P. and Clark, C. E. F. (2018). Object detection for cattle gait tracking. 2018 IEEE International Conference on Robotics and Automation (ICRA), pp. 2206–2213.

Gargiulo, J. I., Eastwood, C. R., Garcia, S. C. and Lyons, N. A. (2018). Dairy farmers with larger herd sizes adopt more precision dairy technologies. *Journal of Dairy Science* 101(6): 5466–5473.

Gargiulo, J. I., Lyons, N. A., Clark, C. E. F. and Garcia, S. C. (2022). The AMS integrated management model: a decision-support system for automatic milking systems. *Computers and Electronics in Agriculture* 196: 106904.

Halachmi, I., Guarino, M., Bewley, J. and Pastell, M. (2019). Smart animal agriculture: application of real-time sensors to improve animal well-being and production. In: Lewin H. A. and Roberts R. M. (Eds) *Annual Review of Animal Biosciences*, Vol. 7, pp. 403–425. Available at: https://www.annualreviews.org/doi/abs/10.1146/annurev-animal-020518-114851?journalCode=animal.

Harris Park Group (2019). TechKISS topic: activity meters. Available at: https://www.harrisparkgroup.com.au/wp-content/uploads/2019/12/TechKISS-Topic-Activity-Meters-2019.pdf (accessed 30 March 2021).

Hendriks, S. J., Phyn, C. V. C., Huzzey, J. M., Mueller, K. R., Turner, S. A., Donaghy, D. J. and Roche, J. R. (2020). Graduate student literature review: evaluating the appropriate use of wearable accelerometers in research to monitor lying behaviors of dairy cows. *Journal of Dairy Science* 103(12): 12140–12157.

Innovation for Agriculture (2019). 4D4F technology warehouse. Available at: https://www.4d4f.eu/content/technology-warehouse (accessed 24 May 2022).

Islam, M. A., Lomax, S., Doughty, A. K., Islam, M. R. and Clark, C. E. F. (2020). Automated monitoring of panting for feedlot cattle: sensor system accuracy and individual variability. *Animals* 10(9): 1518.

Islam, M. A., Lomax, S., Doughty, A. K., Islam, M. R. and Clark, C. E. F. (2021a). Timing of eating during transition impacts feedlot cattle diet and liveweight gain. *Animal* 15(3): 100137.

Islam, N. M. M., Rashid, M. M., Pasandideh, F., Ray, B., Moore, S. and Kadel, R. (2021b). A review of applications and communication technologies for internet of things (IoT) and unmanned aerial vehicle (UAV) based sustainable smart farming. *Sustainability* 13(4): 1821.

Jago, J., Eastwood, C., Kerrisk, K. and Yule, I. (2013). Precision dairy farming in Australasia: adoption, risks and opportunities. *Animal Production Science* 53(9): 907–916.

Kamphuis, C., DelaRue, B., Burke, C. R. and Jago, J. (2012). Field evaluation of 2 collar-mounted activity meters for detecting cows in estrus on a large pasture-grazed dairy farm. *Journal of Dairy Science* 95(6): 3045–3056.

Klevenhusen, F., Pourazad, P., Wetzels, S. U., Qumar, M., Khol-Parisini, A. and Zebeli, Q. (2014). Technical note: evaluation of a real-time wireless pH measurement system relative to intraruminal differences of digesta in dairy cattle. *Journal of Animal Science* 92(12): 5635–5639.

Krieger, S., Oczak, M., Lidauer, L., Berger, A., Kickinger, F., Öhlschuster, M., Auer, W., Drillich, M. and Iwersen, M. (2019). An ear-attached accelerometer as an on-farm device to predict the onset of calving in dairy cows. *Biosystems Engineering* 184: 190–199.

Langworthy, A. D., Verdon, M., Freeman, M. J., Corkrey, R., Hills, J. L. and Rawnsley, R. P. (2021). Virtual fencing technology to intensively graze lactating dairy cattle I: technology efficacy and pasture utilization. *Journal of Dairy Science* 104(6): 7071-7083. DOI: 10.3168/jds.2020-19796.

Lazzarini, B., Baudracco, J., Tuñon, G., Gastaldi, L., Lyons, N., Quattrochi, H. and Lopez-Villalobos, N. (2019). Review: milk production from dairy cows in Argentina: current state and perspectives for the future. *Applied Animal Science* 35(4): 426–432.

Lee, C., Colditz, I. G. and Campbell, D. L. M. (2018). A framework to assess the impact of new animal management technologies on welfare: a case study of virtual fencing. *Frontiers in Veterinary Science* 5: 187.

Li, G., Huang, Y., Chen, Z., Chesser, G. D., Purswell, J. L., Linhoss, J. and Zhao, Y. (2021). Practices and applications of convolutional neural network-based computer vision systems in animal farming: a review. *Sensors* 21(4): 1492.

Livestock Improvement Corporation Limited and DairyNZ Limited (2020). New Zealand dairy statistics 2019–20. Available at: https://www.dairynz.co.nz/media/5794073/nz-dairy-statistics-2019-20-dnz.pdf.

Lomax, S., Colusso, P. and Clark, C. E. F. (2019). Does virtual fencing work for grazing dairy cattle? *Animals* 9(7): 429.

Lovarelli, D., Bacenetti, J. and Guarino, M. (2020). A review on dairy cattle farming: is precision livestock farming the compromise for an environmental, economic and social sustainable production? *Journal of Cleaner Production* 262: 121409.

Lowe, G. L., Sutherland, M. A., Waas, J. R., Schaefer, A. L., Cox, N. R. and Stewart, M. (2019). Physiological and behavioral responses as indicators for early disease detection in dairy calves. *Journal of Dairy Science* 102(6): 5389–5402.

Marchesini, G., Mottaran, D., Contiero, B., Schiavon, E., Segato, S., Garbin, E., Tenti, S. and Andrighetto, I. (2018). Use of rumination and activity data as health status and performance indicators in beef cattle during the early fattening period. *The Veterinary Journal* 231: 41–47.

Maroto-Molina, F., Navarro-García, J., Príncipe-Aguirre, K., Gómez-Maqueda, I., Guerrero-Ginel, J. E., Garrido-Varo, A. and Pérez-Marín, D. C. (2019). A low-cost IoT-based system to monitor the location of a whole herd. *Sensors (Basel, Switzerland)* 19(10): 2298.

Marti, S., Melendez, D. M., Pajor, E. A., Moya, D., Gellatly, D., Janzen, E. D. and Schwartzkopf-Genswein, K. S. (2018). Effect of a single dose of subcutaneous meloxicam prior to band or knife castration in 1-wk-old beef calves: II. Inflammatory response and healing. *Journal of Animal Science* 96(10): 4136–4148.

Mayo, L. M., Silvia, W. J., Ray, D. L., Jones, B. W., Stone, A. E., Tsai, I. C., Clark, J. D., Bewley, J. M. and Heersche, G. (2019). Automated estrous detection using multiple commercial precision dairy monitoring technologies in synchronized dairy cows. *Journal of Dairy Science* 102(3): 2645–2656.

McGavin, S. L., Bishop-Hurley, G. J., Charmley, E., Greenwood, P. L. and Callaghan, M. J. (2018). Effect of GPS sample interval and paddock size on estimates of distance travelled by grazing cattle in rangeland, Australia. *The Rangeland Journal* 40(1): 55.

Merenda, V. R., Figueiredo, C. C., González, T. D. and Chebel, R. C. (2020). Technical note: validation of a system for monitoring individual behavior of Holstein cows. *Journal of Dairy Science* 103(8): 7425–7430.

Merenda, V. R., Marques, O., Miller-Cushon, E. K., Dilorenzo, N., Laporta, J. and Chebel, R. C. (2019). Technical note: validation of a system for monitoring individual behavior in beef heifers. *Journal of Animal Science* 97(12): 4732–4736.

Michie, C., Andonovic, I., Davison, C., Hamilton, A., Tachtatzis, C., Jonsson, N., Duthie, C. A., Bowen, J. and Gilroy, M. (2020). The Internet of things enhancing animal welfare and farm operational efficiency. *Journal of Dairy Research* 87(Suppl. 1): 20–27.

Molfino, J., Clark, C. E. F., Kerrisk, K. L. and García, S. C. (2017). Evaluation of an activity and rumination monitor in dairy cattle grazing two types of forages. *Animal Production Science* 57(7): 1557–1562.

Moore, S. G., Aublet, V. and Butler, S. T. (2021). Monitoring estrous activity in pasture-based dairy cows. *Theriogenology* 160: 90–94.

Neethirajan, S. (2017). Recent advances in wearable sensors for animal health management. *Sensing and Bio-Sensing Research* 12: 15–29.

O'Brien, B., Gleeson, D., O'Donovan, K., Ruane, D., Kinsella, J., Mee, J. F., Boyle, L. and McNamara, J. (2007). Labour efficiency on-farm.Teagasc, End of Project Report 5193, pp. 1–37.

Pereira, G. M., Heins, B. J. and Endres, M. I. (2018). Technical note: validation of an ear-tag accelerometer sensor to determine rumination, eating, and activity behaviors of grazing dairy cattle. *Journal of Dairy Science* 101(3): 2492–2495.

Pereira, G. M., Sharpe, K. T. and Heins, B. J. (2021). Evaluation of the RumiWatch system as a benchmark to monitor feeding and locomotion behaviors of grazing dairy cows. *Journal of Dairy Science* 104(3): 3736–3750.

Qiao, Y., Su, D., Kong, H., Sukkarieh, S., Lomax, S. and Clark, C. (2020). Data augmentation for deep learning based cattle segmentation in precision livestock farming 2020 IEEE 16th International Conference on Automation Science and Engineering (CASE), pp. 979-984.

Ramón-Moragues, A., Carulla, P., Mínguez, C., Villagrá, A. and Estellés, F. (2021). Dairy cows activity under heat stress: a case study in Spain. *Animals* 11(8): 2305.

Reiter, S., Sattlecker, G., Lidauer, L., Kickinger, F., Öhlschuster, M., Auer, W., Schweinzer, V., Klein-Jöbstl, D., Drillich, M. and Iwersen, M. (2018). Evaluation of an ear-tag-based accelerometer for monitoring rumination in dairy cows. *Journal of Dairy Science* 101(4): 3398-3411.

Rutten, C. J., Kamphuis, C., Hogeveen, H., Huijps, K., Nielen, M. and Steeneveld, W. (2017). Sensor data on cow activity, rumination, and ear temperature improve prediction of the start of calving in dairy cows. *Computers and Electronics in Agriculture* 132: 108–118.

Rutten, C. J., Steeneveld, W., Lansink, A. G. J. M. and Hogeveen, H. (2018). Delaying investments in sensor technology: the rationality of dairy farmers' investment decisions illustrated within the framework of real options theory. *Journal of Dairy Science* 101(8): 7650–7660.

Rutten, C. J., Velthuis, A. G. J., Steeneveld, W. and Hogeveen, H. (2013). Invited review: sensors to support health management on dairy farms. *Journal of Dairy Science* 96(4): 1928–1952.

Shalloo, L., O'Donovan, M., Leso, L., Werner, J., Ruelle, E., Geoghegan, A., Delaby, L. and O'Leary, N. (2018). Review: grass-based dairy systems, data and precision technologies. *Animal* 12(s2): s262–s271.

Shepley, E., Lensink, J., Leruste, H. and Vasseur, E. (2020). The effect of free-stall versus strawyard housing and access to pasture on dairy cow locomotor activity and time budget. *Applied Animal Behaviour Science* 224: 104928.

Smid, A. M. C., Weary, D. M. and von Keyserlingk, M. A. G. (2020). Effect of outdoor open pack space allowance on the behavior of freestall-housed dairy cows. *Journal of Dairy Science* 103(4): 3422-3430.

Smith, J. L., Vanzant, E. S., Carter, C. N. and Jackson, C. B. (2015). Discrimination of healthy versus sick steers by means of continuous remote monitoring of animal activity. *American Journal of Veterinary Research* 76(8): 739-744.

Stachowicz, J. and Umstatter, C. (2020). Overview of commercially available digital systems in animal husbandry. *Agroscope Transfer* 294: 1-28.

Stachowicz, J. and Umstatter, C. (2021). Do we automatically detect health- or general welfare-related issues? A framework. *Proceedings of the Royal Society B: Biological Sciences* 288(1950): 20210190.

Stangaferro, M. L., Wijma, R., Caixeta, L. S., Al-Abri, M. A. and Giordano, J. O. (2016a). Use of rumination and activity monitoring for the identification of dairy cows with health disorders: Part I. Metabolic and digestive disorders. *Journal of Dairy Science* 99(9): 7395-7410.

Stangaferro, M. L., Wijma, R., Caixeta, L. S., Al-Abri, M. A. and Giordano, J. O. (2016b). Use of rumination and activity monitoring for the identification of dairy cows with health disorders: Part II. Mastitis. *Journal of Dairy Science* 99(9): 7411-7421.

Stangaferro, M. L., Wijma, R., Caixeta, L. S., Al-Abri, M. A. and Giordano, J. O. (2016c). Use of rumination and activity monitoring for the identification of dairy cows with health disorders: Part III. Metritis. *Journal of Dairy Science* 99(9): 7422-7433.

Steeneveld, W., Rutten, C. J., Oude Lansink, A. G. J. M. and Hogeveen, H. (2017). Why not investing in sensors is logical for dairy farmers. Precision Livestock Farming 2017 - Papers Presented at the 8th European Conference on Precision Livestock Farming, ECPLF 2017, 55-65.

Steinmetz, M., von Soosten, D., Hummel, J., Meyer, U. and Dänicke, S. (2020). Validation of the RumiWatch Converter V0.7.4.5 classification accuracy for the automatic monitoring of behavioural characteristics in dairy cows. *Archives of Animal Nutrition* 74(2): 164-172.

Stygar, A. H., Gómez, Y., Berteselli, G. V., Dalla Costa, E., Canali, E., Niemi, J. K., Llonch, P. and Pastell, M. (2021). A systematic review on commercially available and validated sensor technologies for welfare assessment of dairy cattle. *Frontiers in Veterinary Science* 8: 634338.

Swain, D. L., Friend, M. A., Bishop-Hurley, G. J., Handcock, R. N. and Wark, T. (2011). Tracking livestock using global positioning systems - are we still lost? *Animal Production Science* 51(3): 167-175.

Talukder, S., Kerrisk, K. L., Clark, C. E. F., Garcia, S. C. and Celi, P. (2015). Rumination patterns, locomotion activity and milk yield for a dairy cow diagnosed with a left displaced abomasum. *New Zealand Veterinary Journal* 63(3): 180-181.

Tamas, J., Kiss, N. É., Riczu, P., Gálya, B., Budayné-Bódi, E., Gross, M. and Nagy, A. (2019). Application of lidar technology in cattle grazing areas. *Natural Resources and Sustainable Development* 9(2): 106-112.

Theurer, M. E., White, B. J., Coetzee, J. F., Edwards, L. N., Mosher, R. A. and Cull, C. A. (2012). Assessment of behavioral changes associated with oral meloxicam administration at time of dehorning in calves using a remote triangulation device and accelerometers. *BMC Veterinary Research* 8: 48.

Tucker, C. B., Jensen, M. B., de Passillé, A. M., Hänninen, L. and Rushen, J. (2021). Invited review: lying time and the welfare of dairy cows. *Journal of Dairy Science* 104(1): 20–46.

USDA-NASS (2013). Milk cows: number of operations by year, US.

Vázquez Diosdado, J. A., Barker, Z. E., Hodges, H. R., Amory, J. R., Croft, D. P., Bell, N. J. and Codling, E. A. (2018). Space-use patterns highlight behavioural differences linked to lameness, parity, and days in milk in barn-housed dairy cows. *PLoS ONE* 13(12): e0208424.

Verdon, M., Langworthy, A. and Rawnsley, R. (2021). Virtual fencing technology to intensively graze lactating dairy cattle II: Effects on cow welfare and behavior. *Journal of Dairy Science* 104(6): 7084–7094. DOI: 10.3168/jds.2020-19797.

Wang, D., Shao, Q. and Yue, H. (2019). Surveying wild animals from satellites, manned aircraft and unmanned aerial systems (UASs): a review. *Remote Sensing* 11(11): 1308.

Wang, J., He, Z., Zheng, G., Gao, S. and Zhao, K. (2018). Development and validation of an ensemble classifier for real-time recognition of cow behavior patterns from accelerometer data and location data. *PLoS ONE* 13(9): e0203546.

Werner, J., Umstatter, C., Leso, L., Kennedy, E., Geoghegan, A., Shalloo, L., Schick, M. and O'Brien, B. (2019). Evaluation and application potential of an accelerometer-based collar device for measuring grazing behavior of dairy cows. *Animal* 13(9): 2070–2079.

Wolfger, B., Timsit, E., Pajor, E. A., Cook, N., Barkema, H. W. and Orsel, K. (2015). Technical note: accuracy of an ear tag-attached accelerometer to monitor rumination and feeding behavior in feedlot cattle. *Journal of Animal Science* 93(6): 3164–3168.

Zambelis, A., Wolfe, T. and Vasseur, E. (2019). Technical note: validation of an ear-tag accelerometer to identify feeding and activity behaviors of tiestall-housed dairy cattle. *Journal of Dairy Science* 102(5): 4536–4540.

Chapter 6

Developments in data analysis for decision-making in precision livestock farming systems

Elaine van Erp-van der Kooij, HAS University of Applied Sciences, The Netherlands

1 Introduction

1.1 Big data in smart farming

Because smart machines and sensors have entered many farms, farm data grow in quantity and scope, leading to more data-driven farming decisions. In Precision Livestock Farming, data streams are growing to 25 images per second, 20 000 sound samples per second or 250 accelerometer samples per second. Smart farming is impelled by developments in the Internet of Things and cloud computing. Precision livestock farming (PLF) is based on individual animal data, considering the in-herd variability, and taking into account the context, the situation and real-time events. Smart farming systems offer real-time assistance to the farmer, alerting the farmer in time to take actions in case of changes in operational or animal conditions (e.g. climate or disease alerts). Smart devices can be aware of their environment by using built-in sensors, and are increasingly capable of performing autonomous actions, but the farmer will always be involved in the process. Machines are increasingly capable of making decisions based on data, but they are not 'intelligent' yet, in the sense that they can deal with new or trying situations. So far, PLF might offer support in process decisions, while the farmer takes the business decisions (Rutten et al., 2013).

http://dx.doi.org/10.19103/AS.2021.0090.06

Smart machines use their sensors to measure data in their environment, that is used for the machines' behaviour. This ranges from relatively simple feedback mechanisms, such as the ventilation and heating system, to deep learning algorithms that predict disease outbreaks or a calving event. These data, sometimes combined with external data sources such as weather data, add up to large data sets: sometimes too large or complex to be handled by traditional data processing applications, especially when audio, video, photographs and structured data are combined and the speed of data acquisition is very large. Big data require new techniques to reveal the information from such large data sets: that is where Data Mining and Data Science comes in (Wolfert et al., 2017). However, also for smaller data sets, data mining techniques are a useful tool.

1.2 From business understanding to data understanding

There are large benefits from basing company decisions on data instead of gut feeling or experience. In a study on how data-driven decision-making affects firm performance, it was concluded that the more data a firm uses, the more productive it is. More use of data also leads to a higher return on assets, return on equity, asset utilization and market value (Brynjolfsson et al., 2011). However, it is essential to first start with a business question, and not to start looking at the randomly available data to see what it will bring. A target has to be defined.

For every company that aims to apply data science, the first step in any process is always defining the problem. What is the business problem that data science needs to solve? Only if the business question is understood completely, we can proceed to the data understanding. Business questions that can be answered using data science for a livestock farm could be:

- Who is the most profitable (group of) animal(s)?
  This question can be answered relatively simply, using a database query. We can select the animal or group of animals with the highest milk yield or growth rate, or with the best feed conversion.
- Is there a difference between the more profitable and the average animals?
  This question requires a statistical analysis. Can we find differences between groups of animals?
- Who are these profitable animals, can we characterize them?
  This question can be approached in two ways. First, we can perform a database query, using summary statistics to find the characteristics of the high- and average- (or low-)performing animals. But we can also do a deeper analysis, using pattern recognition techniques. What characteristics differentiate productive animals from less productive animals?
- Which young animals will develop into a high-producing animal?

> This requires more advanced techniques, where we use historical data to find relations, build a prediction model, and apply that model to new data to predict profitability (Provost and Fawcett, 2013). We might find relations between birth weight of piglets and growth rate in the fattening period, and use that to predict growth rates for new litters in the farrowing pen.

Before we can begin with modelling, we must understand the available data and possibly prepare it. This process can be divided in successive stages. First, we evaluate what data we have, and we perform the necessary data preparation. What are the data sources? Do we have enough accurate data or should extra data be acquired? Do the data have to be cleaned, or matched, or otherwise prepared? Beware of 'leaks' in the data: sometimes there is information on the target in the data set, that is not available when the decision must be made (Provost and Fawcett, 2013). An example is a data set with known oestrus lengths, from which we want to predict the optimal insemination moment. Oestrus length is only known when an oestrous event has finished, so this parameter cannot be used to predict the optimal insemination moment in that oestrus. In this stage, or even before this stage, a data architecture had to be built, storing data in a logical and structured way, for example, in a data warehouse. In this chapter we will not go further into the process of data acquisition and data storage, but focus on the data mining process.

Next comes the modelling stage, where data science techniques are used to find relations between the target and the parameters in the data set. Results from this analysis are models or algorithms, for example, to predict an event or to classify an animal. If the model is a regression model or a decision tree, this can be done using a query. Making use of standard query language (SQL) or graphical user interface (GUI), subsets or statistics can be requested from a data set (Provost and Fawcett, 2013). For example, when a relation has been found between high activity on day-1 and calving moment, we can write a query in SQL selecting the cows from the data set showing high activity on day-1 and alert the farmer; those cows have a higher chance of calving the next day than the other cows in the herd.

2 Data science and data mining

Data mining is defined as the extraction of knowledge from data, based on the principles of data science. Data-driven decision-making is based on discoveries in the data, and is especially useful for decisions that must be taken repeatedly. Even a small increase in decision-making accuracy from data can improve decisions.

There are a few fundamental concepts in data science.

1 Extracting useful knowledge from data to solve business problems can be done systematically. There are standards, described in the Cross Industry Standard Process for Data Mining (CRISP-DM).
2 Information technology can be used to find informative descriptive attributes[1] (variables, factors) of entities of interest (individuals or groups). For example, we can use information technology to find relations between behaviour and disease occurrence of individual animals.
3 If your model is based too much on the specific characteristics of the training data, there is a risk of 'overfitting'; you might find relations that are only valid for this specific data set. The solution is using a test data set with data that are independent of the training data.
4 Data mining solutions and evaluating results involves thinking about the context in which they will be used. You have to ask yourself if the knowledge you have gained from the data is useful knowledge, and if the solution will lead to better decisions.

Several techniques can be used in the Data mining process. The most used are classification and regression, and to a lesser degree clustering. An overview is given is in Table 1 (Provost and Fawcett, 2013). Data science techniques can be used for supervised or unsupervised modelling. In a supervised learning model, the algorithm learns on a labelled data set, providing an answer key that the algorithm can use to evaluate its accuracy on training data. An unsupervised model, in contrast, provides unlabelled data that the algorithm tries to make sense of by extracting features and patterns (Salian, 2018).

In this chapter, we will explain relevant techniques in data mining that are being applied most in livestock research: classification, regression and clustering. The difference between supervised and unsupervised modelling is explained further, as well as how to fit the model to the data and how to evaluate the model.

3 Machine learning

Machine learning is the application and science of algorithms that make sense of data. It has evolved as a subfield of artificial intelligence (AI) involving self-learning algorithms that turn data into information, with the aim to make predictions. Machine learning is a more efficient way to make prediction models from the knowledge that is hidden in the data, than the way we used to do that:

1 Some authors also use the term 'features': a feature is a specified attribute, for example, when the attribute is 'breed', then the feature is 'Holstein Friesian' for dairy cows.

Table 1 An overview of data mining techniques for supervised and unsupervised learning (Provost and Fawcett, 2013)

Technique	Description	Application example	Supervised/ unsupervised
Classification	Determining whether an individual belongs to a certain group or cluster; used for prediction	Lameness prediction, or oestrus detection; target variable is binomial	supervised
Regression	Estimating the numeric value of a parameter in an equation that connects variables	Yield prediction; target variable is numeric	supervised
Similarity matching	Identifying similar individuals	Prediction; target variable can be binomial or numeric	supervised/ unsupervised
Clustering	Grouping individuals based on similarities	Exploration of data with no target	unsupervised
Co-occurrence grouping	Finding associations between individuals, based on events or transactions	In marketing: 'what others also bought'	unsupervised
Profiling	Describing behaviour	Descriptive statistics such as daily activity patterns of individual animals, eating or lying time of the herd, and so on	unsupervised
Link prediction	Predicting the connection between data items	In social networking; determining social connections and interactions in a herd or flock	supervised/ unsupervised
Data reduction	Reducing a (big) data set without losing critical information	Reducing data by combining different behaviours or choosing specific animal features for the analysis	supervised/ unsupervised
Causal modelling	Determining what events influence other events	Finding relations between health events on a farm	supervised

with people manually building models based on rules that were derived from large data sets, which took a lot of time and effort (Raschka and Mirjalili, 2019).

Machine learning can be categorized in supervised and unsupervised modelling or learning. As explained above, in a supervised learning model, a labelled data set is used to train the model. In an unsupervised model, unlabelled data are used to find features and patterns. So, if there is a target for grouping your data, and there is a labelled data set, the process is supervised. This is the case when you want to use a certain data mining technique to predict an event: the event is your target, and the process is supervised. Accurate labelling of animal data is a challenging and time-consuming process (Rutten et al., 2013; Nilsson et al., 2015). If there is no target, but you are just interested in whether the individuals fall into different groups, than the process is unsupervised. Clustering or grouping based on similarities is an unsupervised process, which will or will not be meaningful (Caruana and Niculescu-Mizil, 2006).

3.1 Supervised learning

For supervised learning, labelled data sets are needed. Supervised means that the input data should contain a number of known output signals, called labels (Raschka and Mirjalili, 2019). Labelling gives detailed context information that allows to create features in relation to the target that has been set. For livestock, this would mean, for example, an activity data set from a cow herd, with sensor data from accelerometers of individual cows and with known individual calving dates and times.

3.1.1 Classifying objects and predicting events

Mechanistic models have served traditionally as causal pathway analysis and decision-support tools in animal production, but these models need many accurate inputs and values of parameters in the models, user training and accuracy and precision of on-farm predictions. Data-driven modelling methods such as machine learning and deep learning use patterns in data to accurately predict events (forecasting) or to classify objects (Ellis et al., 2020).

If you feed the labelled training data to a machine learning algorithm, a predictive model can be fitted that can make predictions on new data sets. Following the example of calving data for cows, we can use the activity data with the correctly labelled calving events to predict whether in a new data set, certain activity patterns will predict calving in individual cows. This is an example of a classification task. Another subcategory of supervised learning is regression, with a continuous outcome value instead of a binary value (Raschka and Mirjalili, 2019). Another example of supervised learning is classifying objects. This is interesting when animal species or even individuals can be recognized. Counting of animals, distribution of animals in a farm or a pen, and monitoring of individual behavioural patterns become possible with this technique.

In the current PLF systems, sensors monitor animal behaviour and algorithms detect deviations of certain behaviours; when deviations exceed a certain threshold, an alert goes to the farmer. For example, when for a dairy cow the eating time or rumination time differs from an expected value (a fixed threshold, or a prediction based on herd or individual cow data), the farmer is alerted. He can then check on this specific cow. For these algorithms, well-known statistical techniques have been used, such as regression models, to determine the relation between, in this case, disease and activity. However, more companies are using machine learning techniques to analyse the same activity data, detecting anomalies in the behaviours and link them to disease or other events in the animal's life. With these techniques, predictions can be made. Events that are interesting to predict in livestock production are for example occurrence of disease, oestrus or birth. For this, labelled data are needed. When predicting disease from activity data, labelled data sets are needed with known disease events and activity data from the same animals. In that way, a training set can be selected, the model can be trained to recognize patterns in the activity data associated with disease, and in a new data set, disease can be predicted.

3.1.2 Predictive modelling and decision trees

One of the most used techniques in predictive modelling is supervised segmentation using tree induction (Caruana and Niculescu-Mizil, 2006). In this process a predictive model is built based on supervised data mining. The predictive model is a formula for estimating the value of a target. This formula can be a statement or 'rule', a mathematical formula, or a hybrid of both. The supervised data mining model can be a classification model (or class-probability estimation model), or a regression model. The outcome of this model is used to build the predictive model or formula (Provost and Fawcett, 2013).

Formulas in supervised segmentation or tree induction are based on the purity measure or entropy of the segmentation. In information theory, the entropy of a random variable is the average level of information, surprise, or uncertainty in the variable's possible outcome. For example, flipping a coin has a probability p of landing on heads and of $1-p$ of landing on tails: the maximum surprise is for $p=0.5$, which is an unbiased coin (Shannon, 1948). Entropy, used for discrete variables, is similar to variance or standard deviation, used for continuous variables. Supervised segmentation is based on the idea that the data population consists of a group of entities (individuals) with different variables and a mixed target variable. The aim of the segmentation is to find features (specified attributes or variables) that divide the population into groups that are less impure in their target value. The measure for this impurity is entropy: the more mixed ('impure') a group is with regard to the target variable, the higher the entropy. For example, a group of cows with different individual characteristics, and that are either healthy or lame, can be divided into segments of cows based on these individual characteristics; in these segments, there will be a more uniform value of the target, for example, in one segment most cows are lame, in the other, most cows are healthy. If this is the case, then the purity in each segment is higher and the entropy is lower than in the total herd. The splitting criterion for the segments is the information gain (IG), which is the difference between entropy of the total data set minus weighted average of entropy in the segments (Provost and Fawcett, 2013). See textbox for a numeric example.

Numeric example

Entropy can be calculated with the following formula:

$$\text{Entropy}(S) = -p_1 \log_2(p_1) - p_2 \log_2(p_2) - \ldots - p_i \log_2(p_i)$$

where p is the probability of property i within the data set.

When we have 100 cows, of which half is lame, the entropy of the herd is

$$-[0.5*\log_2(0.5) + 0.5*\log_2(0.5)] = -[0.5*(-1)+0.5*(-1)]=1$$

When we divide the herd into segments with high and low milk yield, and in these groups we see a different ratio of lame and healthy cows, we can calculate the entropy in each group that results from the segmentation. If the high-producing group consists of 50 cows and of these cows 35 are lame (p=35/50=0.7), while in the group of 50 low-producing cows only 15 are lame (p=15/50=0.3), we can calculate the new entropies.

Entropy of high producing group: $-[0.7*\log_2(0.7)+0.3*\log(0.3)]=0.88$

Entropy of low producing group: $-[0.3*\log_2(0.3)+0.7*\log(0.7)]=0.88$

Information gain (IG) = $1-[0.5*0.88+0.5*0.88]=0.12$

When we have continuous variables instead of discrete variables, we cannot use entropy to calculate information gain by splitting the data set into segments. For a continuous target variable, variance is used as a measure for impurity. If a subset of data has almost no variation in target value, than the impurity is low and the variance is close to zero. When the variance is high, and target values in the subset differ greatly, then the variance is close to 1. Instead of looking for IG, we are now looking for a reduction in variance. The best segmentation is when the weighted average variance reduction is highest. This means that we have found the variables with the best correlation with the target variable.

Tree structures can be used to build models dividing data into meaningful segments, thus finding the best predictors for values of the target variable. We can simply use the model to classify the entities (animals) into the different groups, this is a classification tree, or we can predict the probability of belonging to a certain group: for discrete variables this is a probability estimation tree, for continuous variables this is a decision tree (Provost and Fawcett, 2013). Trees are understandable and easy to explain, and therefore widely used in the application of data science, for example in classification of behaviours from accelerometer data (Berckmans, 2017), analysing multi-sensor data from sheep (di Virgilio et al., 2018), identifying broiler breeder behaviour (Pereira et al., 2013), classifying stress vocalizations in piglets (da Silva Cordeiro et al., 2013; Cordeiro et al., 2018) and predicting probability of becoming ill in dairy cows (Halachmi, 2015). In Fig. 1, an example of a classification tree is shown.

The outcome of the tree can be translated into a set of rules to select animals at risk from the database. This would look like the following, where Age=Age (in years) of the cow, Milk Yield=Milk Yield per day (in kilogrammes) and Walking Time=Total walking time per day (in hours):

IF (Age>2) AND (Milk Yield>25) AND (Walking Time<AverageWT) THEN CLASS = LAME;
IF (Age>2) AND (Milk Yield>25) AND (Walking Time>AverageWT) THEN CLASS = HEALTHY;
IF (Age>2) AND (Milk Yield<25) THEN CLASS = HEALTHY;
and so on.

The probability in each class can be estimated by using instance counts: a frequency-based estimate of the class membership probability is the number of positive instances divided by the total instances in the segment. If 25 of the 50 cows with age >2 are lame, the probability estimate is 25/50 = 0.5. For small numbers per group, a correction needs to be done. Most used is the Laplace correction. If n=the number of positive instances and m=the number of negative

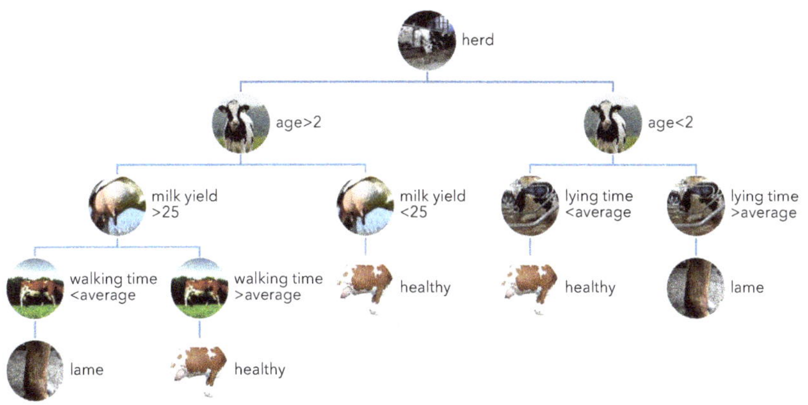

Figure 1 Decision tree for predicting lameness in dairy cows using fictional data.

instances, then the probability estimate is n/(n+m); with the Laplace correction, the probability estimate is (n+1)/(n+m+2) (Provost and Fawcett, 2013).

3.1.3 Fitting a model to data

One way of finding a model for the target variable with the descriptive attributes is building a supervised segmentation model or tree structure. Another way of dividing your data into meaningful segments is using a structure that is described by a function or equation of continuous variables. The values of parameters can be chosen based on domain knowledge or on Information Gain of Variance Reduction techniques, as described earlier. In this case, the data scientist decides what structure to use and which attributes: the goal is finding a model that fits the data best. This is called 'parameter learning' or 'parametric modelling'. Many data mining procedures use this framework, and most use linear models. A well-known example in the scientific world is linear regression. With these linear models, predictions can be made about unknown binary values, the likelihood of events or an unknown numeric value (Provost and Fawcett, 2013).

Classification can be done via a mathematical function. It might be that the combination of two attributes predicts a target value very well. An example for prediction lameness with two attributes, the variables milk yield and lying time, is shown in Fig. 2. In this example, lame cows will be found in the upper right

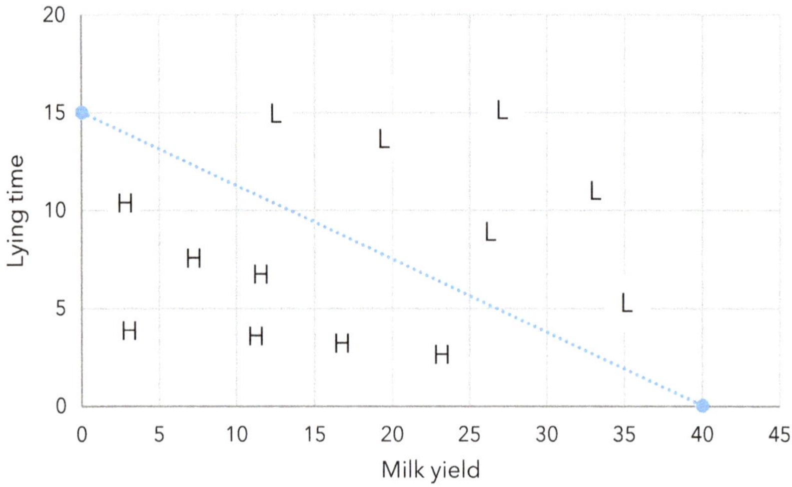

Figure 2 Example of segmentation of a group of cows with binary target lameness: lame (L) or healthy (H), using a linear function of lying time (in hours/day) and milk yield (in kilogrammes/day).

segment, and the linear function of the attributes lying time and milk yield can separate lame cows from healthy ones.

The function of the line can be written as Lying time = –(15/40)*Milk yield + 15, or as:

$$f(x) = -1 * \text{Lying time} - \frac{15}{40} * \text{Milk yield} + 15$$

We can use this equation to write the classification function, if we want to predict lameness:

$$\text{Class}(x) = \begin{cases} \text{Healthy if } -1 * \text{Lying time} - \frac{15}{40} * \text{Milk yield} + 15 > 0 \\ \text{Lame if } -1 * \text{Lying time} - \frac{15}{40} * \text{Milk yield} + 15 \leq 0 \end{cases}$$

The weights of the equation can be interpreted as the importance of each parameter, if the variables are normalized to the same range. The data mining will 'fit' this parameterized model to a particular data set and find a good set of weights for the variables. Therefore we have to optimize the object function. Choosing the weights of the linear equation can be based on linear regression (Montgomery et al., 2012), logistic regression (Hosmer and Lemeshow, 2013) or support vector machines (Statnikov et al., 2011). These are different ways of fitting a linear model to data.

We can use linear discriminant function (Flury, 2013) to classify instances, or to estimate the class membership probability. In some cases, only a rank order is needed for the instances, of how likely it is that they belong to a certain class. This ranking results naturally from the output of the discriminant function. The most uncertainty about which class an instance belongs, is near the decision boundary, when $f(x)$ is close to zero; the uncertainty is low when an instance is far from the boundary, with $f(x)$ being large. In the example above, when $f(x)$ = - 1*Lying time - (15/40)*Milk yield + 15 is large and positive, the probability of that cow being healthy is high.

Measuring how well a model fits the training data, you can give a penalty for each wrongly classified data point, termed 'loss'. Loss functions (Lin, 2002) determine how much penalty should be assigned to an instance based on the error in the model's predicted value. Support vector machines use 'hinge loss' (Statnikov et al., 2011); hinge loss is only positive for instances on the wrong side of the boundary and beyond the margin. Loss increases linearly with the distance from the margin. Zero-one loss assigns zero for a correct and one for an incorrect decision. Squared error is a loss function based on the square of the distance from the boundary. This is used for numeric value prediction as in regression, but that can misjudge points far on the correct side of the decision boundary, and therefore adjusted measures have been created (Provost and Fawcett, 2013).

For numeric target variables, linear regression is a powerful tool to optimize the model's fit to the data. That is done by minimizing the difference between the value of the estimated data by the model and the true values of the training data. Standard linear regression procedures do this by minimizing the sum or means of the squares of the error: the 'linear least squares' regression or the 'total linear least squares' regression in which errors are assumed to exist on both variables in x and y axis.

An example of a linear regression model predicting foot pad lesions in broilers is explained here. In this study (van Erp-van der Kooij et al., 2020b) foot pad lesions were manually scored in six groups of broilers between day 17 and day 38 of age. The data set was simplified to mean percentage of foot pad lesions per day, per group. Per measuring day, birds were divided into three weight classes: average weight birds, light birds (average minus 10% or below) and heavy birds (average plus 10% and above). A repeated measured analysis (Generalized Estimating Equations in SPSS 24)(Greasly, 2007) gave the following output (Table 2).

The Goodness of Fit of this model is reported in the output as well, with the 'Quasi-likelihood under Independence Model Criterion (QIC)' (Montgomery et al., 2012). This can be used to choose between two correlation structures, given a set of model terms. The structure that obtains the smaller QIC is 'better' according to this criterion. The second criterion that is reported is the Corrected Quasi-likelihood under Independence Model Criterion (QICC). This can be used to choose between two sets of model terms, given a correlation structure. The model that obtains the smaller QICC is 'better' according to this criterion.

We can use this output to make a model per weight class to predict the percentage of foot pad lesions during the production round.

For weight class = 1 (light birds):

$$\%\,\text{foot pad lesions} = -10.19 + 3.35 + 0.414 * \text{production day}$$

For weight class = 2 (average birds):

$$\%\,\text{foot pad lesions} = -10.19 + 2.629 + 0.414 * \text{production day}$$

For weight class = 3 (heavy birds):

$$\%\,\text{foot pad lesions} = -10.19 + 0.414 * \text{production day}$$

Table 2 Output of a linear regression model for foot pad lesions with variables weight class and production day measured in 200 broilers. Number of birds per class was calculated for every measuring day, with an average of 55 heavy birds (ranging 46–67), 98 average weight birds (range 86–108) and 48 light birds (range 43–54)

Parameter estimates							
			95% Wald confidence interval		Hypothesis test		
Parameter	B	Std. error	Lower	Upper	Wald chi-square	df	Sig.
Intercept	−10.19	3.0695	−16.206	−4.174	11.021	1	0.001
Weight class=1, light	3.350	1.0475	1.297	5.403	10.229	1	0.001
Weight class=2, average	2.629	1.6516	−0.609	5.866	2.533	1	0.111
Weight class=3, heavy	0[a]	.	.	.	.	.	.
Production day	0.414	0.1175	0.183	0.644	12.383	1	0.000

Dependent variable: foot pad lesions.
Model: intercept, weight class, production day.
[a] Set to zero because this parameter is redundant.

In the graphs the original data (Fig. 3a) and the predictions (Fig. 3b) are shown.

From Figs. 3a and 3b it is clearly visible that this model might not be the best predictor for foot pad lesions in broilers. Only two variables are used and a simple model was built. Using more variables in the model, adding interactions and higher-order functions might improve the model, making better predictions.

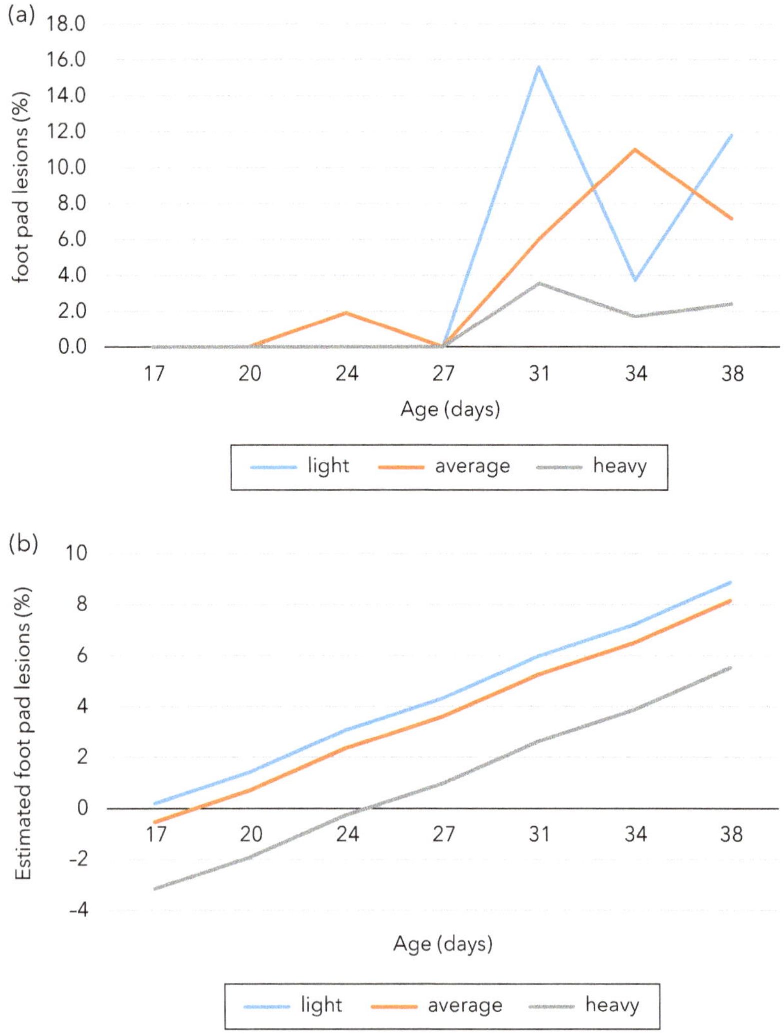

Figure 3 (a) Foot pad lesion percentage in 200 light, average and heavy broilers between 17 and 38 days of age. (b) Estimates of foot pad lesion percentage based on a linear regression model with factors such as age and weight (light, average and heavy).

For binary or class target variables, class probability estimation and logistic regression techniques can be used to fit the model to the data. The odds of an event is the ratio between the probability of the event occurring to the probability of the even not occurring. The logarithm of the odds can be modelled with a logistic regression model. For probability estimation, logistic regression uses the same linear model as the linear discriminants for classification of the linear regression for numeric target value estimation. The output of the logistic regression model is the log-odds of belonging to a certain class. Log-odds can be translated directly into the class membership probability.

If the odds of an event (x) happening are the probability of an event happening (p_+) divided by the probability of an event not happening ($1-p_+$), than the odds can be written as:

$$\text{Odds} = \frac{p_+(x)}{(1-p_+(x))}$$

And the log-odds linear function as:

$$\log\frac{(p_+(x))}{(1-p_+(x))} = f(x) = w_0 + w_1x_1 + w_2x_2 + \cdots$$

where w_i are the variables. The estimated probability of class membership can thus be solved:

$$p_+(x) = \frac{1}{(1+e^{-f(x)})}$$

This is the logistic function. The function is a sigmoid curve, with probabilities between 0 and 1, and the decision boundary being the line at 0.5: the probability of a coin toss (odds of 50/50). When using this equation to separate cows that are lame or healthy, the likelihood that a certain labelled example belongs to the correct class with parameters w, is calculated with the function:

$$g(x,w) = \begin{cases} p_+(x) \text{ if } x = L(\text{lame}) \\ 1-p_+(x) \text{ if } x = H(\text{healthy}) \end{cases}$$

We can sum the *g* values for a labelled data set, for different models; the model that gives the highest sum of *g* values is the model with the best set of weight, or the highest likelihood. This is the maximum likelihood model: giving the highest probabilities to the positive examples and the lowest probabilities to the negative examples. If we apply this technique to a data set with labelled instances (lame or healthy) and other parameters such as milk yield, age, activity and lying time, the intercept and the weights derived from the logistic regression model can be used to predict lameness in the data set. The accuracy of the model is the percentage of instances that the model classifies correctly, and is often used to compare models. However, a better accuracy does not always determine the better model – this is explained further in paragraph 3.1.5.

A linear classifier such as logistic regression can use decision boundaries of any direction or orientation, while a classification tree is a classifier that segments the instance space recursively. It is not easy to determine in advance which will give a better match to your data set. A consideration might be that logistic regression can be difficult to understand for people not schooled in statistics, while decision trees are far more understandable for people without a strong mathematical background (Provost and Fawcett, 2013).

3.1.4 Support vector machines, non-linear functions and neural networks

Support vector machines (SVM) are linear discriminants. SVM's classify instances based on a linear function of the attributes. Instead of separating the data into segments using a line, a wide bar is used to segment the data; after the widest bar has been identified, the linear discriminant is the line through the centre of the bar. To find the best fit for this line, the margin around the linear discriminant is maximized. In several studies, classifiers based on SVM are used to develop models, for example, to identify behavioural patterns in cattle (Behmann et al., 2016), to classify vocalizations in laying hens (Du et al., 2020), to develop a segmentation algorithm to detect sick broilers from posture data (Zhuang et al., 2018) or to classify vocalisations in sheep, cattle and dogs (Bishop et al., 2019).

Non-linear support vector machines are a systematic way of adding more complex terms and fitting a linear function to them. Support vector machines have a 'kernel function' which maps the original attributes (variables) to some other feature space. Then a linear model is fitted to this new feature space. A non-linear vector machine can have a polynomial kernel, which means that higher-order feature combinations can be considered, such as squared variables or products of variables. These algorithms, used to find relations (clusters, rankings, classifications, correlations), are also known as kernel methods (Hofmann et al., 2008). A kernel is a similarity function over pairs of data points.

Neural networks also implement complex non-linear numeric function, represented as a 'stack' of models. The lower layer is the set of original variables, above that are a series of models, each one building on the results of the layer below. The target labels for training are only defined for the last or upper layer. The stack of models can be seen as a large numeric function that can be optimized to find the best parameters. Neural networks are used for classification and regression, but also for other data mining tasks such as clustering, time series analysis and profiling (Provost and Fawcett, 2013). Examples of applying neural networks are weight prediction in broilers (Zhuang et al., 2018), sound analysis in dairy cows, turkeys or broilers (Ikeda and Ishii, 2008; Liu et al., 2018a; Du et al., 2020), estimating height of heifers (Nir et al., 2018), aggression detection in pigs (Oczak et al., 2014) and predicting lameness in sows (Halachmi, 2015).

An appealing example of using neural networks is individual animal identification or facial recognition, which is developed for several livestock species (Hansen et al., 2018; Liu et al., 2018b). In these analyses, usually neural networks are used, building several layers to identify animals, starting with recognizing an object or animal, going through several layers of feature extraction towards a final layer where all individuals are recognized. Although most of the time it is not known exactly what happens in the different layers, it might be explained as follows. We start with pictures from faces. In the first layer, pixels are recognized, then, contrasts and edges, followed by shapes such as curves or circles; next come attributes that combine shapes, such as nose, mouth and eyes, and finally, specific combinations of these attributes are recognized as faces (Fig. 4).

An example of using neural networks in facial recognition is applied in pain recognition in sheep. In this study, sheep faces and expressions are recognized using neural networks, classifying healthy sheep and sheep that are in pain (McLennan and Mahmoud, 2019). In several steps in the process, first sheep faces are detected from photographs, than features in the faces, and finally sheep in pain versus healthy sheep are distinguished (Fig. 5).

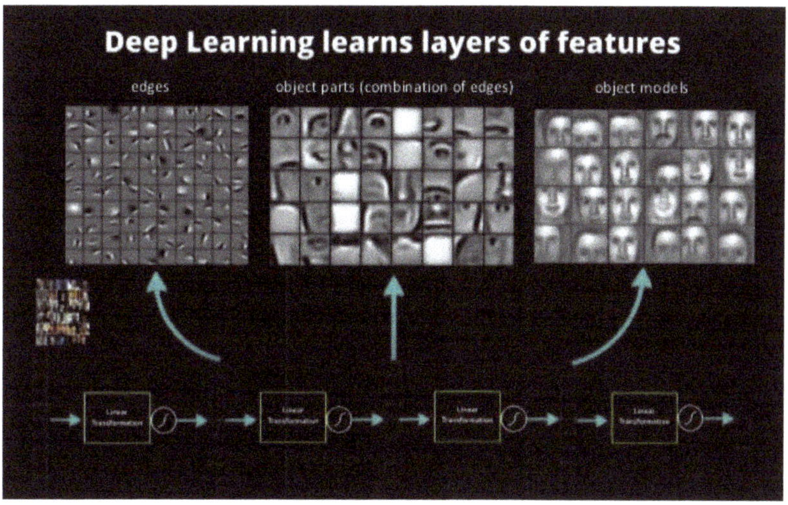

Figure 4 Layers of the neural network in facial recognition (Akagi, 2014).

Figure 5 Detecting painful expressions in sheep using neural networks (McLennan and Mahmoud, 2019).

3.1.5 Overfitting

Models are built based on training data, with the purpose to apply to new data, independent from the population that the data set came from: generalization is the aim. If a model is built to fit the training data exactly, chances are that they will not generalize. This is called overfitting. Evaluation of a model cannot be done on the training data. For testing a model, usually 'holdout data' are used. These are data with known target variables, which are concealed for the model, also called a 'test set'. Comparing the predicted target values with the true target values gives an estimate of the generalization performance of the model.

Using a tree induction procedure to build a model with too many branches and pure leaves (segments with only class A or class B), we will end up with a perfectly accurate model, correctly assigning each instance to a class. However, this model will probably not perform very well on a test set. If the tree model is too big, with too many nodes, it overfits. This means it will perform very well on the training set, but not well on the test set, because it is tuned too specifically to the training set. During the tree induction process, as the decision tree grows in complexity, performance of the model on the training data will increase with every step. Performance on the test data will increase at first, but as the model becomes more complex, performance on the test data will first stabilize and then decrease. The best model (tree size) is where performance on the test set stabilizes and not decreases.

Overfitting in linear models occurs when the models become too large or complicated, with too many variables. Simple real-time models may perform better than complicated neural networks. In a PhD study on behaviour in pigs it was concluded that using simple real-time models to monitor complex animal behaviour gave good results, while it was impossible to estimate parameters in real-time in complex models with many parameters. Models with fewer parameters are more flexible in real-time PLF applications where they must adapt to time-varying behaviour of individuals (Oczak, 2018). In a PhD study on optimizing broiler production, accuracies with neural networks were not high and the author concluded that to improve the models, data reduction should be applied (Johansen, 2019).

The tendency to add extra variables such as squared original variables or ratios between original variables induces the risk of overfitting. If the dimensionality is increased, eventually you can find a perfect fit for your training data. Cross-validation is used to prevent the network from overfitting to the training data by validating its performance on separate test data (Haykin, 2004). The holdout method is the simplest cross-validation method and performs a single split of training and testing data (Devroye and Wagner, 1979). This holdout technique can be used to decide when the number of variables in the model and the complexity is optimal. This attribute selection can be done manually or by using the automatic feature selection option.

A more sophisticated way to evaluate your model than splitting your data into one training and one holdout or test set, is cross-validation. Cross-validation performs multiple splits of the data sets and computes the estimates of the performance of the model over all data. Cross-validation splits the labelled data set usually into 5 or 10 partitions or folds. Training and testing is performed on each set of test and training data. From these results, the average and standard deviation of the performance estimates can be calculated. However, one must be careful, since often the folds do not contain independent data and it is wise to test the model on data from a different case (e.g. livestock house) with the same process.

With more training data, the performance of a model increases, up to a certain point. A plot of the performance relative to the training data is called a learning curve. Learning curves typically start steep, with the model finding the most apparent patterns in the data set, but flatten out when more data cannot increase the accuracy anymore. Learning curves can be used to decide whether it is worthwhile to invest in acquiring more data to improve the model.

Avoiding overfitting in tree induction can be done in two ways. Strategies are:

1 Stop the decision tree before it gets too large
This can be done by agreeing on a fixed minimal number of instances in each segment or 'leaf'. This way, we avoid making decisions based on very small numbers of instances, since they tend to be inaccurate. The number to use can be based on experience, or on a statistical technique. We can use a threshold value (e.g. 5%) based on a hypothesis test on the probability that the observed difference in Information Gain is due to chance.
2 Allow the tree to become too large, then reduce the size and complexity ('pruning')
Removing branches from too large trees can be done based on accuracy. Branches can be replaced with leaves when this does not decrease accuracy.

A general method to avoid overfitting is to split the training data set into a sub-training set and a validation set. The sub-training set is used to build the model with the optimal complexity, and the validation set is used to validate the model, estimating generalized performance. This can also be done using cross-validation: the procedure is then called nested cross-validation. Different sets of variables can be tested this way, using the data to choose the complexity and pick the best model. Examples of selection procedures are sequential forward selection (SFS) (Marcano-Cedeño et al., 2010) and sequential backward elimination (Reeves and Zhe, 1999) of variables. These procedures either start with a simple model with one feature, using nested holdout procedure to pick the best model, then adding features until the accuracy of classification no longer improves (SFS), or work the other way around, starting with a large model and eliminating features one by one until there is a loss of performance by eliminating a variable.

Avoiding models to become too complex can be done with a procedure called regularization, which is actually a combination of optimizing fit of the model and the simplicity. In regularization, a penalty for complexity is given to each model. A regularized model has an extra term with a weight λ that is representing the importance of the penalty. Cross-validation on subsets of training data can determine λ, and this can be used to learn the regularized model on the complete training data set. Optimizing the parameter values this way is known as grid search (Provost and Fawcett, 2013).

Several methods to avoid overfitting can be found in livestock research: using multi-stage SVM, using boosted regression trees, or using k-fold cross-validation to automatically prune the tree model (Berckmans, 2017; Du et al., 2020).

3.1.6 Similarity and clusters

If two animals are similar in some ways, then they often share other characteristics. Data mining procedures are often based on grouping individuals by similarity. Similarity can be used to classification and regression, which is supervised segmentation. In livestock research, similarity is mostly used in studies based on vision techniques, such as facial recognition in pigs (Hansen et al., 2018), estimating height and body mass in heifers (Nir et al., 2018), but also comparing sound patterns (Fontana et al., 2016; Bishop et al., 2019)

A measure for similarity is the distance within a mathematical space. If we visualize data in a two-dimensional space, the distance between data points can be represented by a vector. The Euclidean distance can be computed, which is the overall distance between the data points, by computing the distances (d) of the individual dimensions.

$$\text{Euclidean distance} = \sqrt{(d_{1,A} - d_{1,B})^2 + (d_{2,A} - d_{2,B})^2 + \cdots + (d_{n,A} - d_{n,B})^2}$$

This distance can be used to compare the similarity of one pair to that of another pair of data points. The most similar instances are called the nearest neighbours.

Using similarity for predictive modelling is done by, for each new instance, using the training set to find similar instances. The target value of the new instance is predicted based on the nearest neighbours' target values. For this, a combining function is used (Provost and Fawcett, 2013).

An example of classification of cows using nearest neighbours is given in the text box below.

From a data set of cows with the characteristics age, milk yield and activity level and a binary target variable 'lame', we want to predict whether a new cow, Saskia 27, is likely to be lame or healthy.

Cow	Age (years)	Milk yield (kilogrammes)	Activity level*	Lame	Distance from Saskia 27
Saskia 27	*3*	*25*	*100*	*?*	*0*
Saskia 25	4	20	85	Yes	15.8
Nora 2	3	23	87	Yes	13.2
Nora 15	3	35	100	No	10.0
Liesbeth 4	5	27	90	No	10.4
Carla 1	5	22	105	No	6.2

**from accelerometer data - no unit.*

The distance for each cow from Saskia 27 can be computed with the formula for Euclidean distance. For Saskia 25 this distance is:

$$\sqrt{(4-3)^2+(20-25)^2+(85-100)^2}=\sqrt{251}=15.8;$$

the other distances are calculated accordingly. If we want to use these distances to predict lameness for Saskia 27, the nearest neighbours are Nora 15, Liesbeth 4 and Carla 1. These cows are not lame, so the prediction for Saskia 27 based on the nearest neighbours would be 'healthy'.

Note: the scale of the variables should be roughly in the same range: in this example, it might be problematic that age is in a much smaller range than milk yield and activity level. We could use age in months instead. Otherwise, the distance metric would consider a 1-year age difference as significant as 1 kg difference in milk yield.

Nearest neighbours can also be used to estimate the probability of class membership. Probability estimates can be made from the ratio of target values within the group of nearest neighbours. Usually, more nearest neighbours than just the three from the examples are used, or a correction is made for small samples. Nearest neighbour algorithms are referred to as k-NN where k is the number of nearest neighbours: in the example, this would be 3-NN. In a k-NN classification model, the k is a measure for complexity. If we use $k = n$, which means all the neighbours in the data set are used, we will have a simple model. A model using $k = 1$, will be more complex, defining each training example in its own class. 1-NN models strongly overfit and will predict perfectly on training data; however, they can also perform reasonably well on other data, using the single most similar training example. Choosing the best number for k can be done with similar techniques as to avoid overfitting: using nested cross-validation or other nested holdout testing on the training set for different values of k. The one that gives the best performance will be the chosen k, and this value will be used to build a k-NN model from the entire training set.

Using nearest neighbours for prediction can be done in a similar way as using it for classification. If we use the same data set as in the example above, we can predict milk yield for Saskia 27 using the average (28) or the median (27) of the milk production of her nearest neighbours[2].

Using not only the nearest neighbours but all data in the set, scaled by similarity, we can predict the target probability. A numeric example is in the text box below.

In the Table below, neighbours are ordered by their distance from Saskia 27. For weight of similarity, the following formula is used:

2 The variable milk yield is not used to determine the nearest neighbours, in this case

$$\text{similarity weight} = \frac{1}{d^2}$$

The contributions are the proportion of the weight, adding up to 1. For Carla 1, this is calculated as:

$$\text{Contribution} = \frac{0.026015}{(0.026015 + 0.010000 + 0.009246 + 0.005739 + 0.004006)}$$
$$= 0.474$$

For the other cows, contributions are calculated accordingly.

Cow	Distance	Similarity weight	Contribution	Lame
Carla 1	6.2	0.026015	0.474	No
Nora 15	10.0	0.010000	0.182	No
Liesbeth 4	10.4	0.009246	0.168	No
Nora 2	13.2	0.005739	0.104	Yes
Saskia 25	15.8	0.004006	0.073	Yes

Summing the contributions for the positive and negative classes gives a probability estimate for Saskia 27: 0.18 for lame and 0.82 for healthy.

This weighted scoring procedure for classification can also be applied to regression and class probability estimation.

3.2 Unsupervised segmentation

If we are interested in finding groups of similar animals, without a specified target, unsupervised segmentation or clustering can be used. Clustering is an application of similarity. The aim is finding groups of animals in which the animals are similar, while between groups the animals are not that similar. Such exploratory data analysis can lead to useful and maybe even profitable discoveries. In the clustering process, it is necessary that variables are normalized to the same range, to give the same importance to all variables (Virmani et al., 2015). Clustering techniques are used in studies of social networks in dairy cows (Boyland et al., 2016) or in behavioural studies based on location data (Halachmi, 2015; Meunier et al., 2018).

Two methods of clustering are described in this chapter: hierarchical clustering and clustering around centroids.

3.2.1 Hierarchical clustering

In hierarchical clustering, n data points are grouped based upon (Euclidean) distance. The number of clusters ranges from 1, containing all data points, to n, with each data point being its own cluster. In the process, you start with n clusters, combining data points one by one based on distance. On each level, there is a different number of clusters, growing from few clusters with many data points to many clusters with few data points. Next, a dendrogram can be visualized. In the example (Fig. 6) this is illustrated for a group of 30 cows that are clustered based on the average Euclidian distance of the variables average Body Condition Score (BCS) from 2 measurements, where 2 = very thin and 4 = rather fat, and lactation stage, where 1 = early lactation (0–120 days), 2 = mid lactation (120–200 days) and 3 = late lactation (200–300+ days) (van Erp-van der Kooij et al., 2020a).

The number of clusters must be chosen by the researcher, depending on the goal of the study and the interpretation. A visual representation of the groups is shown in Fig. 7 for three (Fig. 7a) and six (Fig. 7b) clusters.

Many examples of clustering are found in biology (Benigni and Giuliani, 1994), with an appealing visualization in a study on the faecal resistome in slaughter pigs and broilers (Munk et al., 2018), visualizing acquired resistance gene pools in different countries.

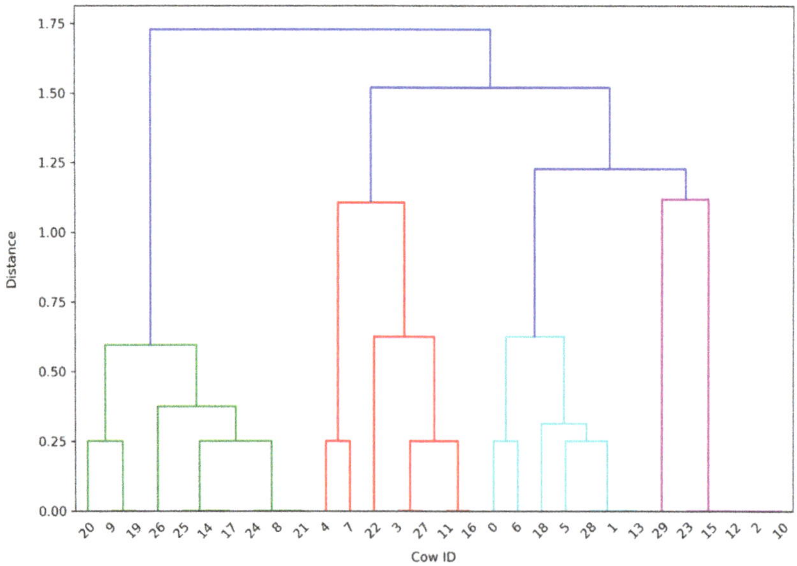

Figure 6 Dendrogram of 30 cows (cow ID 0 to 29) clustered according to average Euclidian distance, based on BCS and lactation stage.

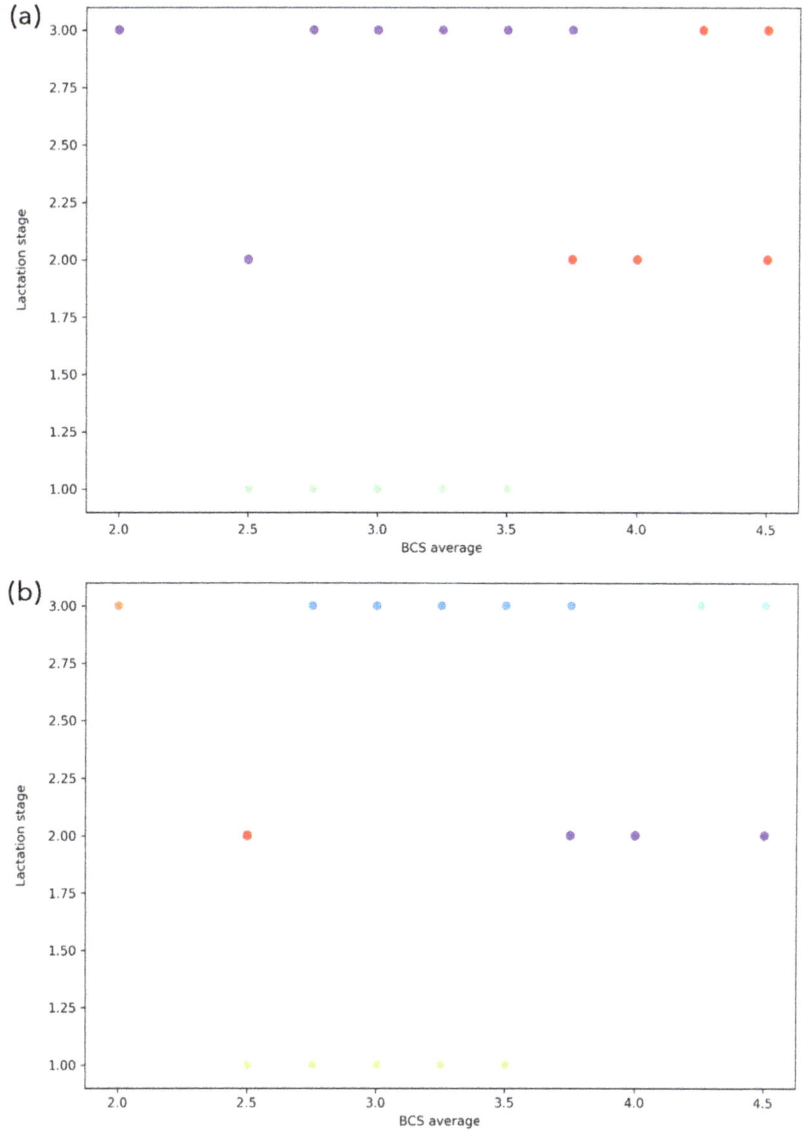

Figure 7 (a) Clustering of 30 cows, based on BCS and lactation stage, with three clusters. (b) Clustering of 30 cows, based on BCS and lactation stage, with six clusters.

3.2.2 Clustering around centroids

While hierarchical clustering focusses on the similarities between the individuals and how they are linked together, clustering can also be done around centroids, focussing on the clusters themselves. Each cluster has a central point or centroid, the geometric centre of each group of instances.

k-means is the most used method for clustering based on centroids. In this algorithm the means are the centroids, calculated as the averages of the values for each dimension for the individuals in the cluster. The centroid is the average of the values for each feature of each example in the cluster. The k in k-means is the desired number of clusters. If the analyst has specified four clusters, then the algorithm would calculate the values of the four centroids and which of the data points belong to each of those clusters. This is also referred to as nearest-neighbour clustering: each cluster contains those data points that are nearest to its centroid.

The method of finding the clusters in a k-means algorithm starts with creating k cluster centres, randomly or chosen by the analyst. Clusters are formed around these cluster centres. Next, the centre is recalculated based on the data appoints around it. This is an iterative process which runs until the centroids no longer shift, or the process is stopped manually. The best clustering can be found by calculating the distortion, a numeric value based on the differences between each data point in a cluster and it centroid. The formula is:

$$\text{distortion} = \sum\left\{(x_1 - c)^2 + (x_2 - c)^2 + \cdots (x_i - c)^2\right\}$$

with x_1-x_i being the data points and c the centroid. The lower the distortion, the better the clustering.

To understand the clustering, it can be helpful to give names to the clusters. These cluster names can be derived from the characteristics of individuals in the cluster, pointing out what makes them similar, or emphasizing the differences with other clusters.

Clustering or unsupervised segmenting of data is a form of data exploration without having a pre-specified target, a process or business problem. Sometimes the goal of the data mining is to explore the data, with only a vague notion of what the process or business problem is. This might lead to unexpected, useful findings, but it might also lead to no result at all and a waste of time. It is important that this risk is understood beforehand, to avoid disappointment. Again the familiarity of the modeller with the process or business is very important since modelling just numbers rarely generates real contributions.

3.3 Evaluation of models

It is important to assess the outcome of the data mining process and to determine whether the target was achieved. There are several evaluation metrics and techniques to determine the 'success' of the data mining process (Caruana and Niculescu-Mizil, 2006).

For evaluation of binary classification models, classification accuracy can be used. Accuracy is defined as the number of correct decisions:

$$\text{accuracy} = \frac{\#\,\text{correct decisions}}{\text{total}\,\#\text{decisions}} = 1 - \text{error rate}$$

This metric is easy to measure, but simplistic, and has some drawbacks. When dealing with a model involving *n* classes, the number of actual and predicted classes can be shown in a *n* * *n* matrix, which is the confusion matrix. In a confusion matrix you can see what classes are 'confused' for other classes. Well-known terms from the field of epidemiology are the false positives (negative instances classified as positive) and the false negatives (positive instances classified as negative). When we have small numbers in certain classes, for example, a rare disease that we want to detect, using accuracy as a measure for model performance is not valid. When the prevalence of a rare disease is 1% and the model predicts every individual to be healthy, the accuracy of the model will be 99% but it will not reach your goal, which is detecting the sick individual. Another problem with accuracy is that when your model is tested on a selected data set, where the prevalence differs from the original population, the accuracy is different for the selected data set than for the whole population. Yet another problem when using accuracy as a measure for performance of classification models is that false positives and false negatives are treated as equally important. In real life, a missed diagnosis might be much worse than a healthy individual being unnecessarily checked and false positives will stop people of using the algorithm.

For regression models, often the R^2, the mean square error (MSE) or the root mean square error (RMSE) is reported as a measure for how well the model performs. This is a good measure for numeric outcomes, but for classification predictions, this is not always the best metric to use: these metrics are not always meaningful or well thought out. It might be better to work with expected values. The expected value is the weighted average of the possible outcome values ($v(o_i)$), where the weight is the probability of that outcome ($p(o_i)$):

$$\text{Expected value} = p(o_1) * v(o_1) + p(o_2) * v(o_2) + \cdots + p(o_i) * v(o_i)$$

With known costs for the different outcomes, we can calculate the expected value of the decision, and from that, derive a decision rule. This is illustrated for the example of lameness prediction.

Assuming we want to evaluate the prediction model for lameness. If the cow is predicted to be lame, and the farmer acts on that assumption, there are two possibilities: (1) she is indeed lame, and she is treated. Let's assume this will cost a fictional amount of 10 euro per treatment and it will save 100 euros in sickness costs for the cow, such as decreased milk yield, other health issues, veterinary costs and so on. The net benefit of the treatment is 90 euros. The other possibility is (2) she is not lame. This will cost you 10 euros for the treatment. If we want to know whether we will make a profit by believing the prediction and acting on it, we calculate whether the expected value is greater than zero, depending on the probability (*p*) of the outcomes L (lame) and the value (*v*) of that outcome:

$$p_L(x) * v_L + [1 - p_L(x)] * v_L > 0$$

The decision rule is then: act upon the model prediction when:

$$p_L(x) > 0.1$$

Or in words: treat the cow when the probability of lameness is higher than 0.1.

There are more sophisticated ways to evaluate the classification models using aggregations of expected values and the confusion matrix. Correct classifications correspond with the benefits while false positives and false negatives correspond with the costs of the model-based decisions. This way, a cost-benefit analysis can be carried out. For these calculations, usually, external information from process knowledge is needed.

For model evaluation, the area under the ROC curve can be used (Hanley and McNeil, 1982). The ROC or receiving operating curve is the true positive rate (TP) or sensitivity plotted against the false positive rate or 100 – specificity (Fig. 8).

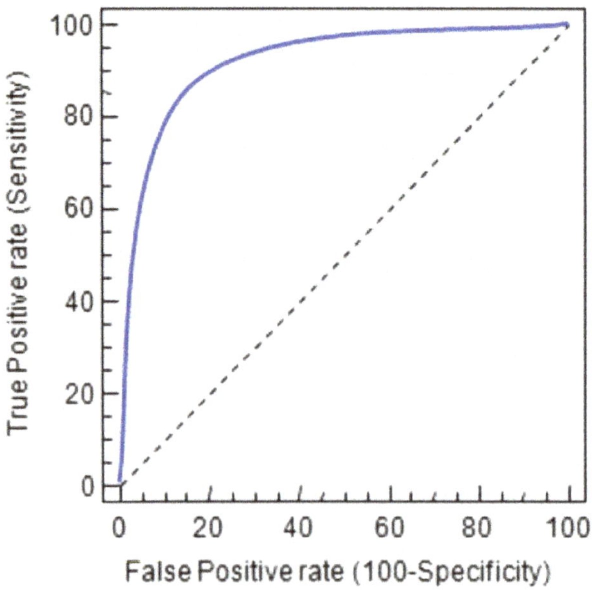

Figure 8 ROC curve (www.medcalc.org).

The area under the ROC Curve, usually abbreviated to AUC, is a good summary statistic for how well the classifier model predicts a true positive. The AUC varies from 0 to 1, with 0.5 meaning the classifier is equal to randomness and does not distinguish between positive and negative, and 1 being a perfect classifier.

4 Conclusion and future trends

Today, more and more businesses gather data from many different processes, store it, analyse it and use the results to support or make data-driven decisions. In livestock farming huge amounts of data are generated from sensors, images and sound devices collected from (individual) animals. This year over 65 billion animals will be slaughtered to fulfil the worldwide demand for animal products. Data mining and analyses have been discovered as a useful tool in making farming more efficient, detecting problems in an early stage so that the farmer is able to prevent losses due to poor animal welfare or events such as sick or badly performing animals. More advanced techniques are used to store data as well as analyse them – large data sets have become less of a problem to analyse, and process or business questions are being translated to data problems that can be solved with data mining techniques. Dashboards with visualized data are built to show farmers the welfare, the health and the performance of their animals, and apps are developed to alert farmers if something is wrong or needs to be checked in the farm. What's next?

4.1 Building data warehouses

As more and more data are being gathered on the farm, the need for on-farm storage systems is growing. Farmers want to see the data combined into one system and visualized in a dashboard, which shows the situation on the farm. This means that data from different sources must be combined and stored into one system. Data warehouses or other systems to store or combine data can be built by any software company; some have specialized into building systems for livestock, other have built a system that combines data but functions not as a data warehouse but as a data hub, channelling data but not bringing it actually in one place. Some companies are specializing in Blockchain systems for the AgriFood sector.

4.2 Combining data from different sources

As more and more data are available for analyses, they are combined and data from more than one source are used to build prediction models. For example, where the first animal-mounted sensors in dairy farming were simple accelerometers counting steps, translating this into oestrous detection, nowadays the accelerometers have become more advanced as are the algorithms behind it, translating results into several different behaviours, monitoring health as well as fertility. Activity information is combined with management data and in some cases body temperature to determine health status of the cow.

4.3 Using more advanced data mining techniques to analyse big data

When at first mainly regression models were used to determine relations between features calculated from animal data and a target variable such as oestrus or disease, now more advanced techniques such as neural networks and support vector machines are used. Unsupervised learning models are used in studies in social networks of animals with the use of automatic location data; image-, sound- and sensor data can be analysed with data mining techniques, and those studies can be done with more animals, faster and with less or no manual or audio-visual observations.

4.4 Sound and vision analysis

Very promising is using sound systems to monitor animals, looking for deviations in behavioural patterns - individual (dairy cows, sows) or per group (fattening pigs, broilers, laying hens); analysing sound and vision with supervised learning methods, using historical data sets to train models and using the algorithms to predict disease or stress in the animals. For pigs, a cough detection system is already on the market, and new systems are being developed for pigs and poultry. Using vision systems to recognize animals and behaviours with the use of machine learning techniques, is also upcoming; again with the aim to monitor health or stress of animals by monitoring behavioural patterns in individuals or groups of animals. Several companies have systems on the market, claiming to be able to monitor animals using cameras and smart software; examples are monitoring systems for dairy cows and for fattening pigs.

The more advanced data mining techniques make these vision and sound analyses possible, including the developments of apps to label sound and vision data in a more efficient, reliable and less time-consuming way.

5 Where to look for further information

A standard introduction to the subject is:

- Provost, F. and Fawcett, T. (2013). 'Data Science for Business - What You Need to Know About Data Mining and Data-Analytic Thinking'. United States of America, O'Reilly Media Inc., 386 pp.

There are many courses on data science, varying from beginner to advanced, cheap or expensive, short or long. A good platform to learn data science online is:https://www.datacamp.com. Here you can do short, online beginner and advanced courses in R, Python and SQL. Free virtual classrooms are available for universities. More intensive data science training can be followed at the Jheronimus Academy of Data Science (JADS) in Den Bosch, the Netherlands, for example 'An introduction to Data Science', a 7-week program to learn the basics of data science:

https://www.jads.nl.

Key journals in this field are:

- Computers and Electronics in Agriculture.
- Biosystems Engineering.

There are a few key conferences in the field of Precision Livestock Farming. Interesting ones are:

- Measuring Behavior (MB) is a conference held every two years, focused on methods, tools and techniques for the study of behavior in humans and animals. Many applications of (sensor) technology and data analysis are discussed and presented. https://measuring behavior.org.
- The European Conference on Precision Livestock Farming (ECPLF) is held every two years and focuses on the application of precision technologies in livestock production systems. https://www.eaplf.eu.

6 Acknowledgements

Many thanks to Diane de Jong, Joshi van Berlo, Larissa van de Klundert and Wilbert van Ettekoven for providing livestock data; to Martijn van Erp for performing a cluster analysis in Python; and to Pieter Rambags and Tim Hanssen for proofreading.

7 References

Akagi, D. (2014). A primer on deep learning, Data Robot. Available at: https://www.datarobot.com/blog/a-primer-on-deep-learning/.

Behmann, J., Hendriksen, K., Müller, U., Büscher, W. and Plümer, L. (2016). Support Vector machine and duration-aware conditional random field for identification of spatio-temporal activity patterns by combined indoor positioning and heart rate sensors, *GeoInformatica* 20(4), 693–714. doi: 10.1007/s10707-016-0260-3.

Benigni, R. and Giuliani, A. (1994). Quantitative modeling and biology: the multivariate approach, *American Journal of Physiology* 266(5 Pt 2), R1697–R1704. doi: 10.1152/ajpregu.1994.266.5.R1697.

Berckmans, D. (2017). Precision livestock farming (PLF). In: Berckmans, D. and Keita, A. (Eds), Precision Livestock Farming 2017. Nantes, France, p. 1. doi: 10.1016/j.compag.2007.09.002.

Bishop, J. C., Falzon, G., Trotter, M., Kwan, P. and Meek, P. D. (2019). Livestock vocalisation classification in farm soundscapes, *Computers and Electronics in Agriculture* 162, 531-542. doi: 10.1016/j.compag.2019.04.020.

Boyland, N. K., Mlynski, D. T., James, R., Brent, L. J. N. and Croft, D. P. (2016). The social network structure of a dynamic group of dairy cows: From individual to group level patterns, *Applied Animal Behaviour Science* 174, 1-10. doi: 10.1016/j.applanim.2015.11.016.

Brynjolfsson, E., Hitt, L. and Kim, H. (2011). Strength in numbers: how does data-driven decision-making affect firm performance?. In: *International Conference on Information Systems 2011, ICIS 2011*, 1, pp. 541-558.

Caruana, R. and Niculescu-Mizil, A. (2006). An empirical comparison of supervised learning algorithms, *ACM International Conference Proceeding Series* 148, 161-168. doi: 10.1145/1143844.1143865.

Cordeiro, A. F. d. S., Nääs, I. d. A., da Silva Leitão, F., de Almeida, A. C. M. and de Moura, D. J. (2018). Use of vocalisation to identify sex, age, and distress in pig production, *Biosystems Engineering* 173(April), 57-63. doi: 10.1016/j.biosystemseng.2018.03.007.

da Silva Cordeiro, A. F., de Alencar Nääs, I., Oliveira, S. R., Violaro, F., de Almeida, A. C. and Neves, D. P. (2013). Understanding vocalization might help to assess stressful conditions in piglets, *Animals* 3(3), 923-934. doi: 10.3390/ani3030923.

Devroye, L. and Wagner, T. (1979). Distribution-free performance bounds for potential function rules, *IEEE Transactions on Information Theory* 25(5), 601-604. doi: 10.1109/TIT.1979.1056087.

di Virgilio, A., Morales, J. M., Lambertucci, S. A., Shepard, E. L. C. and Wilson, R. P. (2018). Multi-dimensional Precision Livestock Farming: A potential toolbox for sustainable rangeland management, *PeerJ* 5, 1-23. doi: 10.7717/peerj.4867.

Du, X., Carpentier, L., Teng, G., Liu, M., Wang, C. and Norton, T. (2020). Assessment of laying hens' thermal comfort using sound technology, *Sensors* 20(2), 1-14. doi: 10.3390/s20020473.

Ellis, J. L., Jacobs, M., Dijkstra, J., van Laar, H., Cant, J. P., Tulpan, D. and Ferguson, N. (2020). Review: synergy between mechanistic modelling and data-driven models for modern animal production systems in the era of big data, *Animal* 14, s223-s237. doi: 10.1017/S1751731120000312.

Flury, B. (2013). *A First Course in Multivariate Statistics*. New York: Springer Science & Business Media.

Fontana, I., Tullo, E., Scrase, A. and Butterworth, A. (2016). Vocalisation sound pattern identification in young broiler chickens, *Animal: An International Journal of Animal Bioscience* 10(9), 1567-1574. doi: 10.1017/S1751731115001408.

Greasly, P. (2007). *Quantitative Data Analysis Using SPSS*. UK: McGraw-Hill Education.

Halachmi, I. (2015). *Precision Livestock Farming 2015, Precision Livestock Farming*. Wageningen: Wageningen Academic Publishers.

Hanley, J. A. and McNeil, B. J. (1982). The meaning and use of the area under a receiver operating characteristic (ROC) curve, *Radiology* 143(1), 29-36.

Hansen, M. F., Smith, M. L., Smith, L. N., Salter, M. G., Baxter, E. M., Farish, M. and Grieve, B. (2018). Towards on-farm pig face recognition using convolutional neural networks, *Computers in Industry* 98, 145-152. doi: 10.1016/j.compind.2018.02.016.

Haykin, S. (2004). *Neural Networks*. New York: Prentice Hall.

Hofmann, T., Schölkopf, B. and Smola, A. J. (2008). Kernel methods in machine learning, *Annals of Statistics* 36(3), 1171-1220. doi: 10.1214/009053607000000677.

Hosmer, D. W. and Lemeshow, S. (2013). *Applied Logistic Regression Analysis, Applied Logistic Regression Analysis*. New York: John Wiley & Sons. doi: 10.4135/9781412983433.

Ikeda, Y. and Ishii, Y. (2008). Recognition of two psychological conditions of a single cow by her voice, *Computers and Electronics in Agriculture* 62(1), 67-72. doi: 10.1016/j.compag.2007.08.012.

Johansen, S. V. (2019). *Automated Optimization of Broiler Production*. Aalborg, Denmark: Aalborg Universitet.

Lin, Y. (2002). A note on margin-based loss functions in classification, *Statistics and Probability Letters* 68(1), 73-82.

Liu, L., Ni, J. Q., Li, Y., Erasmus, M., Stevenson, R. and Shen, M. (2018a). Assessment of heat stress in turkeys using animal vocalization analysis. In: *ASABE 2018 Annual International Meeting*, pp. 1-7. doi: 10.13031/aim.201801743.

Liu, L. S., Ni, J., Zhao, R., Shen, M., He, C. and Lu, M. (2018b). Design and test of a low-power acceleration sensor with Bluetooth Low Energy on ear tags for sow behaviour monitoring, *Biosystems Engineering* 176, 162-171. doi: 10.1016/j.biosystemseng.2018.10.011.

Marcano-Cedeño, A., Quintanilla-Domínguez, J., Cortina-Januchs, M. G. and Andina, D. (2010). Feature selection using sequential forward selection and classification applying artificial metaplasticity neural network. In: *IECON Proceedings (Industrial Electronics Conference)*, pp. 2845-2850. doi: 10.1109/IECON.2010.5675075.

McLennan, K. and Mahmoud, M. (2019). Development of an automated pain facial expression detection system for sheep (Ovis aries), *Animals* 9(4), 1-7. doi: 10.3390/ani9040196.

Meunier, B., Pradel, P., Sloth, K. H., Cirié, C., Delval, E., Mialon, M. M. and Veissier, I. (2018). Image analysis to refine measurements of dairy cow behaviour from a real-time location system, *Biosystems Engineering* 173, 32-44. doi: 10.1016/j.biosystemseng.2017.08.019.

Montgomery, D. C., Peck, E. and Vining, G. G. (2012). *Introduction to Linear Regression Analysis* (4th edn.). New York: John Wiley & Sons.

Munk, P., Knudsen, B. E., Lukjancenko, O., Duarte, A. S. R., Van Gompel, L., Luiken, R. E. C., Smit, L. A. M., Schmitt, H., Garcia, A. D., Hansen, R. B., Petersen, T. N., Bossers, A., Ruppé, E., EFFORT Group, Lund, O., Hald, T., Pamp, S. J., Vigre, H., Heederik, D., Wagenaar, J. A., Mevius, D. and Aarestrup, F. M. (2018). Abundance and diversity of the fecal resistome in slaughter pigs and broilers in nine European countries, *Nature Microbiology* 3(8), 898-908.

Nilsson, M., Herlin, A. H., Ardö, H., Guzhva, O., Åström, K. and Bergsten, C. (2015). Development of automatic surveillance of animal behaviour and welfare using image analysis and machine learned segmentation technique, *Animal: An International Journal of Animal Bioscience* 9(11), 1859-1865. doi: 10.1017/S1751731115001342.

Nir, O., Parmet, Y., Werner, D., Adin, G. and Halachmi, I. (2018). 3D Computer-vision system for automatically estimating heifer height and body mass, *Biosystems Engineering* 173, 4-10. doi: 10.1016/j.biosystemseng.2017.11.014.

Oczak, M. (2018). *Precision Livestock Farming for Sows and Weaner Pigs*. Leuven, Belgium: KU Leuven.

Oczak, M., Viazzi, S., Ismayilova, G., Sonoda, L. T., Roulston, N., Fels, M., Bahr, C., Hartung, J., Guarino, M., Berckmans, D. and Vranken, E. (2014). Classification of aggressive behaviour in pigs by activity index and multilayer feed forward neural network, *Biosystems Engineering* 119, 89–97. doi: 10.1016/j.biosystemseng.2014.01.005.

Pereira, D. F., Miyamoto, B. C. B., Maia, G. D. N., Tatiana Sales, G., Magalhães, M. M. and Gates, R. S. (2013). Machine vision to identify broiler breeder behavior, *Computers and Electronics in Agriculture* 99, 194–199. doi: 10.1016/j.compag.2013.09.012.

Provost, F. and Fawcett, T. (2013). *Data Science for Business* (1st edn.). Edited by Loukides, M. and Blanchette, M. Sebastopol, CA: O'Reilly Media Inc.

Raschka, S. and Mirjalili, V. (2019). *Python Machine Learning* (3rd edn.). Birmingham, UK: Packt Publishing Ltd.

Reeves, S. J. and Zhe, Z. (1999). Sequential algorithms for observation selection, *IEEE Transactions on Signal Processing* 47(1), 123–132. doi: 10.1109/78.738245.

Rutten, C. J., Velthuis, A. G. J., Steeneveld, W. and Hogeveen, H. (2013). Invited review: sensors to support health management on dairy farms, *Journal of Dairy Science* 96(4), 1928–1952. doi: 10.3168/jds.2012-6107.

Salian, I. (2018). What's the Difference *between* Supervised, Unsupervised, Semi-Supervised and Reinforcement Learning?, NVIDIA. Available at: https://blogs.nvidia.com/blog/2018/08/02/supervised-unsupervised-learning.

Shannon, C. E. (1948). A mathematical theory of communication, *Journal of Bell System Technology* 27(3), 379–423; 623–656.

Statnikov, A., Aliferis, C. F., Hardin, D. P. and Guyon, I. (2011). *A Gentle Introduction to Support Vector Machines in Biomedicine*. Singapore: World Scientific Publishing Co.

van Erp-van der Kooij, E., van de Klundert, L., Ettekoven, W. and van Erp, M. J. (2020a). Study on body condition score, lactation stage and lameness in dairy cows. Unpublished raw data.

van Erp-Van Der Kooij, E., de Jong, D. and van Berlo, J. (2020b). Study on foot pad lesions and weight in broilers. Unpublished raw data.

Virmani, D., Taneja, S. and Malhotra, G. (2015). Normalization based K means clustering algorithm. *International Journal of Advanced Engineering Research and Science*, submitted, pp. 1-5. Available at: http://arxiv.org/abs/1503.00900.

Wolfert, S., Ge, L., Verdouw, C. and Bogaardt, M. (2017). Big data in smart farming–a review, *Agricultural Systems* 153, 69–80. doi: 10.1016/j.agsy.2017.01.023.

Zhuang, X., Bi, M., Guo, J., Wu, S. and Zhang, T. (2018). Development of an early warning algorithm to detect sick broilers, *Computers and Electronics in Agriculture* 144, 102–113. doi: 10.1016/j.compag.2017.11.032.

Part 2

Applications

Chapter 7

Monitoring and control of livestock housing conditions using precision livestock farming techniques

Daniela Lovarelli and Marcella Guarino, University of Milan, Italy

1 Introduction

Agriculture, and more specifically livestock farming, has been put under growing pressure in recent decades. The global population is expected to reach 10 billion by 2050 (Wilson, 2005). Hence, producing enough food for everyone will be a huge challenge. It is expected that intensive livestock systems, in particular, will grow rapidly by a factor of 3.0–3.5 for pigs, 4.4–5.0 for broilers, 2.0–2.4 for laying hens (Mikovits et al., 2019). Secondly, increased production requires more inputs, putting pressure on minerals, fossils and other resources, and also has potentially damaging impacts on the environment (International Dairy Federation, 2015 Lovarelli et al., 2020a; Tullo et al., 2019). Livestock (especially cattle) production is characterized by a high carbon footprint per unit of product (de Vries and de Boer, 2010; Lovarelli et al., 2019; Mikovits et al., 2019), due to lower conversion efficiency and enteric methane production. Slurry and manure production and its use also contribute to global warming, soil acidification and freshwater and marine eutrophication (Cao et al., 2019; Fournel et al., 2017; Provolo et al., 2018).

http://dx.doi.org/10.19103/AS.2021.0090.07

There are several ways to increase efficiency in livestock production:

- balancing inputs introduced with outputs produced (e.g. by improving animal feed conversion efficiency);
- adopting a circular-based approach to reduce external inputs by better use or reuse of existing resources in the production chain (e.g. valorizing nutrients in manure and slurry for use in crop production); and
- Monitoring animals and their environment in real-time to assess their health and welfare status in order to avoid production losses due to environmental factors (e.g. extremes of temperature or poor air quality) or health and welfare issues (e.g. illnesses, stress or aggression).

Looking at the first option, livestock feed conversion rates have improved significantly. This is partly due to genetic selection for improved animal performances and partly due to optimizing feed composition, quality and intake to achieve satisfactory growth rates, weight gain or, in the case of dairy cattle farming, the amount of milk produced per kilogram of feed ingested. This more efficient use of resources has both environmental and economic benefits in making farming more sustainable. The environmental benefits of using fewer inputs, including reduced extraction, processing and transport costs, can be measured using Life Cycle Assessment (LCA) methodology, regulated by the ISO Standards 14040 Series (ISO Standards 2006).

A greater focus on more efficient and sustainable use of resources provides a route to a circular-based approach, which prioritizes valorization of resources. In the case of livestock farming, the main resource to be valorized is manure. Manure is rich in nutrients but is also an important component of environmentally damaging emissions to air, soil and water from livestock production (Rong and Aarnink, 2019). Dairy cattle emit methane during enteric fermentation, whilst manure from cattle, swine and poultry releases ammonia and nitrous oxide. These emissions contribute substantially to climate change since methane (CH_4) and nitrous oxide (N_2O) are important greenhouse gases (GHGs). Ammonia (NH_3), nitrates (NO_3^-) and phosphates (PO_4^{3-}) also contribute substantially to acidification and eutrophication (Rojano et al., 2019).

CH_4 emissions are one of the main causes of the environmental impact of livestock production (de Vries and de Boer, 2010; Lovarelli et al., 2019). Methane is not only produced during enteric fermentation but also during manure and slurry storage. These effects can be reduced by storage work at lower ambient temperatures (Im et al., 2020) as well as the use of anaerobic digestion (Aguirre-Villegas et al., 2019; Bacenetti et al., 2016).

It has been estimated that 60% of NH_3 contributing to global warming 60% is associated with livestock production, particularly dairy farming (Sanchis et al., 2019). The total amount of NH_3 released in the atmosphere depends mainly on manure excretion and its contribution to total ammonia nitrogen (TAN). Production of TAN depends on multiple factors, including manure management (Finzi et al., 2020), livestock nutrition and management practices and environmental conditions (Brentrup et al., 2000). These emissions can be reduced in a variety of ways, including through animal breeding, health and diet (Bell et al., 2011), frequency of slurry removal from the barn (Hoff et al., 2006), slurry treatment (Dinuccio et al., 2011), use of covered storages (Guarino et al., 2006), and direct slurry injection on fields as a crop fertilizer (Mattila and Joki-Tokola, 2003).

Turning the focus on the animals themselves, developments in automatic monitoring technologies have laid the foundations for precision livestock farming (PLF). PLF allows real-time monitoring of livestock, providing information on housing conditions, animal health and welfare to make better decisions (Halachmi et al., 2019). Introducing sensors and other monitoring tools in livestock farming provides the potential for continuous, real-time measurement (Fig. 1). This makes it possible for earlier detection of production, welfare and health (including reproductive) issues. This is particularly important given the drive towards large herd sizes (to optimize economies of scale) in intensive farming, which makes monitoring large numbers of animals more challenging. Early detection and corrective action have benefits for production efficiency, animal health and welfare, and environmental sustainability. Measuring these benefits of PLF is challenging given the complexity of the farming system (Kamphuis et al., 2015; Rojo-Gimeno et al., 2019). However, Gunn et al. (2019) and Lovarelli et al. (2020a) have shown the potential economic and other advantages of using PLF techniques.

2 Key variables affecting livestock housing conditions

Housing conditions are fundamental to the health and welfare of livestock, particularly in intensive farming. Animals living close to each other are potentially more susceptible to the spread of diseases and physiological disorders from lack of movement as well as problems such as stress from competition for feed or social conflict (Rojo-Gimeno et al., 2019).

Housing can be affected by the following:

- environmental conditions;
- air quality.

These are discussed in more detail subsequently.

Environmental conditions include temperature, humidity, lighting, CO_2 concentration, air movement/velocity and corresponding sensible and latent energy releases. Both in open barns, typically used for dairy cattle, and

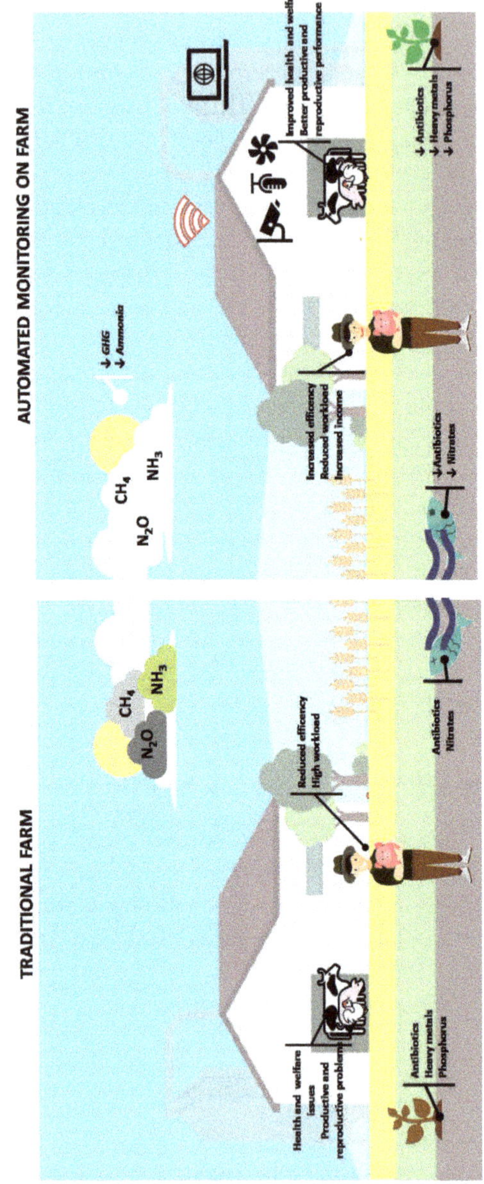

Figure 1 Main aspects of a traditional farm vs a farm equipped with sensors and tools for the automated monitoring of livestock aspects.

closed barns, for poultry and pig, the micro-climate created by the interaction between the housing environment and the animals housed (e.g. the heat animals generate) has a major effect on animal welfare, health, behaviour and production efficiency. Therefore, indoor environmental conditions need to be monitored and kept at satisfactory levels (Lovarelli et al., 2020b; Mikovits et al., 2019). Sensors have an important role to play in effective environmental management (Halachmi et al., 2019).

Air quality is another important factor. Cattle are responsible for methane released from enteric fermentation whilst all livestock species (cattle, pigs and poultry) produce manure. Emissions include ammonia (NH_3) from manure production and CH_4 from its storage. Particulate matter (PM_{10} and $PM_{2.5}$) is also released during feeding (Brown et al., 2019). Bio-aerosols present in the air also contribute to the spread of disease. All these emissions affect animal health, including the spread of pathogens (bacteria and viruses), production and reproductive problems from heat stress, cardiovascular and respiratory disease (Mayo et al., 2019).

Global warming has increased potential risks to livestock from poor housing, such as heat stress (Mikovits et al., 2019). It also affects the release of environmental pollutants, such as ammonia, in the air (Dinuccio et al., 2012). In Central Europe, for example, a +1.4°C mean increase in temperature by 2050 has been predicted, potentially with +14.4 days/year with temperatures > 25°C and +7/+13 days/year with temperatures > 30°C (Mikovits et al., 2019). Mikovits et al. (2019) have studied different methods to optimize environmental conditions in pig housing over the period 1981–2017. They compared four energy-saving air management technologies (direct and indirect cooling systems and a heat exchanger) and three livestock management systems (reduction of stocking intensity, doubling of the rate of summer ventilation, and altering diurnal feeding and resting periods). They found the use of a heat exchanger system to be most effective in reducing heat stress (by 90–100%). The next best solutions were the use of direct and indirect cooling systems. Other important factors are design and construction (e.g. roof slope, ridge height, ridge opening, window position and dimension, and barn orientation).

3 Automated monitoring of animals and their environment

PLF involves the automatic, continuous and real-time monitoring of animals and their housing conditions to support farmers in making decisions (Berckmans, 2017; Berckmans and Guarino, 2017). Modern modelling techniques, combined with vision, images, videos and sound analyses, can be used to assess animal health and behaviour (Wang et al., 2017). Halachmi et al. (2019) have identified the main sensors and other technologies used in PLF. These are summarized in Table 1. PLF involves both monitoring animals and their environment, the

Table 1 Main applications and sensors available for precision livestock farming (PLF). Modified from Halachmi et al. (2019).

Applications	Machine vision, thermal camera	Machine vision, 3D camera	Machine vision, RGB camera	Sound analysis	Response surface	Accelerometers	GPS and positions systems	Loadcell	Bolus	Electronic nose	Other
Body Condition Scoring	Dairy	Dairy	Dairy								Dairy
Body Weight		Pigs Cattle		Broilers				Dairy Broiler	Dairy Beef		
Early detection of diseases or lameness	Dairy Poultry	Dairy	Dairy Poultry	Pigs	Dairy	Dairy Pigs	Dairy	Dairy	Dairy Pigs	Dairy Poultry	
Quantifying pain and stress				Pigs		Dairy					
Feed intake and feeding behavior		Dairy	Dairy	Dairy Beef Poultry			Dairy	Dairy	Cattle		
Rumination time				Dairy		Dairy					
Estrus detection						Dairy Pigs	Dairy				
Milk yield and composition											Dairy
Calving detection						Dairy					
Body temperature	Poultry										Dairy
Aggressive behavior			Pigs								
Quantifying animal welfare				Pigs							
Grazing management							Beef Dairy	Cattle			
Virtual fencing							Beef Dairy				
Heart rate											Beef
Air quality										Dairy Beef Poultry Pigs	

collection of measurement data and its analysis to provide useful information, for example, about variances that require some form of corrective action. Examples include high temperature humidity index (THI) measurements or a high concentration of airborne pollutants in livestock barns, abnormal animal behaviour (e.g. excessive lying or standing times, aggressive behaviour, etc.) or indications of disease (e.g. excessive somatic cells count, parasite numbers in poultry faeces, unhealthy coughs, etc.). A fundamental aspect is that PLF is not designed to replace human decision-making processes but to enhance decision-making by providing better information (Halachmi et al., 2019). Where appropriate, PLF systems can automatically adjust re-conditions when a threshold is exceeded. For example, if THI or pollutants, such as volatile organic compounds (VOCs) or aerosol particles, exceed a defined threshold, ventilation can be automatically initiated or intensified to restore atmospheric conditions to target parameters. If further action is required, the farmer can be alerted to take action based on the information provided by the sensor.

Continuous and real-time monitoring of housing conditions is very important because micro-environmental variables and air molecules concentrations can change rapidly (Weeks, 2008). For example, especially during cooler seasons when ventilation rates are at a minimum level, some gaseous concentrations can increase significantly, reaching dangerous levels in a short time for both animals and workers (O'Neill and Phillips, 1992). This problem can be dealt with by installing sensors that monitor gas concentrations linked to systems for their removal, such as automatically initiating and varying ventilation rates (Rojano et al., 2019; Tabase et al., 2020b), activation of evaporative cooling (Wang et al., 2019a), use of air or chemical scrubbers or biofilters (Dumont, 2018; Van der Heyden et al., 2015).

PLF technologies are also key to manage warm season conditions where heat stress can be a serious problem for animal health and welfare (West, 2003). Sensors can be installed that monitor temperature and relative humidity (THI) continuously in real-time (Zhang et al., 2020), define thermoneutral zones (Amaral et al., 2020) and develop bioenergetic models (Fournel et al., 2017) or spectral entropy models (Herborn et al., 2020) that provide indicators for assessing animal health and welfare. Infrared thermography using thermal imaging cameras and videos can also be used to identify heat stress or pre-clinical signs of disease (Herborn et al., 2020; Zhang et al., 2020) similar to sound analyses (Amaral et al., 2020; Lee et al., 2015). Studies of sound analyses in pigs have found that animals under heat-stress-related conditions produce different sounds with different frequencies, duration and amplitude compared to non-stressed animals (Ferrari et al., 2013). Van Hirtum and Berckmans (2004) and Wang et al. (2019b) found that cough sound analysis can also provide qualitative information about air quality (e.g. ammonia and dust levels) in livestock buildings. Methods have been developed to detect and analyse sneezing in poultry (Carpentier et al., 2019), coughing in pigs (Silva et al., 2008; Silva et al., 2009) and coughing in calves (Carpentier et al., 2018; Ferrari et al., 2010; Vandermeulen et al., 2016) to detect potential health and environmental issues. These methods make it easier to adjust environmental conditions rapidly, to detect health problems early, separate out sick animals and initiate early interventions, which reduce a potentially damaging reliance on antibiotics, given problems such as antibiotic resistance and potential contamination of food products and the environment with veterinary residues (Tullo et al., 2019).

Cameras, videos, sound monitoring and activity sensors have been used to monitor cow behaviour (Arcidiacono et al., 2017; Charlton and Rutter, 2017; Lovarelli et al., 2020b; Meunier et al., 2018; Tsai et al., 2020), pig behaviour (Liu et al., 2020; Norton et al., 2019; Oczak et al., 2013; Oczak et al., 2015), and poultry behaviour (Fontana et al., 2015; Fontana et al., 2017; Li et al., 2019a; Peña Fernández et al., 2018; Rowe et al., 2019; Wang et al., 2019c). Lameness, which is a particular problem in cows, can be evaluated using sensors and algorithms (Halachmi et al., 2013; Jabbar et al., 2017; Pastell et al., 2008; Van Hertem et al., 2013; Van Hertem et al., 2014; Zhao et al., 2018) to detect abnormal movements indicative of early symptoms of lameness. Machine vision can be used to monitor feed intake, both to improve feed efficiency and to identify potential health problems. Body condition scores (BCS) for cows or body weight gain in pigs and poultry can be measured daily using these images (Ferguson et al., 2006; Fontana et al., 2017; Fournel et al., 2017; Frost et al., 1997; Van Hertem et al., 2018).

Another useful tool in PLF is continuous monitoring of the geographical position of animals both indoors and outside using GPS. For example, knowing the position of dairy cows in the barn makes it possible to turn on fans or sprinklers, if needed, only where cows are present or to adjust according to the environmental conditions. This also allows more targeted and efficient electrical and water consumption as well as optimizing the welfare and health status of the monitored animals. GPS technology can also be used when cows are reared outside as a component of virtual fencing (described elsewhere in this book).

4 The use of precision livestock farming technologies to monitor dairy cattle housing

Cows are susceptible to temperatures higher than 25°C, especially when combined with high air relative humidity (Berman, 2019; Klemm and Hall, 1972; Tao et al., 2020; West, 2003). These conditions can be measured using the THI. The THI is an indicator of thermal comfort of animals and plays a very important role in livestock farming since environmental stresses negatively affect production efficiency, health, reproduction and welfare (Mayo et al.,

2019). An excessive THI value is the cause of health problems (reduction of fertility), reduced feed intake, reduced milk production, and could even cause an increase in cardiovascular and respiratory diseases and mortality (Mikovits et al., 2019; Wang et al., 2019a). Moreover, heat stress conditions also affect methanogenesis in the rumen, increasing emission of methane (CH_4) from ruminants, which is widely seen as a significant factor in global warming (Fifth Assessment, 2014; IPCC, 2006). Hempel et al. (2020) found that thermal stress is a potential cause of high CH_4 emissions because of its effect on the digestive system. An optimal temperature of about 10–15°C was found to achieve satisfactory feed intake, digestibility and emission of CH_4.

THI can potentially range between 0 and 100. There are areas where summers are hot and are frequently hit by heat waves. As an example, in Northern Italy, the average temperature in summer is 26°C, with peaks > 35°C. In contrast, winters in the same areas can be quite cold. For example, Northern Italy has an average winter temperature of 1°C with peaks of < −10°C in mountainous areas. Whilst THI values of 30–50 can be maintained in winter, in summer conditions, THI can easily exceed the threshold identified as causing heat stress in cows (THI > 72) if the barn is not properly insulated and managed (Brown-Brandl et al., 2005).

As mentioned previously, an important GHG resulting from livestock production is NH_3. NH_3 is emitted during production, management, treatment, storage and field application of manure (Balsari et al., 2007; Bell et al., 2016; Dinuccio et al., 2011; Finzi et al., 2019; Provolo et al., 2018). A meta-analysis by Sanchis et al. (2019) focused on models to predict NH_3 emissions in dairy cattle housing. It was found that NH_3 emissions were affected by environmental factors (such as temperature, relative humidity, wind speed and ventilation rate) and increased linearly by 1.47 g/day per cow with an increase of air temperature in the barn of 1°C. In addition, when the ventilation rate increased by 100 m^3/h per cow, NH_3 emission increased by 0.007 g/day per cow.

Such research has led to methods that improve barn environments to reduce emissions as well as promote animal welfare, health and productivity. Techniques include the introduction of forced ventilation systems (e.g. using fans or blowers) to circulate fresh air with automatic starting systems when defined THI thresholds are exceeded, cow shower systems to cool animals, mobile shades to protect animals from the sun or reduction of paved areas exposed to the sun to reduce reflected heat (Bar et al., 2019; Berman, 2019; Schütz et al., 2009). Forced ventilation has been widely studied to identify the best solutions to deal with heat stress. Pinto et al. (2019) tested different levels of frequency for use of ventilation systems during the day, showing beneficial effects on animal respiration rates when ventilation was turned on more frequently. Honig et al. (2012) tested up to eight ventilation interventions, finding higher lying and ruminating periods (positive welfare indicators) when interventions were more frequent. Porto et al.

(2017) compared a fogging system with forced ventilation against a sprinkler system with forced ventilation. The fogging system in the resting area was found to encourage lying time, whereas the sprinkler system in the feeding area had no influence on standing time and only a slight influence on feeding activity. Gunn et al. (2019) studied four different levels of intervention to abate heat in a dairy cattle barn, from 'minimal' (open barn-shading) to 'intense' (with air conditioning installed). Despite the additional cost, intense methods were more economic in ensuring a high level of milk production as well as promoting animal welfare and health. Berman (2019) showed the beneficial effect of combining forced ventilation with wetting the surface of cows through the use of showers to increase convective heat loss. Though dependent on barn design and other factors, this is seen as a practical solution to abate heat stress. Similar results were achieved by D'Emilio et al. (2017) and Ruban et al. (2020).

A key element in PLF is the ability to monitor individual animals since each animal is different and has different needs. The most common methods to achieve this are the automatic milking system (AMS) used for dairy cattle and the automatic feeding system (AFS) which can be used for any livestock species. An AMS automatically milks cows without human intervention. The AMS recognizes individual dairy cows when they approach the machine to be milked. The machine automatically analyses milk taken during milking and identifies problems such as excessive somatic cell counts which indicate a health issue (Cabrera et al., 2020; Hogenboom et al., 2019). If a cow approaches the AMS too soon after a previous milking, the machine will not accept it for milking. The system is also designed to alert the farmer if an individual cow is not milked within a defined period, which may indicate a problem with the cow.

An AFS is designed to feed animals at frequent intervals during the day. Most systems do not seek to tailor a different diet to each animal but offer all animals an optimized diet for the group (Bach and Cabrera, 2017; Mattachini et al., 2019). Feeding cows more often during the day has a positive impact on feed intake and conversion ratios. A high feed conversion ratio means cows have better health, higher milk production and better fertility. Frequent feed distribution also means that cows have fresh feed (with fewer leftovers) whilst improved digestibility and feed conversion reduce emissions of CH_4, improving the environmental and economic sustainability of milk production (Lovarelli et al., 2020a).

5 The use of precision livestock farming technologies to monitor poultry housing

Intensive poultry production is often characterized by rearing many animals (sometimes thousands) within single poultry barns at high densities. Major problems include poor air quality, heat stress, risk of disease and its rapid spread within flocks, as well as potential conflict from close proximity of individuals within large groups. All these problems can impact an animal's welfare, health, feed intake, fertility and production efficiency.

Monitoring THI is a key issue for both broilers and laying hens while maintaining the right environmental conditions (Li et al., 2019b). It has been suggested that THI ≤ 70 provides animals with comfortable environmental conditions (El-Tarabany, 2016). Maintaining THI through cooling via adequate ventilation is particularly important. Wang et al. (2019a) found that installing a ventilation system increased layer body mass, egg production and individual egg primarily by reducing heat stress. The new ventilation system reduced heat stress time in the horizontal direction by 29.9% to 64.8% (depending on the housing area) when the alert level (THI in the range 75–78) was reached, and by 0% to 4.5% in the same areas when the danger category (THI in the range 79-83) was reached. Similar values were achieved in the vertical direction: 15.8–47.6% less heat stress in the alert category and 0.9–6.6% less heat stress in the danger category.

The risk of disease can be assessed by monitoring unusual animal behaviour using sound, camera and video tracking. It can also be assessed by monitoring air quality to reduce the spread of airborne bacteria, such as Campylobacter, which pose a food safety risk and can be harmful to poultry (Johannessen et al., 2020; Rasschaert et al., 2020). Thøfner et al. (2019) found a link between the presence of the cocci class of bacteria in housing and footpad lesions in poultry, suggesting footpad lesions may serve as a point of entry for these bacteria. Since cocci are a risk for both poultry and farmworkers, early detection is essential to maintain safe and hygienic conditions for both (Bródka et al., 2012). Borgonovo et al. (2020) tested a data-driven machine learning algorithm to associate critical values of VOCs with abnormal values of oocystis to provide an early alert of coccidiosis risk. Du et al. (2019) analysed the presence of bio-aerosols (i.e. dust, gas particles, antibiotic residues and small particles of a biological origin of 0.5–100 μm), which provide potential media for bacteria. They found higher concentrations of bio-aerosols in the rear of the housing, where the turbulent flow of ventilation inlets increased the diffusion and dispersion of bio-aerosols. Local disturbed flows were found to be responsible for higher local concentration due to the re-suspension of settled bio-aerosols. These results identified priorities for more regular or intense cleaning and disinfection.

One problem affecting indoor air quality is NH_3 emissions from poultry litter, which are both a health risk (in causing respiratory and cardiovascular disease) and an environmental problem. NH_3 emissions vary according to factors, such as breed type, diet, barn design, ventilation and environmental conditions (Rojano et al., 2019; Trokhaniak et al., 2019). Wood et al. (2015) found an average of 600 mg/day of NH_3 per hen within a range of 35–828 mg/day of NH_3 per hen. Dekker et al. (2011) found an average emission of 87.1

mg/day, while Zhao et al. (2013) reported average emissions of 140 mg/day, both within the range found by Wood et al. (2015). Techniques such as manure drying tunnels have been used to minimize NH_3 losses. Rosa et al. (2020) identified the ventilation rate as the main factor influencing the variability of NH_3 concentration in the barn. NH_3 concentrations were lower when the ventilation system was operating, with an annual average equivalent to 93.8 mg NH_3/day. NH_3 emissions were equivalent to 209.3 mg NH_3/day in the manure drying tunnel, mainly influenced by environmental parameters (temperature and relative humidity) and by the dry matter content of manure.

Ammonia is a precursor to the formation of PM_{10} and secondary PM ($PM_{2.5}$), which have detrimental health effects (Li et al., 2020; Peña Fernández et al., 2019). Poultry production is one of the main sectors responsible for high $PM_{2.5}$ emissions, mostly due to feeding operations that release particulate matter, an issue that also affects the feeding of pigs (Cheng et al., 2017). PM emissions range between 3.8 mg/m^3 and 10.4 mg/m^3 in broiler housing and between 0.7 mg/m^3 and 8.7 mg/m^3 in layer hen housing. Guo et al. (2019) introduced a vegetative barrier downwind of the poultry livestock barn to trap PM. They found statistically significant reductions in $PM_{2.5}$ and PM_{10} in the area surrounding the barn, with similar results found by Ajami et al. (2019). Other techniques for abatement of PM include biofilters and scrubbers installed in housing to filter and clean air. Recent solutions include bio scrubbers or bio trickling filters, dry filters, water scrubbers, and wet acid scrubbers that can reduce odours and GHGs, such as CH_4 (Dumont, 2018; Van der Heyden et al., 2015). Mostafa and Buescher (2011) tested different types of scrubber to abate particulate matter and dust. They found a dry scrubber superior to a cyclone and wet scrubber, reducing indoor concentrations of PM and dust by 55% and 72%, respectively.

6 The use of precision livestock farming technologies to monitor pig housing

Like poultry, pigs are reared in large numbers and high stocking densities in intensive systems, with similar potential health, welfare and environmental issues. Kim and Ko (2019) found the highest concentrations of airborne bacteria in housing for growing/fattening pigs and nursery pigs, particularly during the summer. Pig slurry is rich in NH_3, with negative effects on air quality which cause both health issues (e.g. respiratory and cardiovascular diseases) and environmental impacts (Pexas et al., 2020; Tabase et al., 2020a).

Monitoring indoor air quality and environmental parameters can help farmers control the pig environment and increase farm performance (Carbone et al., 2010). Typical methods include installation of air scrubbers, biofilters (to reduce odours and airborne bacteria, though less effective in reducing NH_3), as well as trickle-bed reactors (abatement of dust, airborne bacteria and NH_3), air buffer spaces, inlet ducts, perforated ceilings, roof-chimney exhaust fans and heating equipment (Yeo et al., 2019). Air scrubbers are particularly effective in improving housing air quality. Van der Heyden et al. (2016) compared different scrubber solutions in a pig housing system. They found that the biological air scrubber with an additional nitrification tank performed well in terms of NH_3 removal efficiency (85 ± 6%), while the two-stage biological air scrubber and chemical scrubber were less effective. Lower NH_3 removal efficiency was achieved during the day because the ventilation rate was higher than at night, affecting NH_3 concentrations (Van der Heyden et al., 2020). Mostafa et al. (2017) found that the coupled effect of an air scrubber and the spraying of a water-oil mixture resulted in efficient NH_3 and PM abatement. To abate NH_3 emissions, de Vries and Melse (2017) compared an acid scrubber with two bio-trickling filters, one using nitrification and another employing nitrification and denitrification, finding the acid scrubber most effective. Schauberger et al. (2019) found air exchangers and forced mechanical ventilation were effective in improving air quality and reducing heat stress in pigs. Optimal ventilation system design is challenging since it must take into account seasonal differences in temperature and humidity as well different growth stages, particularly piglets and growing pigs that are especially sensitive to heat stress (Yeo et al., 2019). Measuring and modelling NH_3 fluxes and other aspects of the barn environment remains challenging (Rong and Aarnink, 2019).

Yeo et al. (2019) have analysed the effects of differing environmental conditions on pigs. They found that food intake was about 11% lower in pigs reared at daily temperatures ranging between 20°C and 35°C compared to pigs reared at a constant air temperature of 20°C. Feed intake was found to be 43.5 g/day less and body mass gain 17.6 g/day less. At higher temperatures pigs consumed 8.3 L/day of water compared to the recommended limit of 7.3 L/day, contributing to reduced body mass gain. Whilst setting the ventilation rate at 26°C ensured an adequate temperature, it only achieved half of the recommended rate for humidity, ammonia and dust levels. NH_3 and breathable dust levels (113 ppm and 272.8 $\mu g/m^3$ for NH_3 and breathable dust, respectively) exceeded the permitted exposure levels both for animals and workers (25 ppm for NH_3 and 230 $\mu g/m^3$ for breathable dust). This highlights the importance of accounting for a range of factors, such as NH_3 as well as temperature.

7 Conclusion and future trends

Effective monitoring of livestock housing environmental conditions is becoming more important since it has been shown that micro-environmental variables, such as temperature, relative humidity and air quality, have a significant effect on livestock health and welfare. Variables such as temperature also affect the release of CH_4 and NH_3 from manure and slurry,

CH_4 from enteric fermentation, particulate matter from feeding operations, as well as VOCs and bio-aerosols, which affect both the immediate and wider environment. PLF instrumentation, such as microphones for sound analysis, thermo-cameras and video cameras, can be used to monitor signs of stress from poor environmental conditions. New technologies, such as machine learning, can convert this data into meaningful information for decision-making to meet increasingly challenging targets in such areas as animal welfare and environmental impact whilst meeting growing consumer demand. These techniques allow accurate, continuous and real-time monitoring of environmental conditions, which potentially enables continual, rapid and automated adjustment of the livestock environment. This will optimize feed conversion efficiency, animal growth, prevent disease or detect it much earlier, reduce stress and damaging environmental emissions, making livestock production more efficient and sustainable.

PLF techniques will, therefore, have a role to play in initiatives such as the EU's Green Deal which seeks to make the EU economy more sustainable by setting the following targets: (i) no net emissions of GHGs, (ii) economic growth decoupled from resource use, (iii) no person and no place left behind (European Green Deal, 2020). Part of this initiative concerns transforming agricultural production 'From farm to fork' to assure safe, healthy, affordable and sustainable food for Europeans, tackling climate change, protecting the environment and preserving biodiversity, allowing fair economic returns and increasing organic farming. PLF will play a key role since it allows early recognition of diseases and allowing less use of antibiotics, thus reducing the problem of antimicrobial resistance and environmental contamination. PLF helps to reduce emissions to the environment, improve production efficiency and even make organic farming solutions more adoptable. The motto for the future could be 'sustainable, informed and efficient food production'.

8 References

Aguirre-Villegas, H. A., Larson, R. A. and Sharara, M. A. (2019). Anaerobic digestion, solid-liquid separation, and drying of dairy manure: measuring constituents and modeling emission. *Science of the Total Environment* 696. https://doi.org/10.1016/j.scitotenv.2019.134059.

Ajami, A., Shah, S. B., Wang-Li, L., Kolar, P. and Castillo, M. S. (2019). Windbreak wall-vegetative strip system to reduce air emissions from mechanically ventilated livestock barns–part 3: layer house evaluation. *Water, Air, and Soil Pollution* 230(12). https://doi.org/10.1007/s11270-019-4345-0.

Amaral, P. I. S., Campos, A. T., Yanagi Junior, T., Cecchin, D., Leite, E. M. and Dias e Silva, N. C. (2020). Using sounds produced by pigs to identify thermoneutrality zones for thermal environment assessment ratios. *Engenharia Agricola* 40(3), 266–271. https://doi.org/10.1590/1809-4430-eng.agric.v40n3p266-271/2020.

Arcidiacono, C., Porto, S. M. C., Mancino, M. and Cascone, G. (2017). Development of a threshold-based classifier for real-time recognition of cow feeding and standing behavioural activities from accelerometer data. *Computers and Electronics in Agriculture* 134, 124–134. http://doi.org/10.1016/j.compag.2017.01.021.

Bacenetti, J., Sala, C., Fusi, A. and Fiala, M. (2016). Agricultural anaerobic digestion plants: what LCA studies pointed out and what can be done to make them more environmentally sustainable. *Applied Energy* 179, 669–686. https://doi.org/10.1016/j.apenergy.2016.07.029.

Bach, A. and Cabrera, V. (2017). Robotic milking: feeding strategies and economic returns. *Journal of Dairy Science* 100(9), 7720–7728. https://doi.org/10.3168/jds.2016-11694.

Balsari, P., Airoldi, G., Dinuccio, E. and Gioelli, F. (2007). Ammonia emissions from farmyard manure heaps and slurry stores-effect of environmental conditions and measuring methods. *Biosystems Engineering* 97(4), 456–463. https://doi.org/10.1016/j.biosystemseng.2007.03.033.

Bar, D., Kaim, M., Flamenbaum, I., Hanochi, B. and Toaff-Rosenstein, R. L. (2019). Technical note: accelerometer-based recording of heavy breathing in lactating and dry cows as an automated measure of heat load. *Journal of Dairy Science* 102(4), 3480–3486. https://doi.org/10.3168/jds.2018-15186.

Bell, M. J., Hinton, N. J., Cloy, J. M., Topp, C. F. E., Rees, R. M., Williams, J. R., Misselbrook, T. H. and Chadwick, D. R. (2016). How do emission rates and emission factors for nitrous oxide and ammonia vary with manure type and time of application in a Scottish farmland? *Geoderma* 264, 81–93. https://doi.org/10.1016/j.geoderma.2015.10.007.

Bell, M. J., Wall, E., Russell, G., Simm, G. and Stott, A. W. (2011). The effect of improving cow productivity, fertility, and longevity on the global warming potential of dairy systems. *Journal of Dairy Science* 94(7), 3662–3678. https://doi.org/10.3168/jds.2010-4023.

Berckmans, D. (2017). General introduction to precision livestock farming. *Animal Frontiers* 7(1), 6–11. https://doi.org/10.2527/af.2017.0102.

Berckmans, D. and Guarino, M. (2017). From the editors: precision livestock farming for the global livestock sector. *Animal Frontiers* 7(1), 4–5. https://doi.org/10.2527/af.2017.0101.

Berman, A. (2019). An overview of heat stress relief with global warming in perspective. *International Journal of Biometeorology* 63(4), 493–498. https://doi.org/10.1007/s00484-019-01680-7.

Borgonovo, F., Ferrante, V., Grilli, G., Pascuzzo, R., Vantini, S. and Guarino, M. (2020). A data-driven prediction method for an early warning of coccidiosis in intensive livestock systems: a preliminary study. *Animals: An Open Access Journal from MDPI* 10(4). https://doi.org/10.3390/ani10040747.

Brentrup, F., Küsters, J., Lammel, J. and Kuhlmann, H. (2000). Methods to estimate on-field nitrogen emissions from crop production as an input to LCA studies in the agricultural sector. *The International Journal of Life Cycle Assessment* 5(6), 349–357.

Bródka, K., Kozajda, A., Buczyńska, A. and Szadkowska-Stańczyk, I. (2012). The variability of bacterial aerosol in poultry houses depending on selected factors. *International Journal of Occupational Medicine and Environmental Health* 25(3), 281–293. https://doi.org/10.2478/S13382-012-0032-8.

Brown, M. S., Cole, N. A., Gruber, S., Kube, J. and Teeter, J. S. (2019). Modeling and prediction accuracy of ammonia gas emissions from feedlot cattle. *Applied Animal Science* 35(3), 347-356. https://doi.org/10.15232/aas.2018-01834.

Brown-Brandl, T. M., Eigenberg, R. A., Nienaber, J. A. and Hahn, G. L. (2005). Dynamic response indicators of heat stress in shaded and non-shaded feedlot cattle, part 1: analyses of indicators. *Biosystems Engineering* 90(4), 451-462. https://doi.org/10.1016/j.biosystemseng.2004.12.006.

Cabrera, V. E., Barrientos-Blanco, J. A., Delgado, H. and Fadul-Pacheco, L. (2020). Symposium review: real-time continuous decision making using big data on dairy farms. *Journal of Dairy Science* 103(4), 3856-3866. https://doi.org/10.3168/jds.2019-17145.

Cao, Y., Wu, J., Zhang, X., Niu, B., Li, M., Zhang, Y., Wang, X. and Wang, Z. (2019). Dynamic forage-livestock balance analysis in alpine grasslands on the Northern Tibetan Plateau. *Journal of Environmental Management* 238(March), 352-359. https://doi.org/10.1016/j.jenvman.2019.03.010.

Carbone, C., Decesari, S., Mircea, M., Giulianelli, L., Finessi, E., Rinaldi, M., Fuzzi, S., Marinoni, A., Duchi, R., Perrino, C., Sargolini, T., Vardè, M., Sprovieri, F., Gobbi, G. P., Angelini, F. and Facchini, M. C. (2010). Size-resolved aerosol chemical composition over the Italian Peninsula during typical summer and winter conditions. *Atmospheric Environment* 44(39), 5269-5278. https://doi.org/10.1016/j.atmosenv.2010.08.008.

Carpentier, L., Berckmans, D., Youssef, A., Berckmans, D., van Waterschoot, T., Johnston, D., Ferguson, N., Earley, B., Fontana, I., Tullo, E., Guarino, M., Vranken, E. and Norton, T. (2018). Automatic cough detection for bovine respiratory disease in a calf house. *Biosystems Engineering* 173, 45-56. https://doi.org/10.1016/j.biosystemseng.2018.06.018.

Carpentier, L., Vranken, E., Berckmans, D., Paeshuyse, J. and Norton, T. (2019). Development of sound-based poultry health monitoring tool for automated sneeze detection. *Computers and Electronics in Agriculture* 162, 573-581. https://doi.org/10.1016/j.compag.2019.05.013.

Charlton, G. L. and Rutter, S. M. (2017). The behaviour of housed dairy cattle with and without pasture access: a review. *Applied Animal Behaviour Science* 192, 2-9. https://doi.org/10.1016/j.applanim.2017.05.015.

Cheng, S. S., Li, Y., Geng, S. J., Hu, L. S., Fu, X. F. and Han, X. Y. (2017). Effects of dietary fresh fermented soybean meal on growth performance, ammonia and particulate matter emissions, and nitrogen excretion in nursery piglets. *Journal of Zhejiang University. Science. B* 18(12), 1083-1092. https://doi.org/10.1631/jzus.B1700180.

D'Emilio, A., Porto, S. M. C., Cascone, G., Bella, M. and Gulino, M. (2017). Mitigating heat stress of dairy cows bred in a free-stall barn by sprinkler systems coupled with forced ventilation. *Journal of Agricultural Engineering* 48(4), 190-195. https://doi.org/10.4081/jae.2017.691.

de Vries, J. W. and Melse, R. W. (2017). Comparing environmental impact of air scrubbers for ammonia abatement at pig houses: a life cycle assessment. *Biosystems Engineering* 161, 53-61. https://doi.org/10.1016/j.biosystemseng.2017.06.010.

de Vries, M. and de Boer, I. J. M. (2010). Comparing environmental impacts for livestock products: a review of life cycle assessments. *Livestock Science* 128(1-3), 1-11. https://doi.org/10.1016/j.livsci.2009.11.007.

Dekker, S. E. M., Aarnink, A. J. A., de Boer, I. J. M. and Koerkamp, P. W. G. G. (2011). Emissions of ammonia, nitrous oxide, and methane from aviaries with organic laying hen husbandry. *Biosystems Engineering* 110(2), 123–133. https://doi.org/10.1016/j.biosystemseng.2011.07.006.

Dinuccio, E., Berg, W. and Balsari, P. (2011). Effects of mechanical separation on GHG and ammonia emissions from cattle slurry under winter conditions. *Animal Feed Science and Technology* 166–167, 532–538. https://doi.org/10.1016/j.anifeedsci.2011.04.037.

Dinuccio, E., Gioelli, F., Balsari, P. and Dorno, N. (2012). Ammonia losses from the storage and application of raw and chemo-mechanically separated slurry. *Agriculture, Ecosystems and Environment* 153(2), 16–23. https://doi.org/10.1016/j.agee.2012.02.015.

Du, L., Yang, L., Yang, C., Dominy, R., Hu, C., Du, H., Li, Q., Yu, C., Xie, L. and Jiang, X. (2019). Investigation of bio-aerosol dispersion in a tunnel-ventilated poultry house. *Computers and Electronics in Agriculture* 167. https://doi.org/10.1016/j.compag.2019.105043, 105043.

Dumont, É (2018). Impact of the treatment of NH_3 emissions from pig farms on greenhouse gas emissions. Quantitative assessment from the literature data. *New Biotechnology* 46, 31–37. https://doi.org/10.1016/j.nbt.2018.06.001.

El-Tarabany, M. S. (2016) Impact of temperature-humidity index on egg-laying characteristics and related stress and immunity parameters of Japanese quails. *International Journal of Biometeorology* 60(7), 957–964. https://doi.org/10.1007/s00484-015-1088-5.

European Green Deal. European Commission (2020). Available at: https://ec.europa.eu/info/strategy/priorities-2019-2024/european-green-deal_en. Accessed June 2020.

Ferguson, J. D., Azzaro, G. and Licitra, G. (2006). Body condition assessment using digital images. *Journal of Dairy Science* 89(10), 3833–3841. https://doi.org/10.3168/jds.S0022-0302(06)72425-0.

Ferrari, S., Costa, A. and Guarino, M. (2013). Heat stress assessment by swine related vocalizations. *Livestock Science* 151(1), 29–34. https://doi.org/10.1016/j.livsci.2012.10.013.

Ferrari, S., Piccinini, R., Silva, M., Exadaktylos, V., Berckmans, D. and Guarino, M. (2010). Cough sound description in relation to respiratory diseases in dairy calves. *Preventive Veterinary Medicine* 96(3–4), 276–280. https://doi.org/10.1016/j.prevetmed.2010.06.013.

Fifth Assessment (2014). *Global Warming Potential Values*, 2-5.

Finzi, A., Mattachini, G., Lovarelli, D., Riva, E. and Provolo, G. (2020). Technical, economic, and environmental assessment of a collective integrated treatment system for energy recovery and nutrient removal from livestock manure. *Sustainability* 12(7), 1–18. https://doi.org/10.3390/su12072756.

Finzi, A., Riva, E., Bicoku, A., Guido, V., Shallari, S. and Provolo, G. (2019). Comparison of techniques for ammonia emission mitigation during storage of livestock manure and assessment of their effect in the management chain. *Journal of Agricultural Engineering* 50(1), 12–19. https://doi.org/10.4081/jae.2019.881.

Fontana, I., Tullo, E., Butterworth, A. and Guarino, M. (2015). An innovative approach to predict the growth in intensive poultry farming. *Computers and Electronics in Agriculture* 119, 178–183. https://doi.org/10.1016/j.compag.2015.10.001.

Fontana, I., Tullo, E., Carpentier, L., Berckmans, D., Butterworth, A., Vranken, E., Norton, T., Berckmans, D. and Guarino, M. (2017). Sound analysis to model weight of broiler chickens. *Poultry Science* 96(11), 3938–3943. https://doi.org/10.3382/ps/pex215.

Fournel, S., Rousseau, A. N. and Laberge, B. (2017). Rethinking environment control strategy of confined animal housing systems through precision livestock farming. *Biosystems Engineering* 155, 96–123. https://doi.org/10.1016/j.biosystemseng.2016.12.005.

Frost, A. R., Schofield, C. P., Beaulah, S. A., Mottram, T. T., Lines, J. A. and Wathes, C. M. (1997). A review of livestock monitoring and the need for integrated systems. *Computers and Electronics in Agriculture* 17(2), 139–159. https://doi.org/10.1016/S0168-1699(96)01301-4.

Guarino, M., Fabbri, C., Brambilla, M., Valli, L. and Navarotto, P. (2006). Evaluation of simplified covering systems to reduce gaseous emissions from livestock manure storage. *Transactions of the ASABE* 49(3), 737–747.

Gunn, K. M., Holly, M. A., Veith, T. L., Buda, A. R., Prasad, R., Alan Rotz, C., Soder, K. J. and Stoner, A. M. K. (2019). Projected heat stress challenges and abatement opportunities for U. S. Milk production. *PLoS ONE* 14(3), 1–21. https://doi.org/10.1371/journal.pone.0214665.

Guo, L., Ma, S., Zhao, D., Zhao, B., Xu, B., Wu, J., Tong, J., Chen, D., Ma, Y., Li, M. and Chang, Z. (2019). Experimental investigation of vegetative environment buffers in reducing particulate matters emitted from ventilated poultry house. *Journal of the Air and Waste Management Association* 69(8), 934–943. https://doi.org/10.1080/10962247.2019.1598518.

Halachmi, I., Guarino, M., Bewley, J. and Pastell, M. (2019). Smart animal agriculture: application of real-time sensors to improve animal well-being and production. *Annual Review of Animal Biosciences* 7(1), 403–425. https://doi.org/10.1146/annurev-animal-020518-114851

Halachmi, I., Klopčič, M., Polak, P., Roberts, D. J. and Bewley J. M. (2019). Automatic assessment of dairy cattle body condition score using thermal imaging. *Computers and Electronics in Agriculture*. 99, 35–40. http://dx.doi.org/10.1016/j.compag.2013.08.012

Hempel, S., Willink, D., Janke, D., Ammon, C., Amon, B. and Amon, T. (2020). Methane emission characteristics of naturally ventilated cattle buildings. *Sustainability* 12(10), 1–17. https://doi.org/10.3390/su12104314.

Herborn, K. A., McElligott, A. G., Mitchell, M. A., Sandilands, V., Bradshaw, B. and Asher, L. (2020). Spectral entropy of early-life distress calls as an iceberg indicator of chicken welfare. *Journal of the Royal Society Interface*, 17. https://doi.org/10.1098/rsif.2020.0086.

Hoff, S. J., Bundy, D. S., Nelson, M. A., Zelle, B. C., Jacobson, L. D., Heber, A. J., Ni, J., Zhang, Y., Koziel, J. A. and Beasley, D. B. (2006). Emissions of ammonia, hydrogen sulfide, and odor before, during, and after slurry removal from a deep-pit swine finisher. *Journal of the Air and Waste Management Association* 56(5), 581–590. https://doi.org/10.1080/10473289.2006.10464472.

Hogenboom, J. A., Pellegrino, L., Sandrucci, A., Rosi, V. and D'Incecco, P. (2019). Invited review: hygienic quality, composition, and technological performance of raw milk obtained by robotic milking of cows. *Journal of Dairy Science* 102(9), 7640–7654. https://doi.org/10.3168/jds.2018-16013.

Honig, H., Miron, J., Lehrer, H., Jackoby, S., Zachut, M., Zinou, A., Portnick, Y. and Moallem, U. (2012). Performance and welfare of high-yielding dairy cows subjected to 5 or 8 cooling sessions daily under hot and humid climate. *Journal of Dairy Science* 95(7), 3736–3742. https://doi.org/10.3168/jds.2011-5054.

Im, S., Petersen, S. O., Lee, D. and Kim, D. H. (2020). Effects of storage temperature on CH4 emissions from cattle manure and subsequent biogas production potential. *Waste Management* 101, 35–43. https://doi.org/10.1016/j.wasman.2019.09.036.

International Dairy Federation (2015). *Bulletin of the International Dairy Federation* 479, 1-283. https://doi.org/10.1016/s0958-6946(97)88755-9.

IPCC (Ed.) (2006). Guidelines for national greenhouse gas inventories. Chapter 10 Emissions from livestock and manure management.

ISO 14040 Series (2006). Environmental management - Life Cycle Assessment - Requirements and guidelines. International Organization for Standardization.

Jabbar, A. K., Hansen, M. F., Smith, M. L. and Smith, L. N. (2017). Early and non-intrusive lameness detection in dairy cows using 3-dimensional video. *Biosystems Engineering* 153, 63-69. https://doi.org/10.1016/j.biosystemseng.2016.09.017.

Johannessen, G. S., Garofolo, G., Di Serafino, G., Koláčková, I., Karpíšková, R., Wieczorek, K., Osek, J., Christensen, J., Torp, M. and Hoorfar, J. (2020). Campylobacter in chicken - critical parameters for international, multicentre evaluation of air sampling and detection methods. *Food Microbiology* 90, 103455. https://doi.org/10.1016/j.fm.2020.103455.

Kamphuis, C., Steeneveld, W. and Hogeveen, H. (2015). Economic modelling to evaluate the benefits of precision livestock farming technologies. In: Precision livestock farming applications. Making sense of sensors to support farm management. Edited collection. pp. 87–94. ISBN: 978-90-8686-268-9. https://doi.org/10.3920/978-90-8686-815-5_3.2

Kim, K. Y. and Ko, H. J. (2019). Indoor distribution characteristics of airborne bacteria in pig buildings as influenced by season and housing type. *Asian-Australasian Journal of Animal Sciences* 32(5), 742–747. https://doi.org/10.5713/ajas.18.0415.

Klemm, F. K. and Hall, J. F. (1972). Utility of heat stress indexes and effect of humidity and temperature on single physiologic strains. *Journal of Applied Physiology* 33(4), 436–440.

Lee, J., Noh, B., Jang, S., Park, D., Chung, Y. and Chang, H. H. (2015). Stress detection and classification of laying hens by sound analysis. *Asian-Australasian Journal of Animal Sciences* 28(4), 592–598. https://doi.org/10.5713/ajas.14.0654.

Li, N., Ren, Z., Li, D. and Zeng, L. (2019a). Review: automated techniques for monitoring the behaviour and welfare of broilers and laying hens: towards the goal of precision livestock farming. *Animal* 14(3), 617–625. https://doi.org/10.1017/S1751731119002155.

Li, Z., Wang, H., Zheng, W., Li, B., Wei, Y., Zeng, J. and Lei, C. (2019b). A tracing method of airborne bacteria transmission across built environments. *Building and Environment* 164(17), 106335. https://doi.org/10.1016/j.buildenv.2019.106335.

Li, Z., Wang, C., Li, B., Shi, Z., Zheng, W. and Teng, G. (2020). Concentration and size distribution of particulate matter in a new aviary system for laying hens in China. *Journal of the Air and Waste Management Association* 70(4), 379-392. https://doi.org/10.1080/10962247.2020.1722291.

Liu, D., Oczak, M., Maschat, K., Baumgartner, J., Pletzer, B., He, D. and Norton, T. (2020). A computer vision-based method for spatial-temporal action recognition of tail-biting behaviour in group-housed pigs. *Biosystems Engineering* 195, 27-41. https://doi.org/10.1016/j.biosystemseng.2020.04.007.

Lovarelli, D., Bacenetti, J. and Guarino, M. (2020a). A review on dairy cattle farming : is precision livestock farming the compromise for an environmental, economic and social sustainable production? *Journal of Cleaner Production* 262. https://doi.org/10.1016/j.jclepro.2020.121409, 121409.

Lovarelli, D., Finzi, A., Mattachini, G. and Riva, E. (2020b). A survey of dairy cattle behavior in different barns in Northern Italy. *Animals: An Open Access Journal from MDPI* 10(4), 713.

Lovarelli, D., Bava, L., Zucali, M., D'Imporzano, G., Adani, F., Tamburini, A. and Sandrucci, A. (2019). Improvements to dairy farms for environmental sustainability in Grana Padano and Parmigiano Reggiano production systems. *Italian Journal of Animal Science* 18(1), 1035–1048. https://doi.org/10.1080/1828051X.2019.1611389.

Mattachini, G., Pompe, J., Finzi, A., Tullo, E., Riva, E. and Provolo, G. (2019). Effects of feeding frequency on the lying behavior of dairy cows in a loose housing with automatic feeding and milking system. *Animals: An Open Access Journal from MDPI* 9(4). https://doi.org/10.3390/ani9040121.

Mattila, P. K. and Joki-Tokola, E. (2003). Effect of treatment and application technique of cattle slurry on its utilization by ley: I. Slurry properties and ammonia volatilization. *Nutrient Cycling in Agroecosystems* 65(3), 221–230. https://doi.org/10.1023/A:1022619304798.

Mayo, L. M., Silvia, W. J., Ray, D. L., Jones, B. W., Stone, A. E., Tsai, I. C., Clark, J. D., Bewley, J. M. and Heersche, G. (2019). Automated estrous detection using multiple commercial precision dairy monitoring technologies in synchronized dairy cows. *Journal of Dairy Science* 102(3), 2645–2656. https://doi.org/10.3168/jds.2018-14738.

Meunier, B., Pradel, P., Sloth, K. H., Cirié, C., Delval, E., Mialon, M. M. and Veissier, I. (2018). Image analysis to refine measurements of dairy cow behaviour from a real-time location system. *Biosystems Engineering* 173, 32–44. https://doi.org/10.1016/j.biosystemseng.2017.08.019.

Mikovits, C., Zollitsch, W., Hörtenhuber, S. J., Baumgartner, J., Niebuhr, K., Piringer, M., Anders, I., Andre, K., Hennig-Pauka, I., Schönhart, M. and Schauberger, G. (2019). Global warming impact on confined livestock in buildings: efficacy of adaptation measures to reduce heat stress for growing-fattening pigs. *International Journal of Biometeorology* 63, 221–230. https://doi.org/10.1007/s10584-019-02525-3.

Mostafa, E. and Buescher, W. (2011). Indoor air quality improvement from particle matters for laying hen poultry houses. *Biosystems Engineering* 109(1), 22–36. https://doi.org/10.1016/j.biosystemseng.2011.01.011.

Mostafa, E., Hoelscher, R., Diekmann, B., Ghaly, A. E. and Buescher, W. (2017). Evaluation of two indoor air pollution abatement techniques in forced ventilation fattening pig barns. *Atmospheric Pollution Research* 8(3), 428–438. https://doi.org/10.1016/j.apr.2016.11.003.

Norton, T., Chen, C., Larsen, M. L. V. and Berckmans, D. (2019). Review: precision livestock farming: building "digital representations" to bring the animals closer to the farmer. *Animal* 13(12), 3009–3017. https://doi.org/10.1017/S175173111900199X.

Oczak, M., Ismayilova, G., Costa, A., Viazzi, S., Sonoda, L. T., Fels, M., Bahr, C., Hartung, J., Guarino, M., Berckmans, D. and Vranken, E. (2013). Analysis of aggressive behaviours of pigs by automatic video recordings. *Computers and Electronics in Agriculture* 99, 209–217. https://doi.org/10.1016/j.compag.2013.09.015.

Oczak, M., Maschat, K., Berckmans, D., Vranken, E. and Baumgartner, J. (2015). Classification of nest-building behaviour in sows on the basis of accelerometer data. Precision Livestock Farming 2015–Papers Presented at the 7th European

Conference on Precision Livestock Farming, ECPLF 2015, Milan. https://doi.org/10.1016/j.biosystemseng.2015.09.007.

O'Neill, D. H. and Phillips, V. R. (1992). A review of the control of odour nuisance from livestock buildings: Part 3, properties of the odorous substances which have been identified in livestock wastes or in the air around them. *Journal of Agricultural Engineering Research* 53, 23–50. https://doi.org/10.1016/0021-8634(92)80072-Z.

Pastell, M., Kujala, M., Aisla, A.-M., Hautala, M., Poikalainen, V., Praks, J., Veermäe, I. and Ahokas, J. (2008). Detecting cow's lameness using force sensors. *Computers and Electronics in Agriculture* 64(1), 34–38. https://doi.org/10.1016/j.compag.2008.05.007.

Peña Fernández, A., Demmers, T. G. M., Tong, Q., Youssef, A., Norton, T., Vranken, E. and Berckmans, D. (2019). Real-time modelling of indoor particulate matter concentration in poultry houses using broiler activity and ventilation rate. *Biosystems Engineering* 187, 214–225. https://doi.org/10.1016/j.biosystemseng.2019.09.004.

Peña Fernández, A., Norton, T., Tullo, E., van Hertem, T., Youssef, A., Exadaktylos, V., Vranken, E., Guarino, M. and Berckmans, D. (2018). Real-time monitoring of broiler flock's welfare status using camera-based technology. *Biosystems Engineering* 173, 103–114. https://doi.org/10.1016/j.biosystemseng.2018.05.008.

Pexas, G., Mackenzie, S. G., Wallace, M. and Kyriazakis, I. (2020). Environmental impacts of housing conditions and manure management in European pig production systems through a life cycle perspective: a case study in Denmark. *Journal of Cleaner Production* 253. https://doi.org/10.1016/j.jclepro.2020.120005, 120005.

Pinto, S., Hoffmann, G., Ammon, C., Heuwieser, W., Levit, H., Halachmi, I. and Amon, T. (2019). Effect of two cooling frequencies on respiration rate in lactating dairy cows under hot and humid climate conditions. *Annals of Animal Science* 19(3), 821–834. https://doi.org/10.2478/aoas-2019-0026.

Porto, S. M. C., D'Emilio, A. and Cascone, G. (2017). On the influence of the alternation of two different cooling systems on dairy cow daily activities. *Journal of Agricultural Engineering* 48(1), 21–27. https://doi.org/10.4081/jae.2017.577.

Provolo, G., Mattachini, G., Finzi, A., Cattaneo, M., Guido, V. and Riva, E. (2018). Global warming and acidification potential assessment of a collective manure management system for bioenergy production and nitrogen removal in Northern Italy. *Sustainability* 10(10). https://doi.org/10.3390/su10103653.

Rasschaert, G., De Zutter, L., Herman, L. and Heyndrickx, M. (2020). Campylobacter contamination of broilers: the role of transport and slaughterhouse. *International Journal of Food Microbiology* 322(November 2019), 108564. https://doi.org/10.1016/j.ijfoodmicro.2020.108564.

Rojano, F., Bournet, P. E., Hassouna, M., Robin, P., Kacira, M. and Choi, C. Y. (2019). Modelling the impact of air discharges caused by natural ventilation in a poultry house. *Biosystems Engineering* 180, 168–181. https://doi.org/10.1016/j.biosystemseng.2019.02.001.

Rojo-Gimeno, C., van der Voort, M., Niemi, J. K., Lauwers, L., Kristensen, A. R. and Wauters, E. (2019). Assessment of the value of information of precision livestock farming: a conceptual framework. *NJAS–Wageningen Journal of Life Sciences* 90–91(October). https://doi.org/10.1016/j.njas.2019.100311, 100311.

Rong, L. and Aarnink, A. J. A. (2019). Development of ammonia mass transfer coefficient models for the atmosphere above two types of the slatted floors in a pig house using computational fluid dynamics. *Biosystems Engineering* 183, 13-25. https://doi.org/10.1016/j.biosystemseng.2019.04.011.

Rosa, E., Arriaga, H. and Merino, P. (2020). Ammonia emission from a manure-belt laying hen facility equipped with an external manure drying tunnel. *Journal of Cleaner Production* 251. https://doi.org/10.1016/j.jclepro.2019.119591, 119591.

Rowe, E., Dawkins, M. S. and Gebhardt-Henrich, S. G. (2019). A systematic review of precision livestock farming in the poultry sector: is technology focussed on improving bird welfare? *Animals: An Open Access Journal from MDPI* 9(9), 1-18. https://doi.org/10.3390/ani9090614.

Ruban, S., Borshch, O. O., Borshch, O. V., Orischuk, O., Balatskiy, Y., Fedorchenko, M., Kachan, A. and Zlochevskiy, M. (2020). The impact of high temperatures on respiration rate, breathing condition and productivity of dairy cows in different production systems. *Animal Science Papers and Reports* 38(1), 61-72.

Sanchis, E., Calvet, S., Prado, Ad and Estellés, F. (2019). A meta-analysis of environmental factor effects on ammonia emissions from dairy cattle houses. *Biosystems Engineering* 178, 176-183. https://doi.org/10.1016/j.biosystemseng.2018.11.017.

Schauberger, G., Mikovits, C., Zollitsch, W., Hörtenhuber, S. J., Baumgartner, J., Niebuhr, K., Piringer, M., Knauder, W., Anders, I., Andre, K., Hennig-Pauka, I. and Schönhart, M. (2019). Global warming impact on confined livestock in buildings: efficacy of adaptation measures to reduce heat stress for growing-fattening pigs. *Climatic Change* 156(4), 567-587. https://doi.org/10.1007/s10584-019-02525-3.

Schütz, K. E., Rogers, A. R., Cox, N. R. and Tucker, C. B. (2009). Dairy cows prefer shade that offers greater protection against solar radiation in summer: shade use, behaviour, and body temperature. *Applied Animal Behaviour Science* 116(1), 28-34. https://doi.org/10.1016/j.applanim.2008.07.005.

Silva, M., Exadaktylos, V., Ferrari, S., Guarino, M., Aerts, J. M. and Berckmans, D. (2009). The influence of respiratory disease on the energy envelope dynamics of pig cough sounds. *Computers and Electronics in Agriculture* 64, 286-329. https://doi.org/10.1016/j.compag.2009.07.002.

Silva, M., Ferrari, S., Costa, A., Aerts, J.-M., Guarino, M. and Berckmans, D. (2008). Cough localization for the detection of respiratory diseases in pig houses. *Computers and Electronics in Agriculture* 64(2), 286-292. https://doi.org/10.1016/j.compag.2008.05.024.

Tabase, R. K., Millet, S., Brusselman, E., Ampe, B., De Cuyper, C., Sonck, B. and Demeyer, P. (2020a). Effect of ventilation control settings on ammonia and odour emissions from a pig rearing building. *Biosystems Engineering* 192, 215-231. https://doi.org/10.1016/j.biosystemseng.2020.01.022.

Tabase, R. K., Van linden, V., Bagci, O., De Paepe, M., Aarnink, A. J. A. and Demeyer, P. (2020b). CFD simulation of airflows and ammonia emissions in a pig compartment with underfloor air distribution system: model validation at different ventilation rates. *Computers and Electronics in Agriculture* 171. https://doi.org/10.1016/j.compag.2020.105297, 105297.

Tao, S., Orellana Rivas, R. M., Marins, T. N., Chen, Y. C., Gao, J. and Bernard, J. K. (2020). Impact of heat stress on lactational performance of dairy cows. *Theriogenology* 150, 437-444. https://doi.org/10.1016/j.theriogenology.2020.02.048.

Thøfner, I. C. N., Poulsen, L. L., Bisgaard, M., Christensen, H., Olsen, R. H. and Christensen, J. P. (2019). Correlation between footpad lesions and systemic bacterial infections in broiler breeders. *Veterinary Research* 50(1), 38. https://doi.org/10.1186/s13567-019-0657-8.

Trokhaniak, V. I., Rutylo, M. I., Rogovskii, I. L., Titova, L. L., Luzan, O. R. and Bannyi, O. O. (2019). Experimental studies and numerical simulation of speed modes of air environment in a poultry house. *INMATEH Agricultural Engineering* 59(3), 9–18. https://doi.org/10.35633/INMATEH-59-01.

Tsai, Y. C., Hsu, J. T., Ding, S. T., Rustia, D. J. A. and Lin, T. T. (2020). Assessment of dairy cow heat stress by monitoring drinking behaviour using an embedded imaging system. *Biosystems Engineering* 1, 1–12. https://doi.org/10.1016/j.biosystemseng.2020.03.013.

Tullo, E., Finzi, A. and Guarino, M. (2019). Review: environmental impact of livestock farming and precision livestock farming as a mitigation strategy. *Science of the Total Environment* 650(2), 2751–2760. https://doi.org/10.1016/j.scitotenv.2018.10.018.

Van der Heyden, C., Brusselman, E., Volcke, E. I. P. and Demeyer, P. (2016). Continuous measurements of ammonia, nitrous oxide and methane from air scrubbers at pig housing facilities. *Journal of Environmental Management* 181(3), 163–171. https://doi.org/10.1016/j.jenvman.2016.06.006.

Van der Heyden, C., Demeyer, P. and Volcke, E. I. P. (2015). Mitigating emissions from pig and poultry housing facilities through air scrubbers and biofilters: state-of-the-art and perspectives. *Biosystems Engineering* 134, 74–93. http://doi.org/10.1016/j.biosystemseng.2015.04.002.

Van der Heyden, C., Solon, K., Demeyer, P. and Volcke, E. I. P. (2020). Model-based evaluation of ammonia removal in biological air scrubbers. *Biosystems Engineering* 191, 85–95. https://doi.org/10.1016/j.biosystemseng.2019.12.011.

Van Hertem, T., Maltz, E., Antler, A., Romanini, C. E., Viazzi, S., Bahr, C., Schlageter-Tello, A., Lokhorst, C., Berckmans, D. and Halachmi, I. (2013). Lameness detection based on multivariate continuous sensing of milk yield, rumination and neck activity. *Journal of Dairy Science* 96(7), 4286–4298. http://doi.org/ 10.3168/jds.2012-6188.

Van Hertem, T., Schlageter Tello, A., Viazzi, S., Steensels, M., Bahr, C., Romanini, C. E. B., Lokhorst, K., Maltz, E., Halachmi, I. and Berckmans, D. (2018). Implementation of an automatic 3D vision monitor for dairy cow locomotion in a commercial farm. *Biosystems Engineering* 173, 166–175. https://doi.org/10.1016/j.biosystemseng.2017.08.011.

Van Hertem, T., Viazzi, S., Steensels, M., Maltz, E., Antler, A., Alchanatis, V., Schlageter-Tello, A. A., Lokhorst, K., Romanini, E. C. B., Bahr, C., Berckmans, D. and Halachmi, I. (2014). Automatic lameness detection based on consecutive 3D-video recordings. *Biosystems Engineering* 119, 108–116. https://doi.org/10.1016/j.biosystemseng.2014.01.009.

Van Hirtum, A. and Berckmans, D. (2004). Objective cough-sound recognition as a biomarker for aerial factors. *Transactions of the ASAE* 47(1), 351–356.

Vandermeulen, J., Bahr, C., Johnston, D., Earley, B., Tullo, E., Fontana, I., Guarino, M., Exadaktylos, V. and Berckmans, D. (2016). Early recognition of bovine respiratory disease in calves using automated continuous monitoring of cough sounds. *Computers and Electronics in Agriculture* 129, 15–26. https://doi.org/10.1016/j.compag.2016.07.014.

Wang, K., Zhao, X. and Yong, H. (2017). Review on noninvasive monitoring technology of poultry behavior and physiological information. *Transactions of the Chinese Society of Agricultural Engineering* 33, 197–209.

Wang, Y., Zheng, W., Li, B. and Li, X. (2019a). A new ventilation system to reduce temperature fluctuations in laying hen housing in continental climate. *Biosystems Engineering* 181, 52–62. https://doi.org/10.1016/j.biosystemseng.2019.02.017.

Wang, X. S., Zhao, X. Y., He, Y. and Wang, K. Y. (2019b). Cough sound analysis to assess air quality in commercial weaner barns. *Computers and Electronics in Agriculture* 160, 8–13. https://doi.org/10.1016/j.compag.2019.03.001.

Wang, K., Liu, K., Xin, H., Chai, L., Wang, Y., Fei, T., Oliveira, J., Pan, J. and Ying, Y. (2019c). An RFID-based automated individual perching monitoring system for group-housed poultry. *Transactions of the ASABE* 62(3), 695–704. https://doi.org/10.13031/trans.13105.

Weeks, C. A. (2008). A review of welfare in cattle, sheep and pig lairages, with emphasis on stocking rates, ventilation and noise. *Animal Welfare* 17, 275–284.

West, J. W. (2003). Effects of heat-stress on production in dairy cattle. *Journal of Dairy Science* 86(6), 2131–2144. https://doi.org/10.3168/jds.S0022-0302(03)73803-X.

Wilson, T. (2005). New United Nations world population projections. *People and Place* 13(1), 14–22. https://doi.org/10.1017/CBO9781107415324.004.

Wood, D., Cowherd, S. and Van Heyst, B. (2015). A summary of ammonia emission factors and quality criteria for commercial poultry production in North America. *Atmospheric Environment* 115, 236–245. https://doi.org/10.1016/j.atmosenv.2015.05.069.

Yeo, U. H., Lee, I. B., Kim, R. W., Lee, S. Y. and Kim, J. G. (2019). Computational fluid dynamics evaluation of pig house ventilation systems for improving the internal rearing environment. *Biosystems Engineering* 186, 259–278. https://doi.org/10.1016/j.biosystemseng.2019.08.007.

Zhang, C., Xiao, D., Yang, Q., Wen, Z. and Lv, L. (2020). Review: application of infrared thermography in livestock monitoring. *Transactions of the ASABE* 63(2), 389–399. https://doi.org/10.13031/trans.13068

Zhao, K., Bewley, J. M., He, D. and Jin, X. (2018). Automatic lameness detection in dairy cattle based on leg swing analysis with an image processing technique. *Computers and Electronics in Agriculture* 148, 226–236. https://doi.org/10.1016/j.compag.2018.03.014.

Zhao, Y., Hongwei, X., Shepherd, T., Hayes, M., Stinn, J. and Li, H. (2013). Thermal environment, ammonia concentrations, and ammonia emissions of aviary houses with white laying hens. *Transactions of the ASABE* 56, 1145–1156. http://doi.org/10.13031/trans.56.10097.

Chapter 8

Developments in individual-animal feed efficiency monitoring systems for livestock

Ilan Halachmi and Ran Bezen, The Volcani Centre - Agriculture Research Organization (ARO) and Ben-Gurion University of the Negev, Israel; Assaf Godo, Harel Levit and Victor Bloch, The Volcani Centre - Agriculture Research Organization (ARO), Israel; and Yael Edan, Ben-Gurion University of the Negev, Israel

1 Introduction

The use of advanced low-cost cameras and innovative computer vision algorithms is changing feed-intake monitoring system design. With the help of these technologies, individual feed-intake measurement (IFIM) systems will become available to farmers, replacing expensive and less accurate mechanical weighing systems. IFIM systems offer economic benefits since feed intake is a significant expense in livestock farming such as dairy and beef farm operations. However, there are still difficulties and challenges in implementing these systems. This chapter describes the sensors and algorithms employed in different IFIM systems for dairy cows. The chapter is designed for scientists and R&D staff in the industry interested in the development of feed-intake measurement systems.

Feeding cows accounts for 40–60% of total costs in dairy farming operations, a percentage that can rise to as high as 75% in intensive systems (DeVries et al.,

http://dx.doi.org/10.19103/AS.2021.0090.08

2003; Bach et al., 2004; Buza et al., 2014; Borchers et al., 2016). Individual feed intake is also an essential variable in dairy management (Connor et al., 2019). Farmers can increase feed efficiency in two ways:

- by selecting cows with high feed conversion efficiency (VandeHaar et al., 2016); and
- by appropriate diet design (Chiba, 2009; Connor, 2015).

Proper monitoring of feed intake can significantly improve farm management (Shalloo et al., 2004; Connor, 2015). This may lead to increased overall farm profitability (VandeHaar, 1998; Herd et al., 2003; Buza et al., 2014; Halachmi et al., 2016; Volden, 2011).

Several systems for monitoring individual feed intake have been developed over the years using various sensors and algorithms. Traditional systems utilise individual weighing balances (electronic scales) and radio frequency identification (RFID) antennae in feeding stalls to measure each cow's amount of feed. Recent research has focused on the use of low-cost cameras with computer vision algorithms to design individual feed-intake measuring systems. This chapter reviews individual feed intake systems, advances in technologies and their impact. Based on the insights from this review, the chapter identifies future challenges and possible research directions.

2 Materials and methods

A systematic search was made of the following three literature databases:

- the Institute for Scientific Information;
- the American Society of Agricultural and Biological Engineers; and
- the United States Department of Agriculture National Agricultural Library.

The search resulted in 60 publications. A Boolean phrase search for the term 'individual feed intake' and at least one term referencing 'measure,' 'system,' 'development,' 'modelling,' 'cattle,' etc. was used to search an article's title, abstract or keywords. Books and book chapters, dissertations and theses and practice-oriented articles were beyond the scope of this analysis. A database was constructed which identified key variables such as the number of animals used in a study, the species (dairy or beef cattle, sheep, goats), the method used to measure feed intake and the method used to calculate dry matter intake (DMI). Studies did not include grazing situations.

Of the 60 papers reviewed (listed in the reference section), 52 address operational aspects while only 8 deal with the design, development or validation of measurement systems (Table 1). The measurement systems were developed with different sensors and algorithms and were applied primarily on research dairy farms. In addition to the 60 papers, another 266 research papers focused on statistical modelling to predict daily DMI for various species (dairy cows, beef cattle, pigs, goats and sheep) (Seymour et al., 2019). These models were built based on data collected using mechanical weighing systems (Halachmi et al., 2004, 2016; NRC, 2007; Volden, 2011; Holtenius et al., 2018). However, data based on these technologies have been shown to be a poor indicator of actual DMI (Schirmann et al., 2012). New sensors and other technologies may provide more accurate results (Halachmi et al., 2016; Bezen et al., 2020; Bloch et al., 2019).

3 Wearable sensors and electronic scales

3.1 Wearable sensors

Weighing or camera systems are impracticable in vast grazing fields where cows walk large distances in search of food, water or shelter. Wearable sensors (e.g. sound, jaw-movement and accelerometer sensors) have potential applications in free-ranging grazing cows. Data loggers, which simultaneously record the number of jaw movements and pauses between jaw movements, have been developed and tested for cows (Matsui and Okubo, 1991), sheep and goats (Matsui, 1994). The reported correlation between manual observers and the automatic system was 91% (Rutter et al., 1997). Discrepancies between manual and automatic recording methods probably reflect inaccuracies in manual measurements (the gold reference) rather than in the automated observations (Andriamandroso et al., 2016). If real-time or on-line data are not required, faecal analysis may be used (Lyons and Stuth, 1992). An acoustic monitoring system to quantify ingestion in free-grazing cattle (Clapham et al., 2011) overcomes the drawbacks of faecal analysis and can recognise jaw movements in free-range livestock (Ungar and Rutter, 2006; Navon et al., 2013). These studies suggest that sensors can be a reliable tool to assess eating behaviour in grazing beef cattle. However, to our knowledge, no scientifically approved off-the-shelf feed-intake monitoring product for grazing situations is yet commercially available.

3.2 Electronic scales

The electronic scale is the oldest and the most straightforward sensor for measuring feed intake in group-housing settings (Halachmi et al., 1998). An electronic scale is placed in a feeding station, and it measures the amount of each individual cow's feed intake weight during meals from a feed bin. Farmers can decide how many scales to spread along a feeding lane.

Table 1 Major reviewed research studies

			Evaluation							
What is measured?	Sensor used	Type of algorithm used	Data for test	Weight range	Outdoor/ Lab	Diff food	Performance measures	Accuracy	Feasibility on a farm	Study
Weight of feed heaps	SLI	Volume scanning for regression	272 images	Intervals of 5 pounds in range of 0 to 22–27 kg	Lab	1	Linear regression indicators	72% of the results were within 4 pounds of the difference, 55% of all the results are within 2 pounds of difference, 3% of data are 11 pounds or greater	Influenced by sunlight	Shelley (2013)
Weight of feed heaps	IR	Volume scanning for regression	-	Intervals of 2.27 kg in range of 0–22.68 kg	Lab	2	Linear regression indicators	Error of 0.2 kg	Influenced by sunlight	Shelley et al. (2016)
Weight of feed heaps	Eight cameras	Volume by photogrammetry for regression	125 images in lab, 60 images in cowshed	Range of 0–7 kg in lab and 40 kg in cowshed	Outdoor and lab	1	Linear regression indicators	Error of 0.483 kg for heaps up to 7 kg in lab and 1.32 for heaps up to 40 kg in cowshed	Not effective in a cowshed on a working farm; require at least eight cameras	Bloch et al. (2019)
Weight of individual meals	RGBD camera	Deep learning	576 tensors represent feed intake of individual meals	Individual meals, ranging from 0 to 8 kg per meal	Outdoor	1	Mean absolute error (MAE) and mean square error (MSE)	MAE of 0.127 kg, MSE of 0.034 kg^2	Needed high computing power	Bezen et al. (2019)

Validation of electronic system. Validated daily intake from any bin	Electronic scale with 100 kg weighing capacity with ±25 g of accuracy	-	12 Holstein cows over 5 days. The feed intake was manually measured for 4-h time periods using an external scale	Any range of weight	Cowshed	1	Compared the sum of the feed intake recorded by the monitoring system with the sum of the external scale weight	Differed of 244 g from total intake: 25.635 ± 2.428 g and 25.391 ± 2.428 g estimated using manual weighing and the electronic system, respectively	A reasonable monitoring system for feeding behaviour and feed intake. High priced and frequent cleaning required	Chizzotti et al. (2015)
Validation of a system for monitoring individual feeding. Validated the daily intake from any bin	Electronic scale	-	A total of 42 Holstein cows were tested with access to 24 feed bins. Tested with an external scale over three consecutive 24-h periods	Any range of weight	Cowshed	1	Compared the sum of the feed intake recorded by the monitoring system with the sum of the external scale weight	Differed of 0.68 kg: 29.92 ± 0.90 kg and 29.24 ± 0.90 kg as estimated by manual weighing and by the electronic system, respectively	The electronic system is a useful tool for monitoring intakes. High priced and frequent cleaning required	Chapinal et al. (2007)

(Continued)

Table 1 (*Continued*)

			Evaluation							
What is measured?	Sensor used	Type of algorithm used	Data for test	Weight range	Outdoor/ Lab	Diff food	Performance measures	Accuracy	Feasibility on a farm	Study
Weight of individual meals	Electronic scale with weighing capacity of 100 kg with an accuracy of ±20 g	-	Series of 4-h observations during 5 different days. Test data measured manually with an external scale at the beginning and at the end of 26 cow visits	Any range of weight	Cowshed	1	Difference between the feed consumed by a cow during a visit and the external scale. Also, compared the total feed in 24 h from each scale in 2 separate days	The difference between the feed consumed by a cow during a visit differed only by 52 g (*P* = 0.99) when (3.97 ± 0.68 kg) compared with the manual (3.91 ± 0.69 kg) and differed by 120 g (20.90 ± 1.33 vs. 20.78 ± 1.32 kg) in the sum of the feed consumed	Proved to be a useful tool to provide reasonable estimates of individual feed intake. High priced and frequent cleaning required	Bach et al. (2004)
Weight of individual bin	Electronic scale with weighing capacity of 150 kg, precision of 0.001% and resolution of 0.1 kg	-	-	Any range of weight	Cowshed	-	-	-	Provide a valuable research tool for nutritionists and animal husbandry scientists	Halachmi et al. (1998)

SLI = structured light illumination; IR = infra-red camera; RGBD and RGB-D are depth sensors for 3D mapping and localisation.

Several companies have developed electronic weighing systems including:

- the Calan Broadbent Feeding System;
- the Controlling and Recording Feed Intake System;
- the GrowSafe System; and
- the Roughage Intake Control System (RCI - Hokofarm Group).

The most common way to identify cows in these systems is to utilise RFID. RFID is used to identify individual cows at feeding stations. Each cow's feed intake is measured by the weighing system that can restrict access to feed for a specific animal.

Numerous researchers have used these weighing systems (Halachmi et al., 1998; DeVries et al., 2003; Bach et al., 2004; van der Honing, 2005; Ferris et al., 2006; Wang et al., 2006; Chapinal et al., 2007; Stajnko et al., 2010; Mendes et al., 2011; Chizzotti et al., 2015). The advantage of electronic weight systems is that the weight sensor is accurate and reliable (±52 g, $P = 0.99$, Bach et al., 2004). The weight sensor can also measure intake regardless of feed type or environmental conditions.

However, weight sensors are not often used in commercial farms due to high costs and the frequent cleaning and maintenance required which most farms cannot afford (Wang et al., 2006; Stajnko et al., 2010). Suitable infrastructure is needed, requiring significant changes in the feed distribution system, including the cleaning system and distribution devices. Electronic scales are currently used almost exclusively by researchers rather than commercial farmers.

4 Camera sensors

Camera sensors can provide images from which feed volume-related features are derived in assessing the amount of feed consumed. The camera is typically positioned above the food pile or feeding lane. Several methods are used to represent a three-dimensional (3D) geometrical position of the target surface visible to the camera.

4.1 Structured light illumination

The structured light illumination (SLI) and time of flight (ToF) methods refer to systems composed of a camera and light projector (Borchersen et al., 2018; Lassen et al., 2018). The projector is used to project images of light patterns across the scene being monitored. These images provide data regarding the vertical axis, generating depth and surface information from the deformation of the known projected pattern.

An SLI system has been used in 3D scanning of cow feed to determine the volume and weight of feed in a bin before and after feeding dairy cows

(Shelley, 2013). The SLI system was tested on 272 heaps in a laboratory (Shelley, 2013). The results showed a high variance between the calculated image weight and actual values. Only 72% of the results were within 4 pounds (814 g) of the difference between the calculated image weight value and the scale-measured weight value. This shows SLI requires controlled lighting conditions and shade. This means that current SLI systems work only in indoor conditions when protected from sunlight. Inconsistent lighting in a real-world situation (i.e. a farm) will lead to error. If there is too much light, the image will be oversaturated, and the resulting data will be useless since obtaining the contour of the feed for analysis becomes impossible (Shelley, 2013).

4.2 Calibrated stereo cameras

Multiple cameras in calibrated stereo configuration can be used to extract depth information from objects via triangulation and the analysis of the disparity between corresponding points. Measured feed mass volume has been calculated using a photogrammetry method, which uses several images of the object taken from various perspectives to create a 3D model of the object's surface (Bloch et al., 2019).

Photogrammetry identifies common features in the different images and triangulates them to find their coordinates in the space photographed. These features create a point cloud of the animal feed heap surface (Bloch et al., 2019). Analysis of the relationship between the precision of the volume calculation and the number of images used for the modelling revealed that more than eight images are required to calculate the volume. The method was tested in laboratory and in cowshed conditions, with 125 and 60 feed heaps, respectively. The estimated error for calculating the mass under laboratory conditions was 0.483 kg for heaps up to 7 kg. The standard deviation for the cowshed experiment was 0.44 kg, resulting in a total error of 1.32 kg for heaps up to 40 kg in the cowshed. A major weakness of this approach is that the coloured markers used for the point cloud processing would not be effective in a cowshed on a working farm (dirt can change their colours, and they can be inadvertently detached from the floor and walls by tractors that regularly traverse the barns). Additionally, eight cameras were required for a single heap, making this method currently impractical.

4.3 Red green blue-depth (RGB-D) cameras and infrared sensors

RGB-D cameras provide a combination of images representing RGB colour channels together with the depth (*D*) of objects. These cameras include a depth sensor based on infrared (IR) or near-IR projector and an RGB camera, resulting in in-depth information per pixel in the RGB image. This 3D data acquisition technique has been used in both research and industry to assess the surface conditions of objects (Johnson, 2018). The RGB-D camera has a major advantage in object surface assessment since it is derived directly from the IR sensor. However, the IR sensor is sensitive to and influenced by sunlight.

Several RGB-D feed intake methods and algorithms have been developed for indoor (Shelley et al., 2016), outdoor (Borchersen et al., 2018; Lassen et al., 2018; Bezen et al., 2019) and open cowshed conditions (Lassen et al., 2018; Bezen et al., 2019). Research with an IR sensor with three different distances between the lens and the floor resulted in a minimum error of 0.2 kg for the weight of a single heap (Shelley et al., 2016). This study was limited to laboratory conditions only (and did not deal with sunlight conditions).

In another study, an RGB-D camera and deep learning algorithm were used to overcome the sun's effect on the IR scanner (Bezen et al., 2020). The data tested came from an open cowshed. The system directly measured the feed intake of a single meal, with a mean absolute error of 0.127 kg per meal; each meal was in the range of 0–8 kg. This method looks promising, perhaps in combination with eating behaviour sensing (Halachmi et al., 2016).

4.4 Volume derived from the 3D geometric shape

Most research using computer vision has tried to find a statistical correlation between the 3D geometric shape of the feed surface and the heap's volume (Shelley, 2013; Shelley et al., 2016; Bloch et al., 2019). The 3D representation is produced by a matrix with the size of the image on horizontal x and y axes and an additional z axis for each pixel. The volume of the heap is calculated as the integral under the geometric surface of this shape.

4.5 RGB-D images with subtract tensors

A recent study for measuring feed intake using an RGB-D camera applied a deep convolutional neural network (CNN) model to predict feed intake (Bezen et al., 2020). The convolutional layers were applied to extract the features from the four-dimensional raw image data. The process entails multiple feature maps, each recognising a specific feature. The features include four-dimensional matrices, based on depth and colour images (corresponding to the four RGB-D input channels) of the feed heap. The four-dimensional matrices contain the results of image subtraction of feed heap images before and after the meal.

5 Image analysis algorithms

5.1 Regressions on image features

Most of the vision-based algorithms aim to determine some sort of linear correlation between the image features, e.g. the volume of a feed heap and the heap's actual weight (Shelley, 2013; Shelley et al., 2016; Bloch et al., 2019). These studies use linear (Bloch et al., 2019) or quadratic regression (Shelley et al., 2016) models assessing the calculated volume and the feed mass. The models were evaluated using statistical analysis that measured the distribution of the residuals between the regression line and the volumes. Models of linear regression (Neter et al., 1996) explain the direct contribution of each feature to the model. However, linear regression models assume linearity, constant variance, independence of errors and lack of perfect multicollinearity in the predictors. These assumptions limit the model.

5.2 Deep learning

A recent study has used a deep learning model (Bezen et al., 2020). Deep learning or, more specifically, CNNs are machine-learning techniques which can be applied to complicated computer vision tasks (Zhang et al., 2018). CNNs are inspired by the way neurons in the human brain learn. CNN models have become popular and available in recent years because of the growth of the computation capabilities of computer graphics processing units (GPUs, Tsai et al., 2018) and the quantity of available data (Zhang et al., 2018). CNNs are based on non-linear, end-to-end training, which enables them to learn millions of parameters. Accordingly, they require relatively large amounts of diverse and annotated data (Ros et al., 2016). Precision agriculture (Espejo-Garcia et al., 2019), precision livestock farming (PLF) (Alvarez et al., 2018; Wang et al., 2018), agricultural robotics (Kamilaris and Prenafeta-Boldú, 2018) and other areas (Altuntaş et al., 2019) have seen increased interest in implementing deep learning methods, especially CNN-based algorithms.

A deep learning CNN is a powerful model for computer vision tasks, but it requires a significant amount of learning data. The training process of CNNs requires high GPU computing power. To obtain a reliable model, CNNs for measuring feed intake must also rely on large and varied labelled data collected from the RGB-D camera and the food pile images' corresponding weight. Collecting and labelling the images is a difficult task, especially when large-scale data are needed. For CNN models, the diversity of the data determines the robustness and generic value of the model. This is not required for real-time models. A CNN model for measuring feed intake (Bezen et al., 2019) was trained on RGB-D data, which included a tensor of four-channel matrices of data, each with 480 × 640 pixels. These data were entered as input to the CNN (four-dimensional matrices representing a single meal). The model was evaluated in an open cowshed in variable sunlight (Bezen et al., 2020).

6 Evaluating feed-intake measurement systems

The feed-intake measurement systems developed to date have been quite different in approach and design and are thus hard to compare. Differences include testing parameters such as the setup conditions and the range of feed heap mass. Various setup conditions include environmental conditions, feed (diet) composition, feed fermentation (temperature), moisture content and testing equipment. Environmental conditions include radiation, lighting conditions (i.e. direct sunlight, shaded), location (i.e. indoors or outdoors), testing in laboratory or farm conditions. Equipment issues include differences in camera design and the accuracy of weighing scales utilised to measure feed intake by a system. For example, the number of cameras per feeding station and the number of computers needed are pieces of equipment that must be precisely defined. Additional complementary modules must be defined, such as the RFID devices used to identify cows.

Feed composition (water content and ratio between hay, silage, corn, wheat, grains and other diet ingredients) and changes in the range of heap mass are critical parameters in weight measurement. These parameters affect weight data sensitivity (intervals of 1/0.1/0.01 kg) along with the minimum and maximum weight and the type of feed that the model should address.

Model robustness is crucial in feed-intake measurement. Several computer vision-based models were evaluated using regression indices and linear correlations between heap volume derived from images and actual readings of heap weights (Shelley, 2013; Shelley et al., 2016; Bloch et al., 2019). There is an inherent deficiency in this method. Two measurements are taken of every single meal. Assuming that each measurement has errors, the calculated result will amplify these errors. A computer vision system based on a deep-learning model works differently, directly measuring the intake of a single meal. The loss function of this model considers the difference between images of two heaps, leading to model indices which account for a whole meal. Combining the possibilities of new technologies (cameras, structured light, CNN models) will improve measuring accuracy.

7 Future trends and conclusion

The literature review presented in this chapter suggests several trends that characterise research on feed-intake measurement systems. A frequent method for measuring feed mass in research farms is the use of feed mass weighing systems. However, due to their high costs, cleaning and maintenance requirements, their use is less feasible for commercial farms, and they are primarily used in research or for calibrating other systems or DMI models.

Several studies have used various computer vision systems, including SLI, ToF, depth cameras and calibrated stereo cameras with various processing methods, such as deep learning and analytical algorithms. The use of low-cost cameras with increasing accuracy, equipped with advanced computer vision algorithms, provides a promising direction toward commercial individual intake measurement systems. However, most systems to date have not been validated in longitudinal studies and a range of conditions. Most studies report on an algorithm tested in a controlled experiment under laboratory conditions, which does not reflect the challenges faced by commercial farms. Possible directions for future development should take computer vision systems a step forward, by incorporating advances in sensor technologies, eating behaviour information and computing systems.

In conclusion, as sensor systems improve in quality and computer systems increase in computational power, more innovations and improvements will be introduced into feed-intake monitoring systems. Further information from other sensors, such as eating behaviour, changes in milk yield, changes in body weight and activity changes may be incorporated into machine vision-based DMI-predicting algorithms. The major challenge is implementing a highly reliable, low-cost system that can operate in real time on farms, with low maintenance requirements.

8 Acknowledgements

This study was supported by the Israeli Chief Scientist of Agriculture fund, 'Kandel' PLF centre of expertise, project number 20-12-0029 and partially supported by the Rabbi W. Gunther Plaut Chair in Manufacturing Engineering at Ben-Gurion University of the Negev.

9 Declaration of interest and ethics statement

There are no conflicts of interest. Ethics statement: not relevant to a review paper.

10 References

Altuntaş, Y., Cömert, Z. and Kocamaz, A. F. 2019. Identification of haploid and diploid maize seeds using convolutional neural networks and a transfer learning approach. *Computers and Electronics in Agriculture* 163, 104874.

Alvarez, J. R., Arroqui, M., Mangudo, P., Toloza, J., Jatip, D., Rodríguez, J. M., Teyseyre, A., Sanz, C., Zunino, A., Machado, C. and Mateos, C. 2018. Body condition estimation on cows from depth images using convolutional neural networks. *Computers and Electronics in Agriculture* 155, 12–22.

Andriamandroso, A., Bindelle, J., Mercatoris, B. and Lebeau, F. 2016. A review on the use of sensors to monitor cattle jaw movements and behavior when grazing. *Biotechnologie, Agronomie, Société et Environnement* 20: 273–286.

Bach, A., Iglesias, C. and Busto, I. 2004. A computerised system for monitoring feeding behavior and individual feed intake of dairy cattle. *Journal of Dairy Science* 87(12), 4207–4209.

Ben Meir, Y. A., Nikbachat, M., Portnik, Y., Jacoby, S., Levit, H., Bikel, D., Adin, G., Moallem, U., Miron, J., Mabjeesh, S. J. and Halachmi, I. 2019. Dietary restriction improved feed efficiency of inefficient lactating cows. *Journal of Dairy Science* 102(10), 8898 -8906.

Bezen, R., Edan, Y. and Halachmi, I. 2019. An automatic data acquisition system for acquiring training data for a deep learning algorithm for individual cow intake prediction. In: European Conference on Precision Livestock Farming, O'Brien, B., Hennessy, D. and Shalloo, L. (Eds), pp. 284–291. *The Organising Committee of the 9th European Conference on Precision Livestock Farming (ECPLF)*, Teagasc, Animal & Grassland Research and Innovation Centre, Moorepark, Fermoy, Co. Cork, Cork, Ireland.

Bezen, R., Edan, Y. and Halachmi, I. 2020. Computer vision system for measuring individual cow feed intake using RGB-D camera and deep learning algorithms. *Computers and Electronics in Agriculture* 172, 105345.

Bloch, V., Levit, H. and Halachmi, I. 2019. Assessing the potential of photogrammetry to monitor feed intake of dairy cows. *Journal of Dairy Research* 86(1), 34–39.

Borchers, M. R., Chang, Y. M., Tsai, I. C., Wadsworth, B. A. and Bewley, J. M. 2016. A validation of technologies monitoring dairy cow feeding, ruminating, and lying behaviors. *Journal of Dairy Science* 99(9), 7458–7466.

Borchersen, S., Hansen, N. W. and Borggaard, C. 2018. System for determining feed consumption of at least one animal. Viking Genetics FMBA, Patent Number 10420328.

Buza, M. H., Holden, L. A., White, R. A. and Ishler, V. A. 2014. Evaluating the effect of ration composition on income over feed cost and milk yield. *Journal of Dairy Science* 97(5), 3073–3080.

Chapinal, N., Veira, D. M., Weary, D. M. and Von Keyserlingk, M. A. G. 2007. Technical note: validation of a system for monitoring individual feeding and drinking behavior and intake in group-housed cattle. *Journal of Dairy Science* 90(12), 5732–5736.

Chiba, L. I. 2009. *Animal Nutrition Handbook*. 2nd Revision. Department of Animal Sciences, Auburn University, AL.

Chizzotti, M. L., Machado, F. S., Valente, E. E. L., Pereira, L. G. R., Campos, M. M., Tomich, T. R., Coelho, S. G. and Ribas, M. N. 2015. Technical note: validation of a system for monitoring individual feeding behavior and individual feed intake in dairy cattle. *Journal of Dairy Science* 98(5), 3438–3442.

Clapham, W. M., Fedders, J. M., Beeman, K. and Neel, J. P. S. 2011. Acoustic monitoring system to quantify ingestive behavior of free-grazing cattle. *Computers and Electronics in Agriculture* 76(1), 96–104.

Connor, E. E. 2015. Invited review: improving feed efficiency in dairy production: challenges and possibilities. *Animal: An International Journal of Animal Bioscience* 9(3), 395–408.

Connor, E. E., Hutchison, J. L., Tassell, C. P. and Cole, J. B. 2019. Defining the optimal period length and stage of growth or lactation to estimate residual feed intake in dairy cows. *Journal of Dairy Science* 102(7), 6131-6143.

DeVries, T. J., Von Keyserlingk, M. A. G., Weary, D. M. and Beauchemin, K. A. 2003. Technical note: validation of a system for monitoring feeding behavior of dairy cows. *Journal of Dairy Science* 86(11), 3571-3574.

Espejo-Garcia, B., Lopez-Pellicer, F. J., Lacasta, J., Moreno, R. P. and Zarazaga-Soria, F. J. 2019. End-to-end sequence labeling via deep learning for automatic extraction of agricultural regulations. *Computers and Electronics in Agriculture* 162, 106 -111.

Ferris, C. P., Keady, T. W. J., Gordon, F. J. and Kilpatrick, D. J. 2006. Comparison of a Calan gate and a conventional feed barrier system for dairy cows: feed intake and cow behaviour. *Irish Journal of Agricultural and Food Research* 45, 149-156.

Halachmi, I., Ben, Y., Miron, J. and Maltz, E. 2016. Feeding behavior improves prediction of dairy cow voluntary feed intake but cannot serve as the sole indicator. *Animal: An International Journal of Animal Bioscience* 10(9), 1501-1506.

Halachmi, I., Edan, Y., Maltz, E., Peiper, U. M., Moallem, U. and Brukental, I. 1998. A real-time control system for individual dairy cow food intake. *Computers and Electronics in Agriculture* 20(2), 131-144.

Halachmi, I., Edan, Y., Moallem, U. and Maltz, E. 2004. Predicting feed intake of the individual dairy cow. *Journal of Dairy Science* 87(7), 2254-2267.

Halachmi, I., Guarino, M., Bewley, J. and Pastell, M. 2019. Smart animal agriculture: application of real-time sensors to improve animal well-being and production. *Annual Review of Animal Biosciences* 7, 403-425.

Herd, R. M., Archer, J. A. and Arthur, P. F. 2003. Reducing the cost of beef production through genetic improvement in residual feed intake: opportunity and challenges to application. *Journal of Animal Science* 81, E9-E17.

Hogeveen, H., Rutten, N., Kamphuis, C. and van der Voort, M. 2017. Economics of precision dairy monitoring techniques. In *Conference on Precision Dairy Farming*, pp. 87-97.

Holtenius, K., O'Hara, L. A. and Karlsson, J. 2018. The influence of milk yield, body weight and parity on feed intake by dairy cows. In *Proceedings of the 9th Nordic Feed Science Conference*, Uppsala, Sweden, 12-13 June 2018, pp. 101-105. Swedish University of Agricultural Sciences, Department of Animal Nutrition.

Johnson, J. W. 2018. Adapting mask-RCNN for automatic nucleus segmentation. *Proceedings of the 2019 Computer Vision Conference* 2, 608-618.

Kamilaris, A. and Prenafeta-Boldú, F. X. 2018. Deep learning in agriculture: a survey. *Computers and Electronics in Agriculture* 147, 70-90.

Lassen, J., Thomasen, J. R., Hansen, R. H., Nielsen, G. G. B., Olsen, E., Stentebjerg, P. R. B., Hansen, N. W. and Borchersen, S. 2018. Individual measure of feed intake on in-house commercial dairy cattle using 3D camera system. In: *Proceedings of the 11th World Congress of Genetics Applied to Livestock Production, Auckland, New Zealand*. Accessed July (vol. 2018).

Lyons, R. K. and Stuth, J. W. 1992. Fecal NIRS equations for predicting diet quality of free-ranging cattle. *Journal of Range Management* 45(3), 238-244.

Matsui, K. 1994. A new ambulatory data logger for a long-term determination of grazing and rumination behaviour on free-ranging cattle, sheep and goats. *Applied Animal Behaviour Science* 39(2), 123-130.

Matsui, K. and Okubo, T. 1991. A method for quantification of jaw movements suitable for use on free-ranging cattle. *Applied Animal Behaviour Science* 32(2-3), 107-116.

Mendes, E. D. M. M., Carstens, G. E., Tedeschi, L. O., Pinchak, W. E. and Friend, T. H. 2011. Validation of a system for monitoring feeding behavior in beef cattle. *Journal of Animal Science* 89(9), 2904-2910.

Navon, S., Mizrach, A., Hetzroni, A. and Ungar, E. D. 2013. Automatic recognition of jaw movements in free-ranging cattle, goats and sheep, using acoustic monitoring. *Biosystems Engineering* 114(4), 474-483.

Neter, J., Kutner, M. H., Nachtsheim, C. J. and Wasserman, W. 1996. *Applied Linear Statistical Models*. Irwin Publishing, Chicago.

NRC 2007. Nutrient requirements of small ruminants; sheep, goats, cervids, and New World camelids. *SciTech Book News*, 384 pages.

Ros, G., Sellart, L., Materzynska, J., Vazquez, D. and Lopez, A. M. 2016. The SYNTHIA dataset: a large collection of synthetic images for semantic segmentation of urban scenes. In: *Proceedings of the IEEE Computer Society Conference on Computer Vision and Pattern Recognition*, pp. 3234-3243.

Rutter, S. M., Champion, R. A. and Penning, P. D. 1997. An automatic system to record foraging behaviour in free-ranging ruminants. *Applied Animal Behaviour Science* 54(2-3), 185-195.

Schirmann, K., Chapinal, N., Weary, D. M., Heuwieser, W. and Von Keyserlingk, M. A. G. 2012. Rumination and its relationship to feeding and lying behavior in Holstein dairy cows. *Journal of Dairy Science* 95(6), 3212-3217.

Seymour, D. J., Cánovas, A., Baes, C. F., Chud, T. C. S., Osborne, V. R., Cant, J. P., Brito, L. F., Gredler-Grandl, B., Finocchiaro, R., Veerkamp, R. F., de Haas, Y. and Miglior, F. 2019. Invited review: determination of large-scale individual dry matter intake phenotypes in dairy cattle. *Journal of Dairy Science* 102(9), 7655-7663.

Shalloo, L., Dillon, P., Rath, M. and Wallace, M. 2004. Description and validation of the Moorepark dairy system model. *Journal of Dairy Science* 87(6), 1945-1959.

Shelley, A. N. 2013. *Monitoring Dairy Cow Feed Intake Using Machine Vision*. Theses and Dissertations-Electrical and Computer Engineering 24, 1-99. Available at: https://uknowledge.uky.edu/ece_etds/24.

Shelley, A. N., Lau, D. L., Stone, A. E. and Bewley, J. M. 2016. Short communication: measuring feed volume and weight by machine vision. *Journal of Dairy Science* 99(1), 386-391.

Stajnko, D., Vindiš, P., Janžekovič, M. and Brus, M. 2010. Non invasive estimating of cattle live weight using thermal imaging. *New Trends in Technologies: Control, Management, Computational Intelligence and Network Systems. Sciyo, Rijeka*, 243-256 (open access).

Tsai, H., Ambrogio, S., Narayanan, P., Shelby, R. M. and Burr, G. W. 2018. Recent progress in analog memory-based accelerators for deep learning. *Journal of Physics D: Applied Physics* 51(28).

Ungar, E. D. and Rutter, S. M. 2006. Classifying cattle jaw movements: comparing IGER behaviour recorder and acoustic techniques. *Applied Animal Behaviour Science* 98(1-2), 11-27.

van der Honing, Y. 2005. *Book of Abstracts of the 56th Annual Meeting of the European Association for Animal Production*. Wageningen Academic Publishers.

VandeHaar, M. J. 1998. Efficiency of nutrient use and relationship to profitability on dairy farms. *Journal of Dairy Science* 81(1), 272-282.

VandeHaar, M. J., Armentano, L. E., Weigel, K., Spurlock, D. M., Tempelman, R. J. and Veerkamp, R. 2016. Harnessing the genetics of the modern dairy cow to continue improvements in feed efficiency. *Journal of Dairy Science* 99(6), 4941-4954.

Volden, H. 2011. *NorFor: The Nordic Feed Evaluation System*. Academic Publishers, Wageningen.

Wang, D., Tang, J. L., Zhu, W., Li, H., Xin, J. and He, D. 2018. Dairy goat detection based on Faster R-CNN from surveillance video. *Computers and Electronics in Agriculture* 154, 443-449.

Wang, Z., Nkrumah, J. D., Li, C., Basarab, J. A., Goonewardene, L. A., Okine, E. K., Crews, D. H. and Moore, S. S. 2006. Test duration for growth, feed intake, and feed efficiency in beef cattle using the GrowSafe System. *Journal of Animal Science* 84(9), 2289-2298.

Zhang, Q., Yang, L. T., Chen, Z. and Li, P. 2018. A survey on deep learning for big data. *Information Fusion* 42, 146-157.

Chapter 9

Developments in automated systems for monitoring livestock health: mastitis

M. van der Voort and H. Hogeveen, Wageningen University & Research, The Netherlands

1 Introduction

Mastitis is an inflammatory reaction in a cow's mammary gland. It is regarded as one of the most important dairy farming production diseases worldwide, resulting in significant economic losses from inefficient milk production (Hogeveen et al., 2019c; Hommels et al., 2021). Besides prevention, the timely and accurate detection of mastitis (caused by intramammary infections (IMI) in cows), followed by appropriate remedial action, is an important aspect of mastitis management. Stimulated by the continuing introduction of automatic milking systems (AMS), an increasing number of sensor-based mastitis detection systems are currently available (Rutten et al., 2013; Khatun et al., 2018; Chakraborty et al., 2019). The demand for accurate mastitis detection remains a topical issue, and research and industry are continuously searching for new mastitis indicators and improved detection systems (Hogeveen et al., 2021).

The overall performance of sensor systems and algorithms in automatically detecting an inflammatory response to infection, causing visibly abnormal milk (i.e. clinical mastitis (CM)), varies within the scientific literature. The required detection accuracy of sensor systems for CM has been determined at 80% sensitivity and 99% specificity according to the International Standards Organization ISO/FDIS 20966 (automatic milking installations – requirements

http://dx.doi.org/10.19103/AS.2021.0090.09

and testing). However, to date, the published performance of CM detection models does not meet this target (e.g. Hogeveen et al., 2010; Miekley et al., 2012, 2013b,c; Dominiak and Kristensen, 2016; Dalen et al., 2019; Khatun et al., 2020). Consequently, farmers have to check a large number of cows per day or accept that a lower proportion of cows with CM will be detected in time for early intervention.

Sensor systems have the potential to support farmers' management of udder health and can encourage appropriate farmer interventions. However, there is a need to develop systems for specific mastitis situations to ensure timely detection of CM (Hogeveen et al., 2019a). This chapter describes the current situation regarding automated systems for monitoring mastitis in dairy cattle and potential new developments in mastitis detection.

2 Components of mastitis sensor systems

Automatic mastitis detection systems usually consist of one or more sensors (hardware) and the algorithms to translate sensor data into alerts (software). In some cases, a decision-making system is also part of a mastitis detection system (Rutten et al., 2013). These detection systems use sensors that are attached to the cow or installed in the milking system. Some in-line sensors continuously measure milk from the beginning to the end of milk flow, whereas some online sensors use a fraction of the milk to measure the components of interest (Hogeveen et al., 2010). Mastitis detection sensors can be regarded as a diagnostic test and sample/data collection occurs at every milking over the entire milking period.

Several steps need to be taken to transform the raw sensor data into information of value to be used for decision support, as illustrated in Fig. 1. An algorithm is applied to the raw data to create a useful parameter, often an alert, indicating a deviation from normal that could be of interest to the decision-maker. Most alerts need further diagnosis, consisting of clinical examination (e.g. visual observation, treatment records) or secondary testing (e.g. bacteriological culturing, somatic cell count (SCC)). When combining the confirmed abnormal status with other information, such as milk yield and composition data as well as behavioural data, farmers can be advised which decision to take or provided with a list of several possible actions that can be taken (Jensen et al., 2016; Sørensen et al., 2016; Khatun et al., 2020).

Mastitis sensor systems can be based on data from a single sensor but can also be combined with data from different sensors or sources. The overall accuracy of sensor systems is firstly dependent on the performance of the sensor(s) and the associated algorithm(s). Secondly, the accuracy of cow ID, sample and sampling quality, carryover effects due to residual milk in the system from the previous cow, measurement quality, the handling of outliers in

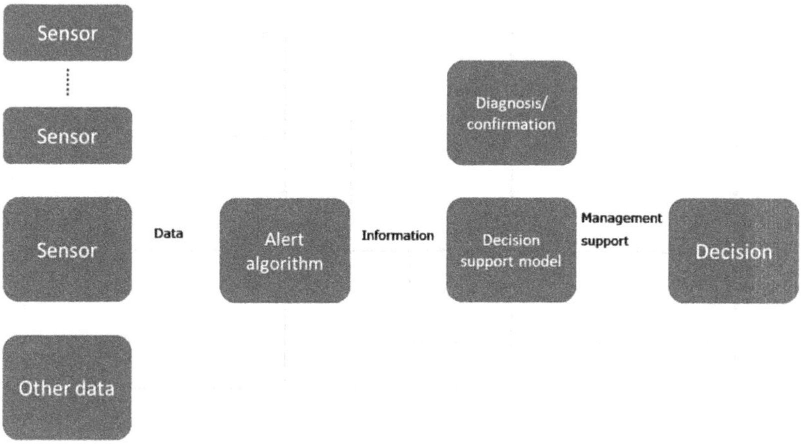

Figure 1 Diagram of pathways from raw (sensor) data to informed management decisions.

data, the variable selected for the algorithm, and the benchmark gold standard (i.e. how sensors are compared with the occurrence of a confirmed mastitis infection) and time window (i.e. the length of time in which mastitis alerts are considered true positive or false negative) contribute to the overall accuracy.

3 Commercially available sensors

Measuring the inflammatory reaction of the udder tissue as a consequence of mastitis was described in the 1970s. Kitchen and Middleton (1976) showed a positive correlation between SCC and milk components, such as *N*-acetyl β-D-glucosaminidase, lactate dehydrogenase and electrical conductivity (EC). Kitchen's research was the basis for using SCC as an indicator of milk quality to improve udder health. Since the 1980s, research and industry have directed effort into developing sensors that automatically detect mastitis using changes in one or more characteristics of milk (Nielen et al., 1992; de Mol et al., 1997). Despite the earlier research on SCC, the first sensor developed for mastitis detection was based on EC, which is the most widely studied sensor technique (Rutten et al., 2013).

3.1 Systems based on EC

EC measures the resistance of a material to an electric current. Mastitis causes changes in the capillary permeability of the blood vessels in the mammary gland, and therefore the ion concentrations in milk change, resulting in a change in EC. EC is a well-known indicator of CM (e.g. Nielen et al., 1992; Norberg et al., 2004), but is also influenced by other factors such as temperature, fat content of the milk and cleaning detergents (Nielen et al., 1992; Kaşikçi et al., 2012), which may decrease the specificity of mastitis detection. Two sensor systems are available to measure EC. The first system measures the conductivity of the whole milk, located for instance in the electronic milk meter or in the long milk tube of a conventional milking machine, and the second system measures the conductivity per udder quarter, located in the claw of the milking cluster or in the long milk tubes of an AMS. As mastitis occurs in udder quarters, quarter-level EC measurements allow the comparison of udder quarters, resulting in better performance (Hogeveen et al., 2010). EC measuring systems are widely available, but their prolonged use in conventional milking systems is limited because they have tended to provide many false alerts. However, improved algorithms, quarter-based measurement and combining EC measures with other data sources could potentially improve detection results (Khatun et al., 2017).

3.2 Systems based on SCC

SCC is an important parameter to monitor udder health and detect (sub)-CM (Dohoo and Leslie, 1991; Satu Pyörälä, 2003). Sub-CM implies inflammation within the udder, but not necessarily infection, in comparison to CM, which causes visible changes in milk or udder appearance. For decades, SCC has been measured at the cow level via milk production recording systems and, in many countries, SCC is measured at the bulk tank level as part of a milk quality payment system. For a more rapid diagnosis of mastitis, online SCC-related measurements are used in some AMSs. These online SCC measurements are good alternatives for the more costly and time-consuming laboratory SCC (L-SCC) measurements (Deng et al., 2020). In addition, the high-frequency screening of high SCC cows within a herd makes it a potentially powerful method to identify episodes of mastitis.

Three sensor systems that measure SCC are commercially available. The first two systems measure SCC indirectly, based on either gel formation of the milk (comparable to the Californian mastitis test; (Deng et al., 2020)) or physical measurements in the milk flow. The outcomes of these sensors are transformed into SCC values (expressed in cells/ml) or SCC classes. A third SCC sensor measures SCC directly, using changes in the cell count to indicate IMI (Sørensen et al., 2016). The repeated online cell counts (OCC) are a proxy for SCCs from every milking at the cow level (Dalen et al., 2019). With the OCC data, a mastitis risk indicator can potentially be developed to detect CM cases (Sørensen et al., 2016).

3.3 Systems based on LDH

LDH is a result of the cow's immune response against infection (presence of udder pathogens) and changes in the cellular membrane (Friggens et al., 2007). An elevated LDH concentration in the milk is a potential indication of CM, because it reflects the response to an IMI rather than the IMI itself (Jørgensen et al., 2016). Since LDH is correlated with SCC, a stronger correlation has been found in cows with CM than in cows without CM (Chagunda et al., 2006; Friggens et al., 2007; Jørgensen et al., 2016; Nyman et al., 2016). An online sensor system based on LDH is commercially available and used as an early indicator of mastitis.

3.4 Systems based on milk colour changes and other parameters

Milk colour changes are a visible aspect of abnormal milk, mostly caused by CM. Colour sensors are based on the reflection of light generated by a light-emitting diode (i.e. the whiter the milk, the more the light is reflected), but sensors that are based on light transmission seem to perform better (Song and Tol, 2010). Although very useful for specific situations (blood in the milk), in general, the diagnostic performance is lower compared with SCC measurement and, therefore, needs to be combined with other sensors (Hogeveen et al., 2010).

Other sensors on the market could support existing mastitis sensor systems (Khatun et al., 2020; Steele et al., 2020), but could also be used to detect other disease signs or improperly functioning farm management. Sensors that have been described and are commercially available include:

- Activity sensors, such as leg, ear-tag or neckstrap-based activity metres or bolus. Deviance in activity patterns or lower than usual activity within a cow can indicate disease (e.g. Khatun et al., 2020).
- Rumination sensors. Changes in rumination, rumen temperature, and pH can be early warning signs of potential systemic health problems (e.g. Stangaferro et al., 2016; Khatun et al., 2018; Kim et al., 2019).
- Temperature sensors. These can be part of, for instance, an ear tag, integrated in a bolus, but also by installing thermal cameras (e.g. Nielen et al., 1995a; Hovinen et al., 2008).
- Body condition scoring (BCS) and body weight (BW) measurements. Changes in cows' BCS values are a good predictor of problems with cow health, production and reproduction (e.g. Berry et al., 2007; Spoliansky et al., 2016). Changes in BW can be associated with changes in energy balance when loss of weight indicates a negative, and weight gain indicates a positive, energy balance. A positive association between BW and SCC has been observed (e.g. Berry et al., 2007; Jensen et al., 2016).

3.5 Performances of mastitis sensors

The automatic collection of sensor data could be used to monitor the health status of the cow to improve mastitis detection while reducing the labour

Table 1 Study characteristics of peer-reviewed published studies conducted from 1990 onwards that used sensor information (in-line and online) for the detection of clinical mastitis (complementing the original of Hogeveen et al. (2010))

Paper	Farms	Definition cases (n)	Definition non-cases (n)	Sensor data	Algorithm	Time window	SE (%)	SP (%)
(Maatje et al., 1992)	1 research farm	Clinical mastitis based on bacteriological culturing and SCC[1] (25)	Based on bacteriological culturing and SCC (200)	EC[2]	Moving average and threshold	14 days	100	-
(Nielen et al., 1995a)	1 research farm	Clinical mastitis based on observing abnormal milk (31)	Based on bacteriological culturing and SCC (25)	EC, milk yield, milk temperature	Artificial neural network and logistic regression	0 day[3]	47.9	91.9
(Nielen et al., 1995b)	1 research farm	Clinical mastitis based on observing abnormal milk or signs of inflammation (13 for training; 13 for testing)	Based on bacteriological culturing and SCC (17 for training; 13 for testing)	EC, milk yield, milk temperature	Artificial neural network	1 day	77.0	69.0
(de Mol et al., 1997)	2 research farms	Clinical mastitis based on clinical signs (52)	- (6495 milkings)	EC, milk yield, milk temperature	Time series with Kalman filter	17 days	90[4]	98.2[5]
(de Mol and Ouweltjes, 2001)[6,7]	1 research farm	Clinical mastitis based on clinical signs (48)	Based on never having clinical mastitis, bacteriological results and SCC (29 033 milkings)	EC, milk yield, milk temperature	Time series with Kalman filter	7 days	100[4]	95.1[5]
(de Mol and Woldt, 2001)	1 research farm	Clinical mastitis based on clinical signs (48)	Based on never having clinical mastitis, bacteriological results and SCC (29 033 milkings)	EC, milk yield, milk temperature	Fuzzy logic	7 days	100[8]	99.8
(de Mol et al., 2001)	4 semi-research farms	Clinical mastitis based on visual observation (95)	Based on not having clinical mastitis in the collection period, SCC and times milked (299 842 milkings)	EC, milk yield, milk temperature	Time series with Kalman filter	4 days	67[4]	97.9[5]

(Norberg et al., 2004)	1 research farm	Clinical mastitis based on treatment by veterinarian after observing clinical signs by staff members (275)	Based on bacteriological culturing and having no treatment for clinical mastitis by veterinarian (1353)	EC	Discriminant function analysis	0 day[3]	47.9	91.9
(Cavero et al., 2006)	1 research farm	Clinical mastitis based on treatment (651 days of mastitis for training; 348 days of mastitis for testing)	Based on not being treated for clinical mastitis (109 690 healthy days for training; 51 588 healthy days for testing)	EC, milk yield, milk flow	Fuzzy logic	5 days	92.9	93.9
(Hovinen et al., 2006)	8 commercial farms	Clinical mastitis based on observing abnormal milk and California mastitis test (21)	Based on SCC and visual observations (301)	EC, milk colour	Threshold	6 days	31.0	97.9
(Cavero et al., 2007)	1 research farm	Clinical mastitis based on treatment for clinical mastitis, visual observations and SCC (236 and 571 mastitis blocks)	Based on SCC and having no treatment for clinical mastitis (–)	EC	Threshold	5 days	85.4	75.1
(Cavero et al., 2008)	1 research farm	Clinical mastitis based on treatment or SCC (58 432 and 10 112 days of mastitis)	Based on not being treated for clinical mastitis (97 887 and 150 247 days of mastitis)	EC, milk yield, milk flow rate and days in milk	Neural network	5 days	80.0	74.9
(Kamphuis et al., 2008)	1 research farm	Treated cases of clinical mastitis (18 cow milkings)	Based on milking's without treatment records (27 699 cow milkings)	EC, SCC	Fuzzy logic	2 days for alert by model, 1 day for observation	80	99.2[9]
(Claycomb et al., 2009)	1 for training and 1 for testing	Clinical mastitis as clots on filter (23 in test set)	-	EC	Threshold	4 days/2 days	83	99.8

(Continued)

Table 1 (Continued)

Paper	Farms	Definition cases (n)	Definition non-cases (n)	Sensor data	Algorithm	Time window	SE (%)	SP (%)
(Mollenhorst et al., 2010)	3 commercial farms	Clinical mastitis based on visual observation of abnormal milk (19 quarter milkings)	Based on visual normal milk (3172 quarter milkings)	EC, SCC	Threshold	0 day[3]	47.4	99
(Kamphuis et al., 2010a)	6 commercial farms	Based on visual observation by farmers (97 quarter milkings)	Based on visual checks of farmers or on random selection (3000 quarter milkings)	EC, colour, milk yield	Decision-tree induction	<1 day	32	98.7
(Kamphuis et al., 2010b)	9 commercial farms	Based on visual observation by farmers (243 for training; 105 for testing)	Training: cases checked for clinical mastitis and SCC (24 960 quarter milkings). Testing: no observation of clinical mastitis case and without a 2-week range of a case (50 000 quarter milkings)	EC, milk colour, milk yield	Decision-tree induction	<1 day	40	99
(Sun et al., 2010)	1 research farm	Clinical mastitis based on visual observation by farm staff or SCC (895 quarter milkings)	Based on SCC and not being treated for clinical mastitis (3235 quarter milkings)	EC, milk yield	Neural network	0 days[3]	86.9	91.4
(Steeneveld et al., 2010)	1 commercial farm	Clinical mastitis based on visual observation by farm staff (11 314)	Based on visual checks of farmers (447 588 milkings)	EC, milk colour, milk yield	Naive Bayesian networks	4 days	70.0	97.8
(Miekley et al., 2012)	1 research farm	Clinical mastitis based on treatment for clinical mastitis and SCC (139 mastitis blocks)	Based on not being treated for clinical mastitis and SCC (377 healthy blocks)	EC	Cumulative sum control chart	2 days before and after alert by model, 1 day for observation	83.6	59.2

(Ankinakatte et al., 2013)	1 research farm	Clinical mastitis based on treatment for clinical mastitis (141 lactations)	Based on not being treated for clinical mastitis (523 lactations)	EC, SCC, LDH[10], milk yield	Neural network	3 days before and after treatment, 1 day for observation	75.0	80.0
(Mammadova and Keskin, 2013)	1 commercial farm	Clinical mastitis based on SCC (61)	Based on SCC (109)	EC, milk yield, milk duration	Support vector machine	1 day	89.0	92.0
(Miekley et al., 2013a)	1 research farm	Clinical mastitis based on treatment for clinical mastitis (35 for training and 15 for testing)	Based on not being treated for clinical mastitis (115 for training and 50 for testing)	EC, milk yield	Support vector machine	4 days before treatment, 1 day for observation	84.6	78.3
(Miekley et al., 2013b)	1 research farm	Clinical mastitis based on treatment for clinical mastitis (100)	Based on not being treated for clinical mastitis (238)	EC, milk yield, feeding pattern	Cumulative sum control chart	4 days before treatment, 1 day for observation	70.2	78.9
(Miekley et al., 2013c)	1 research farm	Clinical mastitis based on treatment for clinical mastitis (0 for training and 138 for testing)	Based on not being treated for clinical mastitis (100 for training and 100 for testing)	EC, milk yield, feeding pattern	Principal component analysis	4 days before treatment, 1 day for observation	83.3	76.7
(Mammadova and Keskin, 2015)	1 commercial farm	Clinical mastitis based on SCC (61)	Based on SCC (109)	EC, milk yield, milk duration	Neural network and neuro-fuzzy inference	0 day[3]	80.0 and 55.0	91.0 and 91.0
(Mikail and Keskin, 2015)	1 commercial farm	Clinical mastitis based on SCC	Based on SCC	EC, milk yield, milk duration	Fuzzy logic	0 day[3]	82.0	74.0

(Continued)

Table 1 (Continued)

Paper	Farms	Definition cases (n)	Definition non-cases (n)	Sensor data	Algorithm	Time window	SE (%)	SP (%)
(Jensen et al., 2016)	1 research farm	Clinical mastitis based on visual observation by farm staff (1455 milkings for training and 1452 milkings for testing)	Based on visual checks of farmer (498 987 milkings for training and 501 313 milkings for testing)	EC, SCC, milk yield, milk composition, body weight	Naive Bayesian networks	14 days	80.0	81.0
(Khatun et al., 2017)	1 research farm	Clinical mastitis based on visual observation by farm staff and California mastitis test (33 quarters)	Based on EC, visual checks of farm staff, and California mastitis test (139 quarters)	EC	Threshold	27 days	92.0	39.0
(Khatun et al., 2018)	2 commercial farms	Clinical mastitis based on changes in EC and visual observation by farm staff (64)	Based on EC and visual checks of farm staff (294)	EC, milk yield, milk flow	Logistic mixed models	3 days	90.0	91.0
(Ebrahimi et al., 2019)	1 commercial farm	Clinical mastitis based on SCC (278 243 milkings)	Based on SCC (68 005 milkings)	EC, milk yield, milk composition, and SCC	Deep learning, naive Bayes, generalized linear model, logistic regression, decision tree, gradient-boosted tree and random forest	0 day[3]	>93.0	39.7

(Dalen et al., 2019)	1 research farm	Clinical mastitis based on bacteriological culturing and veterinary treatment (113)	Based on bacteriological culturing (60)	OCC[11]	Threshold	7 days	80.0	90.0
(Khatun et al., 2020)	1 research farm	Clinical mastitis based on bacteriological culturing (287 quarters)	Based on bacteriological culturing (144 quarters)	EC, activity and rumination	Threshold	3 days before treatment, 1 day for observation	44.4	80.2

Characteristics included are the number of farms, definition of non-cases and cases, and the number of them included in the study, sensors used, the applied methodology, the time window applied for classification, the sensitivity (SE) and the specificity (SP).

[1] Somatic cell count.

[2] Electrical conductivity.

[3] Considers one milking.

[4] Calculated for a mastitis case (cow level).

[5] Calculated for a mastitis-free milking using only cows that never had mastitis.

[6] Based on a model developed for conventional milking and adapted for an automatic milking system.

[7] Records with indeterminable (e.g. due to missing values) were excluded.

[8] A fuzzy logic was used to classify alerts generated by an earlier developed model (de Mol and Ouweltjes, 2001) in order to decrease the number of false positive alerts, not to increase the sensitivity of the detection model.

[9] Approximately, using formula: false alert rate $\approx 10 \times (100 - \text{specificity})$ (Sherlock et al., 2008).

[10] Lactate dehydrogenase

[11] Online cell count.

required for monitoring. A summary of study characteristics and results of the different sensors currently on the market and/or under study are listed in Table 1. This table is based on Hogeveen et al. (2010) and is further expanded with the results of studies carried out after that review.

4 Future developments: biosensors

In 1981, a review on the progress of machine milking research was published in the *Journal of Dairy Science* (Thompson, 1981). The research reviewed was mainly aimed at the important association between milking vacuum and IMI, a topic that received considerable attention at the time. However, the review also emphasized an increasing role for automation in the milking process and predicted the development of sensors that would result in 'further automation not only of milking tasks but also of management data recording and analysis.' In the decades following this review, automated mastitis detection systems have become common, but detection of mastitis and maintaining udder health with the use of sensors remain challenging (Hogeveen et al., 2010; Dominiak and Kristensen, 2016; Ruegg, 2017; Khatun et al., 2018; Cabrera et al., 2020). In addition, the results presented in this chapter (Table 1) do not indicate much progress in the performance of sensor systems to detect CM. In the search for improved automatic mastitis detection, three pathways can be described:

1 Improving existing and developing new sensors;
2 Defining specific mastitis situations based on sensor data; and
3 Improving existing algorithms and combining more data sources within the analysis.

This section discusses the first area of development.

Although sensors are used to detect mastitis, performance is far from optimal. Improved mastitis detection performance can be achieved via improvements in sensors and with sensors that can be tested commercially.

The latest sensor development for mastitis detection is biosensors, that is, devices at the interface of biology and microsystems technology, which combine a biological element (bioreceptor) with a physical transducer (the sensor) (Martins et al., 2019). Advances in the knowledge of proteomics and genomics have led to the development of several biosensors (Martins et al., 2019). Biosensor-based diagnostic methods can efficiently and rapidly detect mastitis pathogens in milk and milk products. Information regarding pathogens is useful to initiate proper treatment, for example, by better targeting of antibiotics to treat particular pathogens (Griffioen et al., 2016, 2020). These new advanced biosensor-based techniques, including microfluidics, sweat and saliva sensing, as well as nanobiosensors, are gaining attention due to various advantages. Together with these techniques, a comprehensive online system can contribute to better farm management (Neethirajan et al., 2017).

In general, when the biological recognition element interacts with a target molecule, the interaction creates a measurable signal that can be converted into data by the integrated transducer. The time between measurement and result is reduced due to less sample preparation and higher sensitivity, and therefore, it is expected to result in earlier treatment decisions and improved herd management. Lower costs and reduced expenses in sample transportation make biosensors attractive to the veterinary market of food-producing animals and particularly the dairy industry due to low profit margins.

Despite these developments, current biosensors are of limited use for mastitis detection, which is mainly related to the complexity of the milk matrix. The heterogeneous composition of milk (e.g. fat, proteins, carbohydrates, inorganic particles, somatic cells) is known to interfere or inhibit most analytical processes. Indeed, the milk matrix of non-healthy animals is even more complex. In theory, biosensors and lab-on-chip devices can provide microbiological and molecular data faster, with more precision and at lower costs (Peedel and Rinken, 2014; Martins et al., 2019). However, these sensors and devices are still in their infancy for mastitis detection and do not yet meet any of the above requirements.

5 Future developments: specific mastitis situations and improving algoritms

This section discusses the second two areas of development.

5.1 Udder inflammation indicators to detect mastitis status

Using sensor data for cow-specific detection and treatment, the Standing Committee of Animal Health and Welfare of the International Dairy Federation (IDF) defined four situations where automated monitoring systems for mastitis can be of added value (Hogeveen et al., 2019b, 2021):

1. The first situation is the identification of cows that require immediate attention, which need to be identified and treated as fast as possible.
2. The second situation relates to cows that do not need immediate attention, but have a risk of progressing into severe CM. These cows should get the chance to recover spontaneously, but interventions are needed if a cow is at risk of developing CM.
3. The third situation describes cows that need attention at drying off. These cows have an IMI at the moment of drying off and therefore need to be identified for a selective dry-cow therapy programme.

4 The last situation is the constant monitoring of udder health at the farm level. This task combines sensor data from all cows in the herd so that any disturbance can be detected by evaluating sensor-based key performance indicators (KPIs). Mastitis is challenging to identify, and neither human nor sensor will be 100% accurate.

In line with these novel definitions, the current focus in mastitis sensor research is aimed at using daily data, collected by sensors, to determine the duration and progression of udder inflammation indicators (Bonestroo et al., 2021). The high frequency with which sensor data can be collected has the potential to routinely establish patterns of an udder inflammation episode much more precisely than measurements on milk production recording test days. This allows, for instance, distinguishing between cows that recover and those that do not recover after an initial inflammation is detected. Farmers can potentially then make the decision whether or not, and when, to intervene. Decisions could be based on the differences between the patterns of a specific udder inflammation episode and typical patterns of recovered udder inflammation cases. Incorporating all available sensor information from lactation should improve the distinction between cows that will benefit from antimicrobial treatment at drying off to cure existing infections and cows that will not.

5.2 Improved algorithms and inclusion of other data sources

Over the last 30 years, we have seen the development of more sophisticated algorithms. However, further improvements in detection for existing commercial and non-commercial sensor systems could arise from improved algorithms (Slob et al., 2020; Fadul-Pacheco et al., 2021). Algorithms in machine learning models in particular have progressed over the last decade. Such improved algorithms can handle more complex problems and are computationally quicker. The question is whether more advanced algorithms will result in better mastitis detection, as the minimal requirements remain to be met. Table 1 highlights there is not much difference between methods used for mastitis detection. The inclusion of other sources of data in the algorithm, however, appears to offer more potential opportunities (Cabrera et al., 2020; Khatun et al., 2020).

With Big Data and the Internet of Things, data-driven decision-making in precision dairy farming has already shown potential benefits in dairy cattle management (Tantalaki et al., 2019). Individual changes in milk-associated and behavioural parameters alone can be too vague to accurately indicate mastitis without an unacceptable number of false alerts. Combining data from different sensors (see Fig. 1: sensor and other data) can help form a much clearer picture and greater predictive ability. Several studies have shown that detection performance can be improved when combining different data sources (Jensen et al., 2016; Khatun et al., 2018, 2020; Steele et al., 2020). Eventually, sensor systems could progress to the point where, in addition to early mastitis detection, we could also identify the type of bacteria, leading to improved treatment and other management decisions. Changes in sensor-based udder health KPIs can be detected earlier in comparison with (bi)monthly collected KPIs based on milk production recording schemes. Therefore, sensor-based udder health KPIs could potentially be a valuable tool for herds not participating in Dairy Herd Improvement programmes.

The involvement of other sources of data also leads to new ways of data processing, for example, the use of data for pattern recognition to improve udder health management. CM is confirmed when clinical signs occur. However, SCC increases prior to clinical signs, which could be used to detect sub-CM (Dalen et al., 2019). The practical use of online SCC measurements for the detection of sub-CM has been studied, but the current system does not operate at the desired level of a sensor system. Dalen et al. (2019) suggest that the online measurement of SCC could still be advantageous for the herd manager and that the usefulness of the system depends on the desired application (see Fig. 1: diagnosis/confirmation). Therefore, high sensitivity is important for the detection of cows for dry-cow treatment, and high specificity is needed for the detection of IMI during lactation.

6 Conclusion

Considerable development has occurred in automatic mastitis detection over the last 30 years. While a number of sensors have been tested and brought on the market, obtaining the demand of 80% sensitivity and 99% specificity in the detection of CM remains challenging and may not be optimal. Most developed detection systems are specifically developed for CM detection, such as EC, CMT and OCC, and the results of these systems are promising. Although improvements in mastitis detection have occurred through more sophisticated algorithms, combining data from more sources in and outside the field of mastitis detection appears to have considerable potential. Future developments in mastitis detection, the subject of the ongoing search, are focused on using biosensor-based diagnostic methods. Research should not only be aimed at supporting treatment decisions of CM but should also be aimed at other aspects of mastitis management, such as the detection of CM, selecting cows to be treated with antibiotics at drying off or evaluating herd mastitis performance.

Recent research has also been aimed at studying udder inflammation patterns to distinguish between cows that recover and those that do not after an initial inflammation. This form of IMI pattern recognition could be of potential interest to farmers to decide when and what type of management actions to take. Finally, the greatest challenge in mastitis detection systems remains reducing the number of false alerts, which causes farmers to lose trust in such systems. To achieve a minimal number of false alerts, developments in sensor technologies, algorithms and using different sources of data seem to be the key.

7 Where to look for further information

The availability of more and improved sensor technologies in combination with advancing data-driven technologies provide a gateway to mastitis specific management. The International Dairy Federation (IDF; Bruxelles, Belgium) recently published a guideline for mastitis specific cases: https://shop.fil-idf.org/collections/publications/products/bulletin-of-idf-b515-guidelines-for-defining-quarter-and-udder-health-status-and-cured-clinical-and-subclinical-mastitis-cases.

A standard introduction to the subject of this chapter is K. Lockhorst, An introduction to smart dairy farming (https://doi.org/10.31715/20181); M. van der Voort, Toward a common language in data-driven mastitis detection research (https://doi.org/10.3168/jds.2021-20311.); and H. Hogeveen, Novel ways to use sensor data to improve mastitis management (https://doi.org/10.3168/jds.2020-19097).

There are a number of current research projects designed to improve automatic mastitis detection and management, including:

- The ClearFarm project which includes better control and management of dairy health (https://www.clearfarm.eu/).
- The Data Driven Dairy Decision For Farmers (4D4F) network to improve decision making on dairy farms based on sensor data (https://www.4d4f.eu/).
- The Biosens4PrecisionMastitis project which support the development of biomarkers to detect mastitis (http://www.biosensingbovinemastitis.com/).

8 References

Ankinakatte, S., Norberg, E., Løvendahl, P., Edwards, D. and Højsgaard, S. 2013. Predicting mastitis in dairy cows using neural networks and generalized additive models: A comparison. *Comput. Electron. Agric.* 99:1–6. doi:10.1016/j.compag.2013.08.024.

Berry, D. P., Lee, J. M., Macdonald, K. A., Stafford, K., Matthews, L. and Roche, J. R. 2007. Associations among body condition score, body weight, somatic cell count, and clinical mastitis in seasonally calving dairy cattle. *J. Dairy Sci.* 90(2):637–648. doi:10.3168/jds.S0022-0302(07)71546-1.

Bonestroo, J., van der Voort, M., Fall, N., Hogeveen, H., Emanuelson, U., Christine Klaas, I. 2021. Progression of different udder inflammation indicators and their episode length after onset of inflammation using automatic milking system sensor data. *J. Dairy Sci.* 104(3): 3458–3473. doi: 10.3168/jds.2019-18054.

Cabrera, V. E., Barrientos-Blanco, J. A., Delgado, H. and Fadul-Pacheco, L. 2020. Symposium review: Real-time continuous decision making using big data on dairy farms. *J. Dairy Sci.* 103(4):3856–3866. doi:10.3168/jds.2019-17145.

Cavero, D., Tölle, K.-H., Buxadé, C. and Krieter, J. 2006. Mastitis detection in dairy cows by application of fuzzy logic. *Livest. Sci.* 105(1–3):207–213. doi:10.1016/j.livsci.2006.06.006.

Cavero, D., Tölle, K.-H., Henze, C., Buxadé, C. and Krieter, J. 2008. Mastitis detection in dairy cows by application of neural networks. *Livest. Sci.* 114(2–3):280–286. doi:10.1016/j.livsci.2007.05.012.

Cavero, D., Tölle, K.-H., Rave, G., Buxadé, C. and Krieter, J. 2007. Analysing serial data for mastitis detection by means of local regression. *Livest. Sci.* 110(1–2):101–110. doi:10.1016/j.livsci.2006.10.006.

Chagunda, M. G., Larsen, T., Bjerring, M. and Ingvartsen, K. L. 2006. L-lactate dehydrogenase and *N*-acetyl-β-D-glucosaminidase activities in bovine milk as indicators of non-specific mastitis. *J. Dairy Res.* 73(4):431–440. doi:10.1017/S0022029906001956.

Chakraborty, S., Dhama, K., Tiwari, R., Iqbal Yatoo, M., Khurana, S. K., Khandia, R., Munjal, A., Munuswamy, P., Kumar, M. A., Singh, M., Singh, R., Gupta, V. K. and Chaicumpa, W. 2019. Technological interventions and advances in the diagnosis of intramammary infections in animals with emphasis on bovine population–A review. *Vet. Q.* 39(1):76–94. doi:10.1080/01652176.2019.1642546.

Claycomb, R. W., Johnstone, P. T., Mein, G. A. and Sherlock, R. A. 2009. An automated in-line clinical mastitis detection system using measurement of conductivity from foremilk of individual udder quarters. *N. Z. Vet. J.* 57(4):208–214. doi:10.1080/00480169.2009.36903.

Dalen, G., Rachah, A., Nørstebø, H., Schukken, Y. H. and Reksen, O. 2019. The detection of intramammary infections using online somatic cell counts. *J. Dairy Sci.* 102(6):5419–5429. doi:10.3168/jds.2018-15295.

de Mol, R. M., Kroeze, G. H., Achten, J. M. F. H., Maatje, K. and Rossing, W. 1997. Results of a multivariate approach to automated oestrus and mastitis detection. *Livest. Prod. Sci.* 48(3):219–227. doi:10.1016/S0301-6226(97)00028-6.

de Mol, R. M. and Ouweltjes, W. 2001. Detection model for mastitis in cows milked in an automatic milking system. *Prev. Vet. Med.* 49(1–2):71–82. doi:10.1016/S0167-5877(01)00176-3.

de Mol, R. M., Ouweltjes, W., Kroeze, G. H. and Hendriks, M. M. W. B. 2001. Detection of estrus and mastitis: Field performance of a model. *Appl. Eng. Agric.* 17(3):399–407. doi:10.13031/2013.6201.

de Mol, R. M. and Woldt, W. E. 2001. Application of fuzzy logic in automated cow status monitoring. *J. Dairy Sci.* 84(2):400–410. doi:10.3168/jds.S0022-0302(01)74490-6.

Deng, Z., Hogeveen, H., Lam, T. J. G. M., van der Tol, R. and Koop, G. 2020. Performance of online somatic cell count estimation in automatic milking systems. *Front. Vet. Sci.* 7:221. doi:10.3389/fvets.2020.00221.

Dohoo, I. R. and Leslie, K. E. 1991. Evaluation of changes in somatic cell counts as indicators of new intramammary infections. *Prev. Vet. Med.* 10(3):225–237. doi:10.1016/0167-5877(91)90006-N.

Dominiak, K. N. and Kristensen, A. R. 2016. Prioritizing alarms from sensor-based detection models in livestock production - A review on model performance and alarm reducing methods. *Comput. Electron. Agric.* 133:46–67. doi:10.1016/j.compag.2016.12.008.

Ebrahimi, M., Mohammadi-Dehcheshmeh, M., Ebrahimie, E. and Petrovski, K. R. 2019. Comprehensive analysis of machine learning models for prediction of sub-clinical mastitis: Deep Learning and Gradient-Boosted Trees outperform other models. *Comput. Biol. Med.* 114:103456. doi:10.1016/j.compbiomed.2019.103456.

Fadul-Pacheco, L., Delgado, H. and Cabrera, V. E. 2021. Exploring machine learning algorithms for early prediction of clinical mastitis. *Int. Dairy J.* 119:105051. doi:10.1016/j.idairyj.2021.105051:105051.

Friggens, N. C., Chagunda, M. G. G., Bjerring, M., Ridder, C., Højsgaard, S. and Larsen, T. 2007. Estimating degree of mastitis from time-series measurements in milk: A test of a model based on lactate dehydrogenase measurements. *J. Dairy Sci.* 90(12):5415–5427. doi:10.3168/jds.2007-0148.

Griffioen, K., Cornelissen, J., Heuvelink, A., Adusei, D., Mevius, D. and van der Wal, F. J. 2020. Development and evaluation of 4 loop-mediated isothermal amplification assays to detect mastitis-causing bacteria in bovine milk samples. *J. Dairy Sci.* 103(9):8407–8420. doi:10.3168/jds.2019-18035.

Griffioen, K., Hop, G. E., Holstege, M. M. C., Velthuis, A. G. J. and Lam, T. J. G. M. 2016. Dutch dairy farmers' need for microbiological mastitis diagnostics. *J. Dairy Sci.* 99(7):5551–5561. doi:10.3168/jds.2015-10816.

Hogeveen, H., Kamphuis, C., Steeneveld, W. and Mollenhorst, H. 2010. Sensors and clinical mastitis-the quest for the perfect alert. *Sensors (Basel)* 10(9):7991–8009. doi:10.3390/s100907991.

Hogeveen, H., Klaas, I. C., Dalen, G., Honig, H., Zecconi, A., Kelton, D. F. and Mainar, M. S. 2021. Novel ways to use sensor data to improve mastitis management. *J. Dairy Sci.* 104(10):11317–11332. doi:10.3168/jds.2020-19097.

Hogeveen, H., Klaas, I., Dalen, G., Honig, H., Zecconi, A., Mainar, M. S. and Kelton, D. 2019a. Encouraging adequate farmer interventions through the use of sensors for mastitis management. *Page in the IDF International Mastitis Conference*, Copenhagen, Denmark.

Hogeveen, H., Klaas, I. C., Dalen, G., Honig, H., Zecconi, A., Mainar, M. S. and Kelton, D. F. 2019b. Encouraging adequate farmer interventions through the use of sensors for mastitis management. *Page in IDF International Mastitis Conference*, Copenhagen, Denmark.

Hogeveen, H., Steeneveld, W. and Wolf, C. A. 2019c. Production diseases reduce the efficiency of dairy production: A review of the results, methods, and approaches Regarding the economics of mastitis. *Annu. Rev. Resour. Econ.* 11(1):289–312. doi:10.1146/annurev-resource-100518-093954.

Hommels, N. M. C., Ferreira, F. C., van den Borne, B. H. P. and Hogeveen, H. 2021. Antibiotic use and potential economic impact of implementing selective dry cow therapy in large US dairies. *J. Dairy Sci.* 104(8):8931–8946. doi:10.3168/jds.2020-20016.

Hovinen, M., Aisla, A.-M. and Pyörälä, S. 2006. Accuracy and reliability of mastitis detection with electrical conductivity and milk colour measurement in automatic milking. *Acta Agric. Scand. A* 56(3–4):121–127. doi:10.1080/09064700701216888.

Hovinen, M., Siivonen, J., Taponen, S., Hänninen, L., Pastell, M., Aisla, A. M. and Pyörälä, S. 2008. Detection of clinical mastitis with the help of a thermal camera. *J. Dairy Sci.* 91(12):4592–4598. doi:10.3168/jds.2008-1218.

Jensen, D. B., Hogeveen, H. and De Vries, A. 2016. Bayesian integration of sensor information and a multivariate dynamic linear model for prediction of dairy cow mastitis. *J. Dairy Sci.* 99(9):7344–7361. doi:10.3168/jds.2015-10060.

Jørgensen, C. H., Kristensen, A. R., Østergaard, S. and Bennedsgaard, T. W. 2016. Use of inline measures of lactate dehydrogenase for classification of posttreatment mammary Staphylococcus aureus infection status in dairy cows. *J. Dairy Sci.* 99(10):8375–8383. doi:10.3168/jds.2016-10858.

Kamphuis, C., Mollenhorst, H., Feelders, A., Pietersma, D. and Hogeveen, H. 2010a. Decision-tree induction to detect clinical mastitis with automatic milking. *Comput. Electron. Agric.* 70(1):60–68. doi:10.1016/j.compag.2009.08.012.

Kamphuis, C., Mollenhorst, H., Heesterbeek, J. A. P. and Hogeveen, H. 2010b. Detection of clinical mastitis with sensor data from automatic milking systems is improved by using decision-tree induction. *J. Dairy Sci.* 93(8):3616–3627. doi:10.3168/jds.2010-3228.

Kamphuis, C., Sherlock, R., Jago, J., Mein, G. and Hogeveen, H. 2008. Automatic detection of clinical mastitis is improved by in-line monitoring of somatic cell count. *J. Dairy Sci.* 91(12):4560–4570. doi:10.3168/jds.2008-1160.

Kaşikçi, G., Çetin, Ö., Bingöl, E. B. and Gündüz, M. C. 2012. Relations between electrical conductivity, somatic cell count, California mastitis test and some quality parameters in the diagnosis of subclinical mastitis in dairy cows. *Turk. J. Vet. Anim. Sci.* 36:49–55. doi:10.3906/vet-1103-4.

Khatun, M., Clark, C. E. F., Lyons, N. A., Thomson, P. C., Kerrisk, K. L. and Garciá, S. C. 2017. Early detection of clinical mastitis from electrical conductivity data in an automatic milking system. *Anim. Prod. Sci.* 57(7):1226–1232. doi:10.1071/AN16707.

Khatun, M., Thomson, P. C., Clark, C. E. F. and García, S. C. 2020. Prediction of quarter level subclinical mastitis by combining in-line and on-animal sensor data. *Anim. Prod. Sci.* 60(1):180–186.

Khatun, M., Thomson, P. C., Kerrisk, K. L., Lyons, N. A., Clark, C. E. F., Molfino, J. and García, S. C. 2018. Development of a new clinical mastitis detection method for automatic milking systems. *J. Dairy Sci.* 101(10):9385–9395. doi:10.3168/jds.2017-14310.

Kim, H., Min, Y. and Choi, B. 2019. Real-time temperature monitoring for the early detection of mastitis in dairy cattle: Methods and case researches. *Comput. Electron. Agric.* 162:119–125. doi:10.1016/j.compag.2019.04.004.

Kitchen, B. J. and Middleton, G. 1976. Enzymic methods for the estimation of the somatic cell count in bovine milk: II. N-acetyl-β-D-glucosaminidase test for routine estimation of the somatic cell count in milk. *J. Dairy Res.* 43(3):491–494. doi:10.1017/S0022029900016071.

Maatje, K., Huijsmans, P. J. M., Rossing, W. and Hogewerf, P. H. 1992. The efficacy of in-line measurement of quarter milk electrical conductivity, milk yield and milk temperature for the detection of clinical and subclinical mastitis. *Livest. Prod. Sci.* 30(3):239–249. doi:10.1016/S0301-6226(06)80013-8.

Mammadova, N. and Keskin, I. 2013. Application of the support vector machine to predict subclinical mastitis in dairy cattle. *Sci. World J.I* 2013:603897. doi:10.1155/2013/603897.

Mammadova, N. M. and Keskin, I. 2015. Application of neural network and adaptive neuro-fuzzy inference system to predict subclinical mastitis in dairy cattle. *Indian J. Anim. Res.* 49(OF):671–679. doi:10.18805/ijar.5581.

Martins, S. A. M., Martins, V. C., Cardoso, F. A., Germano, J., Rodrigues, M., Duarte, C., Bexiga, R., Cardoso, S. and Freitas, P. P. 2019. Biosensors for on-farm diagnosis of mastitis. *Front. Bioeng. Biotechnol.* 7:186. doi:10.3389/fbioe.2019.00186.

Miekley, B., Stamer, E., Traulsen, I. and Krieter, J. 2013a. Implementation of multivariate cumulative sum control charts in mastitis and lameness monitoring. *J. Dairy Sci.* 96(9):5723–5733. doi:10.3168/jds.2012-6460.

Miekley, B., Traulsen, I. and Krieter, J. 2013b. Mastitis detection in dairy cows: The application of support vector machines. *J. Agric. Sci.* 151(6):889–897. doi:10.1017/S0021859613000178.

Miekley, B., Traulsen, I. and Krieter, J. 2013c. Principal component analysis for the early detection of mastitis and lameness in dairy cows. *J. Dairy Res.* 80(3):335–343. doi:10.1017/S0022029913000290.

Miekley, B., Traulsen, I. and Krieter, J. 2012. Detection of mastitis and lameness in dairy cows using wavelet analysis. *Livest. Sci.* 148(3):227–236. doi:10.1016/j.livsci.2012.06.010.

Mikail, N. and Keskin, I. 2015. Subclinical mastitis prediction in dairy cattle by application of fuzzy logic. *Pak. J. Agric. Sci.* 52:1101–1107.

Mollenhorst, H., van der Tol, P. P. J. and Hogeveen, H. 2010. Somatic cell count assessment at the quarter or cow milking level. *J. Dairy Sci.* 93(7):3358–3364. doi:10.3168/jds.2009-2842.

Neethirajan, S., Tuteja, S. K., Huang, S-T., Kelton, D. 2017. Recent advancement in biosensors technology for animal and livestock health management. *Biosens. Bioelectron.* 98:398–407. doi: 10.1016/j.bios.2017.07.015.

Nielen, M., Deluyker, H., Schukken, Y. H. and Brand, A. 1992. Electrical conductivity of milk: Measurement, modifiers, and meta analysis of mastitis detection performance. *J. Dairy Sci.* 75(2):606–614. doi:10.3168/jds.S0022-0302(92)77798-4.

Nielen, M., Schukken, Y. H., Brand, A., Haring, S. and Ferwerda-Van Zonneveld, R. T. 1995a. Comparison of analysis techniques for on-line detection of clinical mastitis. *J. Dairy Sci.* 78(5):1050–1061. doi:10.3168/jds.S0022-0302(95)76721-2.

Nielen, M., Spigt, M. H., Schukken, Y. H., Deluyker, H. A., Maatje, K. and Brand, A. 1995b. Application of a neural network to analyse on-line milking parlour data for the detection of clinical mastitis in dairy cows. *Prev. Vet. Med.* 22(1–2):15–28. doi:10.1016/0167-5877(94)00405-8.

Norberg, E., Hogeveen, H., Korsgaard, I. R., Friggens, N. C., Sloth, K. H. M. N. and Løvendahl, P. 2004. Electrical conductivity of milk: Ability to predict mastitis status. *J. Dairy Sci.* 87(4):1099–1107. doi:10.3168/jds.S0022-0302(04)73256-7.

Nyman, A. K., Emanuelson, U. and Waller, K. P. 2016. Diagnostic test performance of somatic cell count, lactate dehydrogenase, and *N*-acetyl-β-D-glucosaminidase for detecting dairy cows with intramammary infection. *J. Dairy Sci.* 99(2):1440–1448. doi:10.3168/jds.2015-9808.

Peedel, D. and Rinken, T. 2014. Rapid biosensing of Staphylococcus aureus bacteria in milk. *Anal. Methods* 6(8):2642–2647. doi:10.1039/C3AY42036A.

Ruegg, P. L. 2017. A 100-year Review: Mastitis detection, management, and prevention. *J. Dairy Sci.* 100(12):10381–10397. doi:10.3168/jds.2017-13023.

Rutten, C. J., Velthuis, A. G. J., Steeneveld, W. and Hogeveen, H. 2013. Invited review: Sensors to support health management on dairy farms. *J. Dairy Sci.* 96(4):1928–1952. doi:10.3168/jds.2012-6107.

Pyörälä, S. 2003. Indicators of inflammation in the diagnosis of mastitis. *Vet. Res.* 34(5):565–578. doi:10.1051/vetres:2003026.

Sherlock, R., Hogeveen, H., Mein, G. and Rasmussen, M. 2008. Performance evaluation of systems for automated monitoring of udder health: Analytical issues and guidelines. In *Mastitis Control–From Science to Practice*. Proceedings of international conference 30 September–2 October 2008, The Hague, the Netherlands

Slob, N., Catal, C. and Kassahun, A. 2020. Application of machine learning to improve dairy farm management: A systematic literature review. *Prev. Vet. Med.* 187:105237. doi:10.1016/j.prevetmed.2020.105237.

Song, X. and Tol, R. 2010. Automatic detection of clinical mastitis in astronaut A3 TM milking robot. *Proceedings of the First North American Conference on Precision Dairy Management.*

Sørensen, L. P., Bjerring, M. and Løvendahl, P. 2016. Monitoring individual cow udder health in automated milking systems using online somatic cell counts. *J. Dairy Sci.* 99(1):608–620. doi:10.3168/jds.2014-8823.

Spoliansky, R., Edan, Y., Parmet, Y. and Halachmi, I. 2016. Development of automatic body condition scoring using a low-cost 3-dimensional Kinect camera. *J. Dairy Sci.* 99(9):7714–7725. doi:10.3168/jds.2015-10607.

Stangaferro, M. L., Wijma, R., Caixeta, L. S., Al-Abri, M. A. and Giordano, J. O. 2016. Use of rumination and activity monitoring for the identification of dairy cows with health disorders: Part II. Mastitis. *J. Dairy Sci.* 99(9):7411–7421. doi:10.3168/jds.2016-10908.

Steele, N. M., Dicke, A., De Vries, A., Lacy-Hulbert, S. J., Liebe, D., White, R. R. and Petersson-Wolfe, C. S. 2020. Identifying gram-negative and gram-positive clinical mastitis using daily milk component and behavioral sensor data. *J. Dairy Sci.* 103(3):2602–2614. doi:10.3168/jds.2019-16742.

Steeneveld, W., van der Gaag, L. C., Ouweltjes, W., Mollenhorst, H. and Hogeveen, H. 2010. Discriminating between true-positive and false-positive clinical mastitis alerts from automatic milking systems. *J. Dairy Sci.* 93(6):2559–2568. doi:10.3168/jds.2009-3020.

Sun, Z., Samarasinghe, S. and Jago, J. 2010. Detection of mastitis and its stage of progression by automatic milking systems using artificial neural networks. *J. Dairy Res.* 77(2):168–175. doi:10.1017/S0022029909990550.

Tantalaki, N., Souravlas, S. and Roumeliotis, M. 2019. Data-driven decision making in precision agriculture: The rise of big data in agricultural systems. *J. Agric. Food Inf.* 20(4):344–380. doi:10.1080/10496505.2019.1638264.

Thompson, P. D. 1981. Milking machines - The past twenty-five years. *J. Dairy Sci.* 64(6):1344–1357. doi:10.3168/jds.S0022-0302(81)82707-5.

Chapter 10

Developments in automated systems for monitoring livestock health: lameness

Zoe E. Barker, University of Reading, UK; Nick J. Bell, University of Nottingham, UK; Jonathan R. Amory, Writtle University College, UK; and Edward A. Codling, University of Essex, UK

1 Introduction

Lameness is the behavioural manifestation of a range of locomotor system disorders in livestock such as sheep, pigs and cattle. Lameness can cause ongoing health and welfare problems for affected animals and can lead to significant economic losses for farmers, and hence early detection and treatment is essential. However, lameness can be difficult or time-consuming to detect manually, and so there is a need for novel automated approaches for on-farm monitoring and detection of lameness conditions across livestock species, particularly where herd sizes are increasing.

In dairy cows, lesions of the claw horn and the surrounding skin account for the vast majority of lameness cases (Solano et al., 2016). In sheep, infectious foot lesions, such as interdigital dermatitis, foot rot and contagious ovine digital dermatitis (CODD), are the most prevalent causes of lameness (Winter et al., 2015). In pigs, infectious arthritis and injuries to the legs and feet are common causes of lameness (KilBride et al., 2008; Gillman et al., 2009).

http://dx.doi.org/10.19103/AS.2021.0090.10

The prevalence of lameness reported in dairy cattle across the world is alarmingly high compared with other livestock species, with reported rates of 29.5–32% in the UK (Griffiths et al., 2018; Afonso et al., 2020), 28% in Chile (Tadich et al., 2010), 30% in Canada and 30-55% in the USA (von Keyserlingk et al., 2012). Estimates of lameness on beef farms are limited, ranging from 1.1% in Norwegian suckler herds recorded by locomotion scoring (Fjeldaas et al., 2007) to 4.5% in Canadian feed lots, based on farmer records (Davis-Unger et al., 2019). Farm-level prevalence of lameness in sheep in the UK was recently estimated at 4.1% (Prosser et al., 2019), while the prevalence of lameness on UK pig farms, recorded as part of the 'real welfare' initiative (where over 5 million individual pigs were included in the assessments from all stages of production), was 0.18% (Pandolfi et al., 2017). In a study investigating the prevalence of lameness in sows, the prevalence was 4.5% (Willgert et al., 2014), reflecting that sows are heavier and longer-lived, and therefore exposed to risk factors for longer, than slaughter animals.

Lameness is known to cause behavioural changes in affected animals, including changes in gait and movement, lying and feeding behaviour and social interactions. For this reason, most manual and automated lameness detection methods are based on observing and detecting changes or differences in movement, gait and behaviour. Sensor-based technological tools that have been used to collect data on movement and behaviour in the context of lameness include:

- ground reaction force sensors, load sensors, and pressure-sensitive walkways,
- thermography,
- kinematics (video and acoustic),
- animal-mounted accelerometers,
- proximity sensors, and
- spatial positioning systems.

In this chapter, we give an overview of lameness and its causes in commercial livestock, outline existing manual approaches for detection, and highlight new automated approaches that are being developed to aid in detection and prediction. Throughout the chapter, we focus on dairy cows (particularly those fully housed) as the main case study species since (as outlined above) they are the commercial species with the highest lameness prevalence and impact. We explain how lameness may affect behaviour and how this may allow both direct and indirect automated approaches for lameness detection. We review the current technology available and highlight novel methods for the analysis of behavioural (and other) data that is collected. We finish with a discussion of how automated methods for lameness detection may form a complementary element of more general systems for monitoring multiple behaviours and diseases as part of an integrated and holistic precision livestock farming approach.

2 Lameness and its impacts

2.1 Conditions causing lameness

The first step towards developing automated technologies which will support farmers in reducing the prevalence of lameness on farms is to understand the underlying biology of the disease in different livestock species. For a detailed overview of the causes of lameness, and its prevention and management, in dairy cattle, readers are referred to Bell (2017).

Of the skin conditions causing lameness in cattle probably the most widespread is digital dermatitis, a polymicrobial infection of the soft tissue around and between the claws. Although until recently New Zealand has remained free of digital dermatitis, the remainder of the major dairy-producing countries have experienced widespread endemic digital dermatitis for many years. Foot trimmer records from Canadian dairy herds identified digital dermatitis in 15% of cows and 94% of herds (Solano et al., 2016) which is probably reflective of the situation across much of the northern hemisphere. Once endemic on a farm, treatment often takes the form of whole herd foot bathing rather than individual cow treatment, so the true prevalence on farms is often under-recorded. The impact of digital dermatitis on locomotion score can be variable with only severe lesions associated with an increased likelihood of lameness (Frankena et al., 2009) leading to challenges for early detection of digital dermatitis lesions. While raw open lesions (similar to that shown in Fig. 1a) can be acutely painful, the impacts of digital dermatitis do not appear to be as prolonged as for other causes of lameness. Cows with digital dermatitis had lower nociceptive thresholds on the day of treatment, indicative of a pain response, but no such response when tested again 28 days later (Whay et al., 1998) and milk yields were not significantly different for cows with farmer-recorded cases of digital dermatitis when compared with cows that were never lame (Amory et al., 2008). In sheep, the same causative agent can infect the digital skin but the condition is known as CODD. The number of infected farms has increased significantly over a 10-year period and the condition contributes to an increased prevalence of lameness on infected farms (Dickins et al., 2016) suggesting similar painful outcomes as for cattle.

After digital dermatitis, the most common causes of lameness in cattle are claw horn disruption lesions (CHDL) of which sole bruising (Fig. 1b), sole ulcer (Fig. 1c), white line disease (Fig. 1d) and toe necrosis (Fig. 1e) are the predominant lesions (Solano et al., 2016). Historically, these lesions were

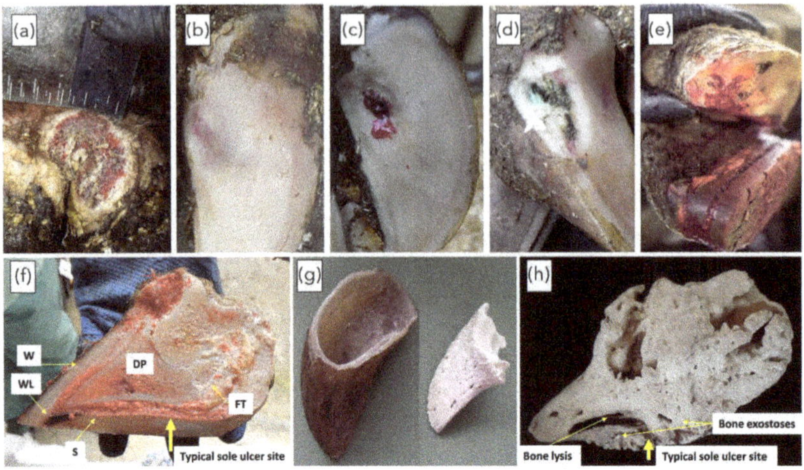

Figure 1 (a–e) Examples of common conditions causing lameness in dairy cows. (a) M2 stage digital dermatitis lesions with characteristic acute and painful red or red-grey ulcerative tissue (see Kofler et al., 2020); (b) Sole bruising haemorrhage; (c) sole ulcer at the typical site on the claw; (d) white line disease lesion; (e) toe necrosis, chronic infection of the corium and hoof horn. (f–h) Examples of the internal structure of the cow foot that are associated with claw horn disruption lesions (CHDLs). (f) Section through a healthy claw showing the wall horn (W), the distal phalanx (DP), the flexor tendon (FT), the white line, which is the anatomical junction between sole and wall horn (WL), the horn of the sole (S), and the typical site of sole lesions; (g) the claw capsule and distal phalanx of a healthy hoof; (h) the distal phalanx of a diseased hoof with significant changes to the bone capable of exerting abnormal forces on germinal epithelium resulting in further lesions.

considered to have a nutritional component, but a recent review of the statistical evidence for such a relationship (Randall et al., 2018) found the evidence to be weak in comparison to the evidence for environmental (e.g. increased time on poor floor surfaces and lack of resting) and individual animal risk factors (e.g. milk yield, weight and body condition score). The increasing risk of lameness with age is well documented (Barker et al., 2009; Solano et al., 2016) as is the increased risk of future lameness associated with a previous CHDL (Hirst et al., 2002; Randall et al., 2016; Newsome et al., 2017). There is now compelling evidence that increasing age and previous CHDLs are associated with irreversible pathological changes to the distal phalanx (Newsome et al., 2016; Fiore et al., 2019; Fabbri et al., 2020), which may predispose cows to future lameness incidents (Fig. 1f–h). Early detection and prompt treatment of lameness have successfully reduced both the severity and duration of CHDLs (Leach et al., 2012; Groenevelt et al., 2014; Miguel-Pacheco et al., 2017) and the associated inflammatory responses hypothesised to be involved with the bone pathologies (Newsome et al., 2016). These actions are essential if we are to increase the longevity of dairy cows as suggested by De Vries (2020), whose modelling indicates that we should be reconsidering the average productive lifespan of dairy cattle to take into account wider societal concerns including animal welfare and environmental sustainability.

As with cattle, there is evidence of environmental risks for lameness in pigs. Suboptimal flooring surfaces have been associated with sole bruising in piglets (Quinn et al., 2015) and bursitis in weaned and finishing pigs (Gillman et al., 2008). Pathological changes to the bones of pig feet are yet to be reported, largely due to the relatively short lifespan of meat pigs, but the increased prevalence of lameness in sows mentioned above indicate that this warrants further study in breeding animals.

2.2 Lameness impacts on health and welfare

There is compelling evidence from around the world that lameness results in significant reductions in milk yield (Amory et al., 2008; Bicalho et al., 2008; Archer et al., 2010) and that higher-yielding dairy cows have an increased risk of becoming lame (Amory et al., 2008; Archer et al., 2010). It is also clear that lameness has a negative impact on a range of measures of fertility including a greater number of services per conception (Vacek et al., 2007), reduced conception (Machado et al., 2010; Alawneh et al., 2011), extended calving interval (Sogstad et al., 2006), and increased calving to conception (Alawneh et al., 2011) and calving to first service intervals (Vacek et al., 2007). Lameness is associated with an increased risk of culling in many studies (Bicalho et al., 2007a; McConnel et al., 2008; Cramer et al., 2009; Machado et al., 2010) but not all (Barkema et al., 1994; Hultgren et al., 2004), reflecting the fact that culling decisions are complicated and the underlying reasons are not always well recorded. As lameness is a disease of the high yielding cow, she may be retained in the herd despite her lameness while her milk yield remains higher than lower-yielding herd mates. These production losses coupled with costs associated with the treatment of lame cows were estimated to cost a typical UK dairy farmer £323 for the average lameness case (Willshire and Bell, 2009). In the Netherlands, the cost per cow in the herd per year was estimated as €53 (Bruijnis et al., 2012).

These economically costly impacts on individual animals occur as a result of pain and also generalised 'sickness behaviours' such as reduced activity, inappetence, and reduced interest in their environment and conspecifics. The application of mechanical threshold testing to measure nociceptive thresholds in cattle has been used to demonstrate the pain associated with lameness. Tadich et al. (2013) measured reduced nociceptive thresholds and therefore hyperalgesia, with increasing lameness severity and hyperalgesia persisted to at least 28 days following treatment (Whay et al., 1998). Haptoglobin, an acute-phase protein indicative of an inflammatory response, was also elevated with increasing lameness severity (Tadich et al., 2013), indicating a systemic inflammatory response to lameness. Invading pathogens and tissue damage can both result in the release of cytokines associated with inflammation and fever, which can result in reduced activity and a loss of appetite (Broom, 2006).

Despite being clear that lameness impairs welfare, causes significant economic losses and leads to an increased susceptibility to further lameness for the lifetime of the animal, underestimation of lameness by farmers remains a problem that impairs prompt treatment (Leach et al., 2010; Tunstall et al., 2019). One promising route to detecting lameness early is the identification of behavioural changes linked to the condition.

3 Lameness and behaviour

3.1 Individual behaviour

It has been demonstrated that lameness is associated with alterations in dairy cattle behaviours such as lying, feeding and social interactions. The majority of recent studies report increased lying times associated with lameness (Blackie et al., 2011; Sepúlveda-Varas et al., 2014; Westin et al., 2016; Weigele et al., 2018; Grimm et al., 2019) along with the decreased number of lying bouts and increased bout durations (Sepúlveda-Varas et al., 2014; Westin et al., 2016), although such findings are not fully unanimous (see e.g. Blackie and Maclaurin (2019), who highlight the methodological challenges in investigating the relationship between lameness and lying). An increase in asymmetric lying by lame cows (Weigele et al., 2018) indicates the lame limb affects the process of lying down and therefore the side the cow is able to lie on. Lying behaviour is not only affected in the milking herd but also in transition and calving periods (Calderon and Cook, 2011; Sepúlveda-Varas et al., 2018). There is also evidence that reduced lying times in the period immediately after calving, which may be associated with increased risk of lameness later in lactation (Sepúlveda-Varas et al., 2018).

Lameness is associated with reduced time spent feeding (González et al., 2008; Barker et al., 2018; Weigele et al., 2018; Grimm et al., 2019), although feed intake may not be affected by these reductions as feeding pace increases (González et al., 2008; Grimm et al., 2019), perhaps as a result of reduced jaw movements before swallowing (Weigele et al., 2018). Lameness has also been associated with reduced reactivity to fresh feed or feed push-ups by lame cows (Blackie et al., 2011; Yunta et al., 2012; Miguel-Pacheco et al., 2016), perhaps due to lame cows being unable or unwilling to compete for food at busy times. This may explain why differences in feeding behaviour have been observed during the day, when there is more competition at the feed face, but not overnight (Barker et al., 2018). Reduced visits to a concentrate feeder by lame cows were also reported by Weigele et al. (2018).

In addition to essential activities, such as feeding and resting, reductions have been reported for behaviours and activities that might be considered as non-essential or for pleasure, such as visiting grooming brushes (Weigele et al., 2018), particularly when further from the feed face (Mandel et al., 2017). Similar or related behavioural changes have also been reported in individual cows for a number of other diseases including reduced activity in cows with a high parasite burden (Szyszka et al., 2013) and reduced intake with clinical mastitis (Sepúlveda-Varas et al., 2016) and ketosis (Goldhawk et al., 2009), supporting the notion of qualitatively different 'sickness behaviours' in dairy cattle.

In the context of other livestock species, there are very limited studies looking at behavioural activity changes associated with lameness in pigs or sheep. Lame sows in farrowing crates are more likely to show uncontrolled lying-down behaviour compared with non-lame sows (Bonde et al., 2004) and induced lameness causes sows to stand less and lie down more (Pairis-Garcia et al., 2015). There are no current studies reporting the direct observation of sheep behaviour activity changes beyond gait alterations, but a recent study using accelerometer and gyroscope-based ear sensors (discussed in more detail below) reports that while standing and lying times are not different between lame and non-lame sheep, lame animals appear to 'lie differently' (Kaler et al., 2020).

3.2 Social behaviour

Social behaviour and interactions can also be influenced by sickness. Lame cows were shown to instigate fewer aggressive interactions with herd mates despite still receiving a similar number of aggressive interactions (Galindo and Broom, 2002); lame cows were also licked more frequently than non-lame cows. The exact role licking plays in the social dynamics of dairy herds is complex and is yet to be fully understood (de Freslon et al., 2019), but it is thought to have a calming effect on the receiver (Laister et al., 2011). Whether lame cows solicited the social licking from herd mates to benefit from these calming effects, or if the actions of non-lame herd mates are spontaneous, is unclear.

More generally, there appear to be links between sickness and social isolation. For example, it is well recognised that cows and many other herd animals opt to isolate themselves from the herd to give birth, but it has now been demonstrated that cows experiencing prolonged calvings, which we might assume to be difficult, and therefore more painful, seek out greater isolation than cows experiencing normal calvings (Rørvang et al., 2018). Social isolation has also been observed in cows with high rectal temperatures post-calving which is indicative of infectious diseases such as mastitis and metritis (Proudfoot et al., 2014). Sick cows may also exhibit a reluctance to undertake social interactions which allow them to access favoured resources such as feed. Cows carried out fewer displacements of healthy herd mates from the feeder in the days preceding diagnosis with mastitis (Sepúlveda-Varas et al., 2016) or ketosis (Goldhawk et al., 2009). Chopra et al. (2020) found no significant differences between lame and non-lame dairy cows when considering individual-level social interactions across a large barn-housed herd over a month-long study, although high stocking rates in this herd may have reduced the opportunities for social isolation.

It is increasingly clear that changes in both the individual and social activities of dairy cows, and other livestock species, are influenced by health status and therefore offer opportunities for us to identify early indicators of disease onset, including lameness.

4 Non-automated detection of lameness

4.1 Ad hoc *and visual detection by farmers*

The manual visual identification of lame cows occurs in an *ad hoc* and informal manner on many dairy farms. Farmers identify changes in the gait of the cows as they herd them between different locations, or as they move amongst them in the sheds or pasture. This approach can be effective, especially where herd sizes are relatively small, the prevalence of lameness is low or the herdsperson is highly attentive. However, in a relatively short number of years, both milk yields and herd sizes have increased significantly, making manual identification of disease harder. Over the past 20 years, the UK average dairy herd size and annual yield per cow have increased from 77 cows producing 5789 L in 1998 to 148 cows producing 7960 L in 2018 (AHDB, 2021a,b). Lameness is typically a disease of high yielding cows, which means many farms are now likely to have an increased risk of greater numbers of lame cows in their herds. As herd size increases, a conscious effort must be made to view all cows and not just those who show clear behavioural differences in the eyes of the farmer. While severely lame cows are more likely to be found towards the end of the milking order, mild and moderately lame cows are distributed throughout the milking order (Main et al., 2010), and it is these cows that are a priority to treat promptly in order to prevent the irreversible changes to the claw horn that have such negative impacts on cow longevity. The underestimation of lameness by farmers asked to estimate the number of lame cows in their herd on a given day, indicates that these *ad hoc* and informal approaches are not entirely successful (Leach et al., 2010; Higginson Cutler et al., 2017).

4.2 Formalised mobility and locomotion scoring

Formalising the visual identification of lame cows using a defined scoring system to assess the entire herd at regular intervals, enables the identification of milder lameness cases (e.g. Chapinal et al. 2009). To support farmers in the UK, the national levy board AHDB Dairy supported the development of a mobility scoring system (MSS) aimed at farmers, with descriptors not only of the visual identifiers (see Fig. 2a) but also suggested actions for each score (AHDB, 2020). A systematic review and meta-analysis of lameness assessment methods in dairy research in the UK were undertaken by Afonso et al. (2020), who reported that although the AHDB MSS was the second most common scoring system, it was only used in 10% of the research studies considered, although its more general

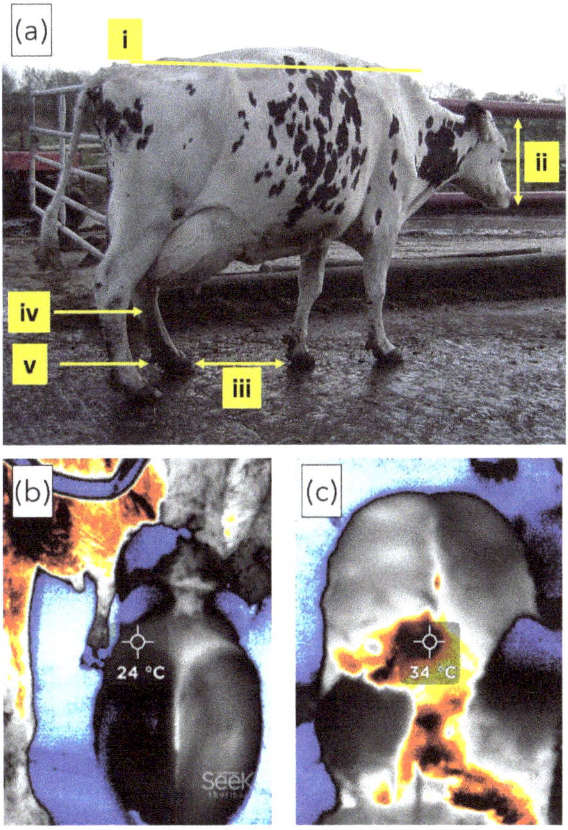

Figure 2 (a) Examples of deviations from normal walking posture and gait in a dairy cow which may be detected by manual observation and mobility scoring, or through automated lameness monitoring systems. This cow was scored as mobility score 3 (AHDB dairy mobility score) so displayed clear deviations from normal walking behaviour: (i) arched spine, (ii) head low or head bobbing, (iii) shortened stride length, (iv) limp, (v) stiffness in the joint. Cows with mild lesions may not show all signs of lameness and those displayed are often less pronounced. (b) and (c) Examples of thermographic images for (b) a non-lame cow and (c) a cow with claw horn lesions in the lateral claw (dark red colour indicating higher relative heat to surrounding foot). Images captured using the Reveal Pro System by Seek Thermal (USA).

on-farm use outside of research may be higher. While farmers are encouraged to use MSS tools to score cows regularly, the task is time-consuming especially if it were to be carried out every month, or more frequently, which would be most ideal to ensure the prompt treatment of lame cows (Groenevelt et al., 2014). It is also worth considering who carries out the scoring. Someone external to the farm may provide an unbiased approach to the locomotion scoring, but this scorer needs to be trusted and the scores believed by the farmer to be acted upon. During discussions with farmers participating in a lameness study (Leach et al., 2010), a range of reasons were offered for why a cow might walk in a certain way, such as 'she had a bad calving' or 'she has always walked like that'. This application of dispensations to specific cows can affect the treatment decisions made for those individuals.

As an alternative to applying a somewhat subjective gait score, researchers have demonstrated a strong relationship between walking speed and locomotion score (Zillner et al., 2018). While this would improve objectivity, the time taken to record the speed of each individual would be prohibitive and would require additional staff to ensure cows continued to move in front of the observer. The presence of milder lesions is not always accompanied by a corresponding increase in locomotion score, indicating that locomotion scoring even by trained observers may not be sensitive enough to detect all lameness cases (Tadich et al., 2010). To be accepted by farmers, technology-based lameness detection methods need to be sensitive enough to detect early lesions, remove the subjectivity of the human observer, and also specific enough to save the farmer time, especially those with large herd sizes.

Although MSSs have also been developed for sheep (e.g. Kaler et al., 2009), sows (e.g. D'eath, 2012) and finishing pigs (e.g. Main et al., 2000), none are in regular industry use due to concerns regarding inter-observer variation and subjectivity.

5 Automated detection of lameness using non-wearable technology

5.1 Ground reaction force (GRF) systems/load sensors - walk over

Initial use of one dimensional, and later three-dimensional, force plates were aimed at understanding better the forces exerted within the claws of cows, and have contributed to our understanding of factors, such as overgrowth of claw horn, foot trimming and floor surface, all of which may affect these forces (Scott, 1989; van der Tol et al., 2002, 2003, 2005; Walker et al., 2010). More recently Ouweltjes et al. (2019) used a three-dimensional ground force reaction system in combination with computerised tomography scanning to investigate the strains occurring within the soft tissues of the claw while under load, finding higher strains in the typical sole ulcer site (Fig. 1c), adding to our understanding of the aetiology of CHDLs. The first GRF system used to detect lameness in individual cow limbs was that of Rajkondawar et al. (2002), which comprised two parallel platforms each with four load cells over which cows would walk, and vertical GRFs were measured for left and right hind limbs.

Leading on from the work of Rajkondawar et al. (2002), the commercial Stepmetrix (Kruegers Bou-Matic, USA) lameness detection system was launched. Field trials by Bicalho et al. (2007b) compared the accuracy of the Stepmetrix system using the commercial settings with a visual lameness score for their ability to detect painful claw lesions at foot trimming. While the specificity was high (93.8%) the sensitivity of the automated system was very low (22.2%) and was outperformed by the visual lameness score. Modelling of data from this system by Lui et al. (2011) led to improvements in lameness prediction, but sensitivity was only improved to 51.92%. The ability of the lame cow to partially redistribute loading through the non-lame claw on the lame limb may have contributed to the low sensitivity.

Three-dimensional GRF systems have overcome some of the sensitivity issues and Dunthorn et al. (2015) reported 90% sensitivity and 93% specificity for the detection of lameness with painful lesions. Thorup et al. (2014) were able to identify mild lameness cases, albeit in a small number of cows, raising the possibility of the early detection of mild lameness cases. An alternative to the parallel platforms described above was investigated by Pastell et al. (2008b). A flexible electromechanical film sandwiched between robust rubber mats, which detects dynamic forces as a cow moves over it, measures a number of possible gait-related variables with the capability of differentiating between lame and non-lame cows.

5.2 Load sensors – weight distribution

There has been significant progress in the development of lameness detection technology designed for automatic milking systems. Using a 4-balance system which measured distribution of weight across individual limbs, Pastell and co-workers have identified asymmetry of weight-bearing in lame cows (Pastell et al., 2006), developed a probabilistic neural network approach to lameness prediction (Pastell and Kujala, 2007), and then produced a real-time monitoring system which alerts the user to a lame cow (Pastell et al., 2008a). This group have also reported on the possibility of using cumulative sum (CUSUM) charts to identify changes in individual cow weight-bearing over time (Pastell and Madsen, 2008). An experimental four-balance system was also used to investigate the effect of flooring surface (Neveux et al., 2006) and the impact of treatment with a non-steroidal anti-inflammatory drug for pain relief in lame and non-lame cows (Chapinal et al., 2010). Using the same balance system Pastell et al. (2010) reported on the accuracy of lameness detection for two thresholds of lameness severity (NRS > 3 and NRS > 3.5 (Flower and Weary, 2006)) and for three lesions causing lameness, sole ulcer, sole bruising and digital dermatitis, compared with no lesions. One might argue that, for early detection, it would be necessary to assess the ability of the system to identify cows with mild abnormalities of gait, in this case, NRS > 2. The authors did however find a clear relationship between sole ulcer and the leg weight ratio and a weaker relationship for sole bruising, indicating there may be some scope for the early detection of lesions. More generally, Garcia et al. (2014) used discriminant analysis to link the lameness status of individual dairy cows to a range of milk production variables collected automatically by a robotic automatic milking systems (AMS) during milking events. This type of approach has great potential for automated monitoring approaches given the wide implementation of AMS on larger farms, particularly if production data can be automatically combined with more general behavioural data as part of a combined monitoring system.

Load sensor systems have also been developed for sheep with differences observed between healthy hooves and those with infection (Byrne et al., 2019), although the value of this technology for early detection may be limited in a species predominantly managed in extensive grazing settings. A potential commercial application for forceplate technology in pigs is the Sow Stance Information System (SowSIS). Initially developed by Pluym et al. (2013) it has been incorporated into an electronic sow feeding system with recent tests showing promising results for on-farm lameness detection (Briene et al., 2021).

5.3 Pressure-sensitive walkway

The GaitWise system was first described by Van Nuffel et al. (2009). This system records location, time and pressure on sensors embedded in a walkway, as individual cows walk over them. These are used to calculate a range of gait attributes, such as stride length and asymmetry in relative pressure. Asymmetry and speed variables provided the greatest accuracy for identifying lame cows (sensitivity 76–90% and specificity 86–100%) (Maertens et al., 2011). Further developments of the system to measure such variables demonstrated the possibility to detect the lame limb (Van Nuffel et al., 2013) and mild lameness cases (Van Nuffel et al., 2015). The size and cost of this and other walk over systems have to date hindered the widespread uptake of such technologies on commercial farms. Despite efforts to optimise the size of the pressure mat in the GaitWise system, the minimum length is 3.28 m to accommodate a full gait cycle (Van De Gucht et al., 2017), which must be accommodated within the existing farm layout.

Pressure mats have also been used for lameness detection in pigs (Mohling et al., 2014), but with the same logistical difficulties requiring the animals to walk evenly down a walkway.

5.4 Thermography

Infrared thermography (IRT) has been successfully validated for the identification of claw lesions (Fig. 2b,c). Stokes et al. (2012) reported IRT sensitivity and specificity of 93% and 54% for clean feet and 80% and 73% for dirty feet when measuring from the plantar aspect of the foot. In this study, it was not possible to distinguish between CHDLs and digital dermatitis, both leading to elevated claw temperature as a result of inflammatory responses to infection. Claw lesions were also identified by measuring the temperature difference between the coronary band and adjacent skin, with a sensitivity and specificity of 85.7% and 55.9% with a threshold of 0.64°C before foot trimming, and 80.0% and 82.9% with a threshold of 1.09°C after foot trimming (Alsaaod and Büscher, 2012). Similar performance was reported for the detection of digital dermatitis, also measuring at the coronary band and adjacent skin (sensitivity 89.1% and specificity 66.6%) with a cut off of a 0.99°C difference between skin and coronary band (Alsaaod et al., 2014). Harris-Bridge et al. (2018) compared the accuracy of measurements at both the coronary band and the plantar aspect of the heel bulb and concluded that the coronary band provided the most accurate predictions. In a 6-month study on a 990-cow farm, Lin et al. (2018) found that cows with lameness had significantly elevated foot temperatures and this correlated significantly with the observed AHDB mobility score.

While IRT techniques have great potential to be accurate tools for lameness prediction they have not yet been developed into fully automated monitoring systems. Such techniques would probably be best suited to AMS where cameras could be focused on the coronary band of each cow entering the robotic milker. Finding suitable locations within non-AMS farms either in stalls or in the milking parlour could prove challenging.

5.5 Kinematics

In early studies, researchers used video images to look in greater detail at alterations in the gait of cows, both with and without lameness (see Fig. 2a). Cows lame with sole ulcers were reported to walk slower, take shorter strides and spend longer with three limbs in contact with the ground in each gait cycle (Flower et al., 2005). However, these authors completed all assessments of the slow-motion videos manually and the cows used were all fitted with reflective markers on the key leg joints to provide contrast in the video. Song et al. (2008) took the first steps towards continuous automatic recording by converting videos to a binary image from which the location of each hoof was determined from the position of the pixels it occupied. They obtained a positive correlation between visual locomotion scores and a measure of tracking up, step overlap. However, further studies revealed that step overlap alone would not be sufficient for automatic lameness detection (Pluk et al., 2010). In a follow-on study, Pluk et al. (2012) were able to synchronise continuous videos with data from a pressure mat, and investigated changes in gait characteristics associated with increases in visual lameness scores, indicating either onset or deterioration in lameness status).

In addition to foot placement, a number of visual lameness scores also incorporate observations of back arch (Fig. 2a), which is indicative of a shift in weight towards the front feet to relieve pressure and pain from the hind feet. Researchers have therefore applied video analysis technologies to the curvature of the spine of cows while walking. Poursaberi et al. (2010) achieved an accuracy of 94.7% (sensitivity 100% and specificity 97.6%) by fitting a circle through specific points on the spine, to estimate a curve using images of the side of the cow from a 2D camera and comparing with a visual locomotion score. These methods were adapted by Viazzi et al. (2013) to take into account the movement of the spine and head of the cow. On a binary scale of lame versus non-lame, the sensitivity of lameness detection was 76%; this was improved where thresholds for individual cows were used rather than a group-based threshold. However, low sensitivity of 25% was reported for the prediction of lame cows where the results were validated against a 3-point scale (non-lame, lame and severely lame), suggesting there may still be challenges for the early detection of lameness due to the subtle changes in posture seen at this stage. Despite some promising results, the use of a 2D side view presented a number of challenges for successful on-farm deployment, such as alterations to the background and variable lighting conditions (Viazzi et al., 2014). To address these issues, the authors compared the existing system with a top view 3D camera system and reported accuracy, sensitivity and specificity values of 91%, 76% and 95% for the 2D side view system and 90%, 82% and 91% for the 3D top view system. The 3D cameras were less susceptible to the background changes and shadows than the 2D camera allowing the segmentation of images to be carried out automatically. In addition, some analysis of cows side by side in the image was possible. The main drawback to the 3D system was the oversensitivity of the camera to natural light which forced the researchers to record the videos during the hours of darkness.

Three-dimensional cameras were also used by Abdul Jabbar et al. (2017) to track the movements of the spine and hook bones in a study of 22 cows. To assess the possibility of detecting lameness early, a threshold for the lameness of locomotion score 2 and above (Sprecher et al., 1997) was used and accuracy, sensitivity and specificity values of 100%, 75% and 95.7% were reported. The applicability of a 3D camera system for automatic detection of lameness on a commercial farm was assessed by Van Hertem et al. (2018). The authors developed a fully automated process that took the video images and allocated a lameness score which was tested at the exit of a rotary parlour on a commercial dairy farm. The system was only able to automatically score around 49% of cows from each milking session and this was attributed to the high rates of flow within the raceways leading to short times between radio-frequency identification (RFID) readings for many cows. These issues are likely to be greater in farms with parlour designs where cows exit in batches. However, due to the continuous nature of the video recordings, 79.1% of cows were automatically scored at least five times a week, which would certainly be frequent enough for early detection of lameness. Jiang et al. (2020) also used a deep learning neural network method on video frames of cow motion. Their optical flow convolution model was able to recognise actions characteristic of lame cows with an accuracy of 98.24% at a high frame rate. Piette et al. (2020) used historical back posture data for individual cows to determine lameness using an automated camera system. For individual thresholds, deviations from the healthy reference were identified with a specificity of 82.3%, a sensitivity of 79%, an accuracy of 82% and a precision of 36.1% when the length of the healthy reference window was not limited. When the length of the healthy reference window was varied the algorithm performance was maximised with a reference window of 200 days.

As an alternative to video analysis, identification of gait changes and lameness in cattle has also been achieved using acoustic analysis of cows walking (Volkmann et al., 2021). Camera-based kinematic systems have also been tested in pigs with gait measures including step-to-stride length ratio have been used to identify lameness with some success (Conte et al., 2014; Stavrakakis et al., 2015).

6 Automated detection of lameness using animal-mounted accelerometers

The use of animal-mounted sensors to automatically collect behavioural data has dramatically increased over the past 10–15 years, mainly because of the accessibility and affordability of new technology including accelerometers and spatial positioning systems (as discussed in Section 7). In general, there are two main approaches for analysing and interpreting the data output from these sensors. In the first approach, classification methods (statistical and/or machine-learning tools) are used to process data signals and patterns into known behaviours, such as feeding, lying and standing (see Fig. 3a), and differences in these behaviours are then used to detect or predict lameness (c.f. Section 3). In the second type of approach, sensor data is analysed (e.g. patterns and signals

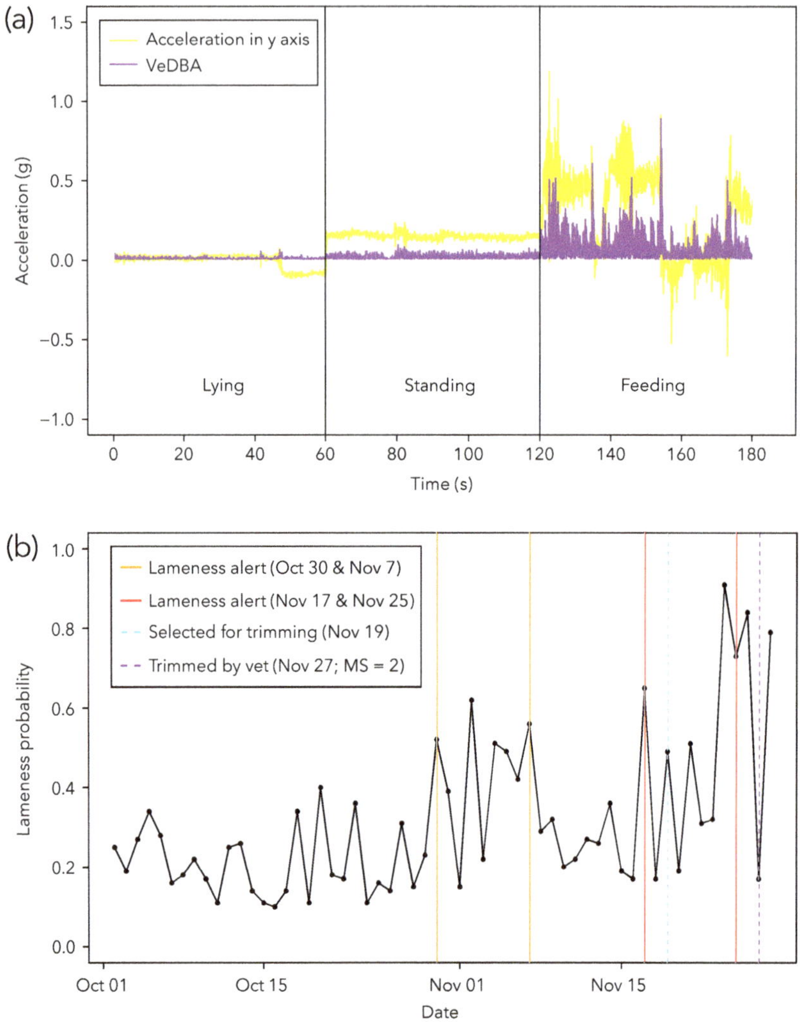

Figure 3 Illustrative examples of how accelerometers can be used in an automated way to determine behaviour and predict lameness. (a) Example output of VeDBA (vector of dynamic body acceleration, a measure of the total acceleration) and acceleration in the vertical (y) axis from a 50 Hz triaxial neck-mounted accelerometer (recreated using data originally published in Vázquez Diosdado et al., 2015) highlighting how lying, standing, and feeding behaviours in dairy cows can be classified (using a decision-tree in this example). By monitoring differences in these behaviours over time it may be possible to predict lameness status. (b) Example output from a commercial leg-mounted accelerometer system for predicting lameness in dairy cows (CowAlert by IceRobotics Ltd, UK). The system determines patterns of behaviours, such as lying and standing, and combines these with data on gait and other farm production data, to produce a daily probability of individual cow lameness; automatic lameness alerts are generated once a certain probability threshold is exceeded. In this example, prior to the vet trimming, the cow was observed to have an AHDB mobility score (MS) of 2 (impaired mobility: likely to be lame and benefit from treatment).

in acceleration during gait analysis) and used directly to detect or predict lameness; the intermediate behavioural classification stage is not undertaken (this may reduce the potential for classification errors being introduced, but means results may be more difficult to interpret or explain to users). Many commercially available systems use a combination of these approaches. For example, the CowAlert system (IceTag and IceQube) by IceRobotics Ltd, UK, uses a leg-mounted accelerometer to classify patterns of lying and standing, and combines these with other available production data, to predict individual cow lameness (Fig. 3b).

6.1 Accelerometers: behavioural classification and monitoring

The concept of wearable technologies in cattle began with the development of pedometers, during the 1980s and early 1990s, which measured changes in step counts to detect oestrous. Accuracy of oestrus detection with these early technologies varied from 22% to 100%, mostly impaired by high numbers of false positives resulting from either technical limitations or adverse on-farm conditions (Lehrer et al., 1992). More recently, the development of tri-axial accelerometers has provided the opportunity for more accurate measurement of activity and the opportunity to infer a number of other behaviours. No longer restrained to the legs, sensors have also been mounted on neck collars and ear tags on both cattle, sheep and pigs, although the position of the sensors affects the type of behaviours that can be inferred. When using leg-mounted accelerometers researchers are constrained to activity level and lying and standing-related variables, unless the pedometers are used in combination with other technologies. Conversely, neck-mounted collars on cattle are able to report feeding and rumination behaviours and overall activity (Fig. 3a) but are poor at identifying lying behaviour (Martiskainen et al., 2009; Vázquez Diosdado et al., 2015). More recent validations of ear-tag-based accelerometers indicate that a wider range of behaviours can be captured including ruminating, standing and lying (Bikker et al., 2014; Roland et al., 2018). In a study of extensively grazed sheep in Australia, neck- and ear-mounted accelerometers were effective at detecting grazing behaviour (Barwick et al., 2018), with the latter also being applied in extensively grazed sheep in New Zealand (Fogarty et al., 2020).

Despite a large number of studies reporting associations between lameness and lying behaviour recorded with leg-mounted accelerometers, relatively few have considered the predictive capabilities of these data. Byabazaire et al. (2019) reported predicting the lameness status of 91% of cows 1 day before and 87% of cows 3 days before being visually identified by the farmer or researcher, using the commercially available Track-a-Cow Long-Range Pedometer (LRP, ENGS Systems, Israel). Alsaaod et al. (2012) reported a 76% accuracy for predicting lameness when using six variables related to lying bouts derived from a leg-mounted pedometer. To achieve this, the authors considered the deviation of the animal from its normal behaviour rather than using an absolute threshold value, due to the wide variation in lying behaviours between cows. This variation was also noted by Westin et al. (2016), who suggested that lameness did not account for all of the variations in lying behaviour, with parity and stage of lactation also influencing behaviour. Variation of behaviour with weather conditions for cows at pasture has also been reported (Thompson et al., 2019), which might explain why most studies using pedometers for lameness prediction do so in combination with other technologies. Accelerometer activity was one of three variables, used in combination with liveweight and milking order, to predict lameness in a model by Kamphuis et al. (2013). However, this only gave a sensitivity of 50% with a specificity of 80% when a mild lameness threshold was used. Using data from automatic feeding troughs, AMS and leg-mounted accelerometers, Grimm et al. (2019) reported lameness prediction sensitivity and specificity of 94% and 81%, with the number of meals, average feed intake per meal, and average duration of a meal reported as the most important variables for lameness detection. de Mol et al. (2013) also combined data from AMS with leg-mounted accelerometers and reported sensitivity of 85.5% and specificity of 88.8%. Nechanitzky et al. (2016) collected data from multiple sources, including feeding behaviours from a noseband with sensors, heart rate variability, leg weight distribution from a pressure mat and lying behaviours from a leg-mounted accelerometer. While a number of the variables had significant associations with lameness, the optimal model included the standard deviation of the weight being taken by the limb bearing the least weight and lying time, achieving a sensitivity of 94%, and a specificity of 80%.

A range of neck-mounted accelerometers is commercially available which offer to predict, oestrous, rumination, and other health issues. However, there are limited studies validating their use in lameness detection. Where this has occurred they have typically been used in combination with other sensors. Van Hertem et al. (2013) included inline milk records, and leg and neck-mounted accelerometer data. Their optimal model contained four milk-related variables and three activity and rumination variables from the neck-mounted accelerometers, giving a sensitivity of 89% and a specificity of 85%.

Kaler et al. (2020) used an accelerometer- and gyroscope-based ear sensor with a sampling frequency of 16 Hz to develop and compare machine-learning algorithms that can differentiate between lame and non-lame sheep during three different activities including walking (76.83% accuracy), standing (81.15% accuracy) and lying (84.91% accuracy), with an overall accuracy of 80% for detecting lameness. Leg-mounted accelerometers have also been used in pigs to measure lying, standing and stepping behaviours, which have subsequently been used to identify lameness (Grégoire et al., 2013; Conte et al., 2014). Traulsen et al. (2016) used an ear-tag solution to address possible husbandry concerns from other pigs biting the sensor mounted on a neck or leg position. This device combined accelerometers with local position technology (discussed in Section 7) to assess lameness in sows but concluded that only positional data was useful for lameness detection, although this research group subsequently used a wavelet transform method on the acceleration data to classify six out of seven lame sows in a group of 14 (Scheel et al., 2017).

6.2 Accelerometers: gait analysis

In addition to generating activity data from which behaviours can be classified or inferred, the data from tri-axial accelerometers can also be adapted to directly measure forces exerted by a cows limb during the process of walking. Beer et al. (2016) measured elements of the gait cycle using a single low-frequency accelerometer attached to a hind leg; stride number, distance and duration, along with walking speed, all had a significant association with lameness but were not included in the optimal model. Haladjian et al. (2018) also applied a single accelerometer to a hind leg but used a higher frequency (100 Hz) and were able to measure the elements of the complete gait cycle. Here the author's simulated unilateral lameness by applying a plastic block to a single claw. Abnormal gait was predicted with an average sensitivity of 74.2% and specificity of 91.6%. Using two high frequency (400 Hz) accelerometers attached to each hind leg, Alsaaod et al. (2017) were able to measure the elements of a complete gait cycle. Using a moderate change in the stance phase duration between each limb the authors were able to predict both lameness (NRS > 3) and the presence of unilateral lesions, including one cow with a lesion but not detectable lameness, with a sensitivity and specificity of 100%. A similar analysis of the full gait cycle was undertaken by Jarchi et al. (2021), who used 16 Hz accelerometers mounted on all four legs of individual cows. Using hierarchical deep learning and syncro-squeezed wavelet transforms enabled the authors to determine periods of movement characteristic of lameness with sensitivity and specificity both over 90%. While such systems show great potential, the requirement to add multiple sensors to each animal may not be favoured by farmers.

In their review of accelerometer-based lameness detection methods, O'Leary et al. (2020) argue that the use of accelerometers to measure alterations in gait offers the greatest promise for the prediction of lameness when compared with behaviour monitoring, due to the higher degrees of accuracy reported in the literature to that point. However, systems that provide multiple services in parallel, from the detection of oestrous to the detection of additional diseases such as lameness and ketosis, may be more highly valued by farmers. In addition, the opportunities to integrate either multiple sensor types into one device, or to integrate available farm production and health data with sensor data are being investigated.

7 Automated detection of lameness using spatial positioning systems

Spatial positioning systems are seeing increasing use as part of on-farm automated monitoring for both research and commercial purposes. These can range from very simple sensors that only record proximity encounters, to full tracking systems using video or local and global positioning systems.

7.1 Proximity sensors

The simplest type of spatial positioning system uses RFID, where an electromagnetic field (radio-wave) automatically identifies nearby tags within given proximity and hence positional information is relative rather than absolute. RFID sensors are cheap and simple to set up and use, and hence these systems are the most common to currently be deployed on farms and to track animals in other contexts (Floyd, 2015). There are typically two main uses for RFID systems for monitoring livestock (Voulodimos et al., 2010). In the first type of system an active (powered) sensor known as a 'reader' is deployed at a fixed location (e.g. the entrance to a milking station, or near a feed area or water trough) and this records when passive (unpowered) animal-mounted tags are within a defined proximity. For example, RFID and ultra-high frequency RFID ear tags have been used to monitor feeding and drinking behaviour in group-housed pigs (Maselyne et al., 2016; Adrion et al., 2018). The second type of system uses powered tags mounted on all animals in a group (and on relevant environmental features); each sensor can then record when other sensors are within defined proximity allowing for detection and monitoring of social interactions between individual animals (e.g. Boyland et al., 2016) as well as key environmental interactions.

While proximity sensors have a number of advantages (the primary being low cost and simplicity), the obvious disadvantage is that they are unable to provide full spatial location information and hence it is not possible to fully track movement and space-use continuously for individuals and herd groups within their local environment.

7.2 Spatial location systems

Outdoor location tracking systems, such as the global positioning system (GPS), have been used to collect movement and space-use data on a wide range of wild animals (Kays et al., 2015) and for animals grazing at pasture (Williams et al., 2016; Schieltz et al., 2017). However, GPS and related systems require strong satellite signals; mean average location errors for GPS are typically around 5 m in commercial systems and can be as high as 19.6 m (Duncan et al., 2013). Crucially, GPS does not function well (or at all) in the indoor barn environments typical of most commercial dairy farms.

Instead of GPS, real-time location systems (RTLS) that use wireless radio networks to triangulate spatial position across a network of fixed and animal-mounted sensors can be used indoors to create a 'local positioning system' (LPS). Validation studies across a range of indoor LPS have reported mean errors better than GPS (typically 2–3 m, although 0.5 m mean error may be achievable with the most recent systems) (Broom, 2006; Gygax et al., 2007; Huhtala et al., 2007; Asikainen et al., 2013; Porto et al., 2014; Tullo et al., 2016; Wolfger et al., 2017; Pastell et al., 2018; Rice et al., 2020). Commercially available LPS systems usually combine location tracking with an accelerometer to provide additional activity data (see Section 6), and include Smartbow (Zoetis, USA; Perisho and Hajnal, 2021), CowView (GEA Technologies, Germany; Veissier et al., 2017), Ubisense indoor positioning RTLS (Ubisense, France; Pastell and Frondelius, 2018), while in our own research we have used the Series 500 LPS sensor system produced by Omnisense Ltd, UK (Barker et al., 2018). While combined LPS and accelerometer systems allow much richer data to be collected than simple proximity sensors (or accelerometer-only systems), they typically require costly fixed infrastructure and regular management input from expert technicians. Hence, to date, they have had less commercial uptake than simpler sensor systems, particularly on smaller farms.

LPS (and GPS) provide detailed data on animal movement and space-use. At a basic level, this takes the form of a time series of (x, y) (and possibly z) location coordinates collected at a regular frequency (typically around 0.1–1 Hz, as a higher frequency is not usually necessary given the low movement speed of livestock). Once location data for individual animals is linked to maps of the local (barn or pasture) environment, and to the time series of locations of other individuals in the herd, then it is possible to determine a range of additional metrics. These include movement metrics such as total distance moved, movement bouts (stop-start), steps taken, movement speed, turning angles and turning rate, movement direction, etc., which can be analysed using tools borrowed from the fields of random walk theory and movement ecology including hidden Markov Models (HMMs), state-space models, and other movement analysis techniques (Codling et al., 2008; Patterson et al., 2008; Langrock et al., 2012). Patterns and signals in the movement data can subsequently be explored and linked to health conditions and more general behavioural patterns (as described in Section 3). Space-use metrics such as range size, site preference, site fidelity, etc., allow further insights into behaviour that can be linked to health or other individual characteristics (Vázquez Diosdado et al., 2018). Location data can also be directly mapped to specific features of interest within the local environment (e.g. feeding area and cubicles area) and hence can be used to directly classify behaviour such as feeding (Pastell and Frondelius, 2018), particularly if combined with accelerometer signals (Fig. 4a; Barker et al., 2018). Social interactions can also be determined: proximity interactions can easily be calculated by comparing locations of animals within the herd over time, and these can be used to determine social networks and group dynamics (Chopra et al., 2020; Rocha et al., 2020). Location data can also provide additional information around social interactions, such as approach avoidance, or the direction and facing of animals during a proximity interaction (Nagy et al., 2013).

It should be noted that similar spatial location data could also potentially be collected and analysed using a video tracking system (e.g. Guzhva et al., 2018; Salau and Krieter, 2020). Such systems can provide valuable additional information about animal behaviour and social interactions, but typically also require large, and potentially expensive, fixed networks of mounted cameras to cover the movement range of animals being monitored. Automated monitoring and tracking of individuals within each video frame requires sophisticated image recognition analysis tools and can be error-prone with misrecognition of individuals a problem (Ter-Sarkisov et al., 2017; Guzhva et al., 2018). In addition, analysing and storing video data can require enormous amounts of processing and storage compared to LPS systems that only record location coordinates.

7.3 Automated monitoring of behaviour and health using spatial data

Given the detailed behavioural information available from combined LPS and accelerometer systems, and the number of recent validation studies of indoor farm-based RTLS (see references above), it is perhaps surprising that there have not been more studies in the literature that directly link observed movement and space-use metrics with behaviour and health in managed livestock.

Veissier et al. (2017) used the CowView RTLS (GEA Farm Technologies, Germany) to monitor the location of 350 cows in a Danish dairy farm for 5 months, with a reported accuracy of <0.5 m deviance. Cow activity was inferred from the location data using a simple decision rule: if the cow was observed in a cubicle, it was classified as resting, if the cow was observed within the feeding area, it was classified as feeding, otherwise, it was classified as in alleys. Factorial correspondence analysis was then used to determine patterns of activity over time allowing the authors to chart a circadian rhythm of activity. This was then used to highlight differences in activity patterns for cows in oestrus, or with

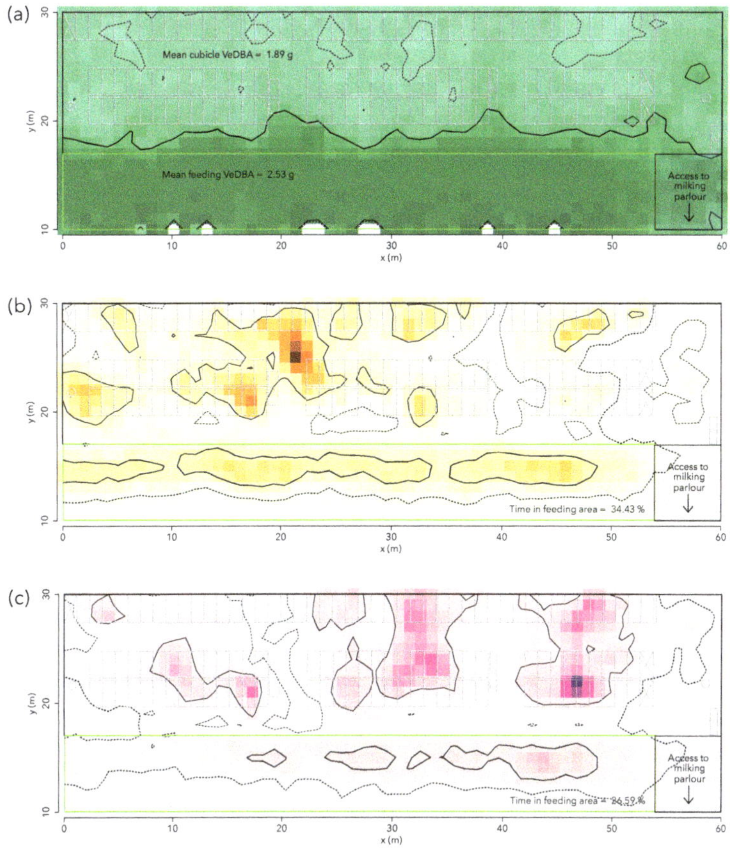

Figure 4 Illustrative examples of dairy cow space-use and activity collected using a local positioning sensor system (Omnisense Ltd, UK). (a) Spatial heat-map of mean activity (measured using a triaxial accelerometer) for a fully barn-housed group of 20 cows over 5 days. Although activity is clearly higher in the feeding area, there are also areas of high activity in other parts of the barn (e.g. in some passageways and near the brush in the upper-right of the barn). Combining information on activity and spatial position can give a more accurate classification of behaviours such as feeding, differences in which can then be used to predict lameness status. (b) and (c) Heat-maps of aggregated space use for (b) 9 non-lame cows and (c) 11 lame cows over 5 days. The lame cows clearly spend less time in the feeding area than non-lame cows and are also more likely to be found on the right-hand side of the barn. All plots have been reproduced using data originally published in Barker et al. (2018) and Vázquez Diosdado et al. (2018). In all plots, the feeding area is indicated with a bright green border (10 m < y < 17 m); cubicles are marked with a grey outline; B = brush, W = water trough. The right-hand side of the barn (x > 54 m, y > 17 m) is used as a 'loafing' area; the area in the lower-right (x > 54 m, y < 17 m) is the passageway providing access to the milking parlour and is not accessible to cows outside milking periods. In all plots, the solid contours highlight the 'core' range of the group (the cells which cumulatively total to the highest 50% of activity or space use), while the dashed contours highlight the 'full' range (cumulative total of 95% of activity or space use), see Vázquez Diosdado et al. (2018) for details.

specific health conditions such as mastitis or lameness. Importantly, variations in circadian activity were observed 1–2 days before conditions were noticed by the farmer, which suggests that this methodology may allow early identification of disease states.

Pastell and Frondelius (2018) used the Ubisense (France) indoor positioning RTLS to collect positional data from a group of 50 cows over 7 days. An HMM was then used to classify feeding behaviour with reported accuracy 97.6%, sensitivity 95.3% and specificity 97.9%. Their HMM was also able to estimate the mean bout duration and the number of feeding bouts. Although their study did not directly consider health, the fact that feeding behaviour is known to change with lameness (see Section 3), suggests this could also be a useful approach that could be further developed as an automated lameness monitoring tool.

In our own research, we have used the Omnisense Ltd (UK) Series 500 RTLS to track the behaviour and activity of groups of dairy cows. In Barker et al. (2018), we undertook a cross-sectional trial with two groups of dairy cows (10 lame and 10 non-lame; although later work highlighted that one cow incorrectly included in the non-lame group was actually lame). High resolution (0.1 Hz) data on location and activity (using the inbuilt accelerometer that forms part of the Omnisense sensors) were collected over 5 days, and a simple decision tree was developed, based on the level of activity and spatial location, to identify behavioural bouts of feeding, non-feeding, and periods out of the barn for milking. Subsequently, for each behaviour, we determined the mean number of bouts, the mean bout duration, and the mean total duration across all bouts on a daily basis, and also separately for the time periods in between milking (morning, afternoon and night). Importantly, we found that the mean total daily feeding duration was significantly lower for lame cows compared with non-lame cows. Behaviour was also affected by the time of day, with significantly lower mean total duration of feeding and higher total duration of non-feeding in the afternoons for lame cows, compared with non-lame cows. Figure 4a highlights the value of combining location and activity data in order to improve the accuracy of behavioural classification, and these types of 'activityscapes' may be a useful tool to determine behavioural changes linked to lameness (and other diseases) moving forwards.

In a follow-on study (Vázquez Diosdado et al., 2018), we analysed the same data set with more focus on the fine-scale space-use metrics of each individual animal. By creating space-use heat-maps for each individual cow over each day of the study we were able to determine the 'core' and 'full' range sizes (i.e. the size and location(s) within the barn where the animal spends the most time), the site fidelity (how often individuals revisit particular areas), and the time spent in areas of interest (such as the feeding or cubicle areas). Although there were no significant differences in the total distance moved between the lame and non-lame groups, there were clear differences identified in the nature of their movement and space-use. In particular, lame cows had a smaller range size, spent less time in the feeding area, and were more likely to be found on the right-hand side of the barn (near to the access passageway to the milking parlour), see Fig. 4a and b. Using these results we were able to demonstrate how a very simple predictive model (based only on observed range size and time spent in the feeding area) was able to accurately predict the lameness status of all individuals in the study (once the lame cow in the incorrect group is accounted for).

The RTLS technology and related methods and approaches described above are directly transferable to other livestock species where animal-mounted sensors are practical. Although there have been few RTLS indoor studies reported for species other than cattle, a recent feasibility study by Zhuang et al. (2020) demonstrated how an ultra-wideband indoor positioning RTLS can accurately track the location of group-housed sows. Perisho and Hajnal (2021) used the Smartbow (Zoetis, USA) indoor positioning system to track the location of swine during and after social reintroduction events, and were able to demonstrate how a range of movement metrics could be used to predict reproduction success.

8 Conclusion and future trends

Lameness continues to be a high prevalence disease in sheep, pigs, and particularly dairy cows, with major economic impact and a highly detrimental effect on animal welfare. New practical automated monitoring approaches are needed on farms where early interventions are essential. Advances in technology and scientific research are the keys to new solutions; we have reviewed load sensors, pressure-sensitive walkways, thermography, kinematics, video, accelerometers and RTLS position technologies that have all contributed to a better scientific understanding of lameness. There are several factors to address in order to see progress towards the use of these technologies in an automated way on the farm. Systems need to be robust and practical for the challenging farm environment of hard metallic edges, destructive animals, and limited staff time and technological expertise. This can be achieved by integrating lameness detection technologies into existing farm settings, for example by taking advantage of a milking return passage such as the Stepmetrix for dairy cattle (Bicalho et al., 2007b), or adding load sensors to an electronic sow feeding station like the SowSIS system (Briene et al., 2021). For a long time dairy cows have worn collars for electronic identification; combining animal-mounted sensors with developments in battery technology and energy-saving algorithms will allow for long term use of accelerometers and spatial positioning technologies. In addition to being practical, for wide industry uptake, the technology has to be relatively inexpensive or, at least, value for money with clear evidence of useful benefits that can only be supported through detailed research in real farm environments. A key element moving forwards is to train, educate and support farmers in the use of such technology, and the data it generates.

Future trends lie in the development of both automated lameness detection methods and, more generally, automated approaches for simultaneous multiple-disease (and behaviour) detection. In addition, systems that work effectively in both the housed environment and at pasture, switching seamlessly between the two, will be required to meet consumer demands. While there have been several studies that have differentiated the activities of lame and non-lame animals in case-controlled or cross-sectional trials, there is a need to move more towards longitudinal trials where more detailed information regarding the onset of the disease can be elucidated, leading to more potential for earlier and more precise diagnosis. In the context of scientific research, combining automatic monitoring approaches with new mathematical, statistical and machine-learning techniques will also provide opportunities to develop a better fundamental understanding of the links between animal health, including lameness, and behaviour. For example, how factorial correspondence analysis can be used to show how lameness affects circadian rhythms (Veissier et al., 2017), or how fine-scale space-use metrics highlight the differences in range size (Vázquez Diosdado et al., 2018) between lame and non-lame dairy cows. Ultimately we expect future developments in precision livestock farming will enable the move towards holistic monitoring and management systems that combine different sensor technologies with new predictive algorithms as part of a wider farm-based decision-support system. Such systems will use a range of behavioural and other metrics to differentiate between different diseases that may affect livestock production and welfare, predicting and flagging those individual animals most at risk, and becoming an essential tool in farm management and animal production research.

9 Where to look for further information

9.1 Further reading

- Healthy Feet Website is an output from a large UK research project and contains numerous links to lameness related materials: https://www.cattle-lameness.org.uk/.
- AHDB dairy has have a library of useful information for farmers and practitioners as part of the Healthy Feet Programme: https://ahdb.org.uk/knowledge-library/lameness-in-cows-the-healthy-feet-programme.

9.2 Key journals and conferences

- EA-PLF: European Association of Precision Livestock Farming, host the European Conference on Precision Livestock Farming every two years: www.eaplf.eu/.
- AC-PLF: Asian Conference on Precision Livestock Farming; 2020 website: http://acplf2020.csp.escience.cn/.
- Lameness in Ruminants – International Symposium and Conference is held every two years; 2022 website: https://lamenessinruminants2022.com/.
- The British Society for Animal Science (BSAS) and European Federation of Animal Science (EAAP) also host regular meeting which increasingly embrace research on PLF and lameness.

9.3 Major international research projects

- SMARTCOW is an EU framework project which scientific and technical skills in animal nutrition, health and welfare and ethics and included a work package focused on sensors technologies and automatic recordings of physiological and behavioural traits: www.smartcow.eu.
- CLEARFARM is an EU funded in which PLF will be used to increase animal welfare in the whole production chain of dairy cows and pigs. It moves beyond the farm and produce a technology driven platform which can support decision making for both farmers and consumers: www.clearfarm.eu.
- TECHCARE is an EU funded project aiming to develop early warning systems for welfare issues in small ruminants: www.techcare-project.eu/.
- The DairyLand Initiative Lifestep Lameness Model puts research and clinical best practice into practice supporting farmers with lameness prevention. The programme includes research on camera-based technologies for lameness detection and improving lameness treatment: https://thedairylandinitiative.vetmed.wisc.edu/home/lifestep-lameness-module/.

10 References

Abdul Jabbar, K., Hansen, M. F., Smith, M. L. and Smith, L. N. 2017. Early and non-intrusive lameness detection in dairy cows using 3-dimensional video. *Biosyst. Eng.* 153, 63–69. https://doi.org/10.1016/j.biosystemseng.2016.09.017.

Adrion, F., Kapun, A., Eckert, F., Holland, E.-M., Staiger, M., Götz, S. and Gallmann, E. 2018. Monitoring trough visits of growing-finishing pigs with UHF-RFID. *Comput. Electron. Agric.* 144, 144–153. https://doi.org/10.1016/j.compag.2017.11.036.

Afonso, J. S., Bruce, M., Keating, P., Raboisson, D., Clough, H., Oikonomou, G. and Rushton, J. 2020. Profiling detection and classification of lameness methods in British dairy cattle research: A systematic review and meta-analysis. *Front. Vet. Sci.* 7, 542. https://doi.org/10.3389/fvets.2020.00542.

AHDB. 2020. Mobility scoring: How to score your cows. AHDB [WWW Document]. Available at: https://ahdb.org.uk/knowledge-library/mobility-scoring-how-to-score-your-cows (accessed 4 November 2021).

AHDB. 2021a. UK milk yield. AHDB [WWW Document]. Available at: https://ahdb.org.uk/dairy/uk-milk-yield (accessed 4 November 2021).

AHDB. 2021b. UK and EU cow numbers. AHDB [WWW Document]. Available at: https://ahdb.org.uk/dairy/uk-and-eu-cow-numbers (accessed 4 November 2021).

Alawneh, J. I., Laven, R. A. and Stevenson, M. A. 2011. The effect of lameness on the fertility of dairy cattle in a seasonally breeding pasture-based system. *J. Dairy Sci.* 94(11), 5487–5493. https://doi.org/10.3168/jds.2011-4395.

Alsaaod, M. and Büscher, W. 2012. Detection of hoof lesions using digital infrared thermography in dairy cows. *J. Dairy Sci.* 95(2), 735–742. https://doi.org/10.3168/jds.2011-4762.

Alsaaod, M., Luternauer, M., Hausegger, T., Kredel, R. and Steiner, A. 2017. The cow pedogram–Analysis of gait cycle variables allows the detection of lameness and foot pathologies. *J. Dairy Sci.* 100(2), 1417–1426. https://doi.org/10.3168/jds.2016-11678.

Alsaaod, M., Römer, C., Kleinmanns, J., Hendriksen, K., Rose-Meierhöfer, S., Plümer, L. and Büscher, W. 2012. Electronic detection of lameness in dairy cows through measuring pedometric activity and lying behavior. *Appl. Anim. Behav. Sci.* 142(3–4), 134–141. https://doi.org/10.1016/j.applanim.2012.10.001.

Alsaaod, M., Syring, C., Dietrich, J., Doherr, M. G., Gujan, T. and Steiner, A. 2014. A field trial of infrared thermography as a non-invasive diagnostic tool for early detection of digital dermatitis in dairy cows. *Vet. J.* 199(2), 281–285. https://doi.org/10.1016/j.tvjl.2013.11.028.

Amory, J. R., Barker, Z. E., Wright, J. L., Mason, S. A., Blowey, R. W. and Green, L. E. 2008. Associations between sole ulcer, white line disease and digital dermatitis and the milk yield of 1824 dairy cows on 30 dairy cow farms in England and Wales from February 2003–November 2004. *Prev. Vet. Med.* 83(3–4), 381–391. https://doi.org/10.1016/j.prevetmed.2007.09.007.

Archer, S. C., Green, M. J. and Huxley, J. N. 2010. Association between milk yield and serial locomotion score assessments in UK dairy cows. *J. Dairy Sci.* 93(9), 4045–4053. https://doi.org/10.3168/jds.2010-3062.

Asikainen, M., Haataja, K. and Toivanen, P. 2013. *Wireless indoor tracking of livestock for behavioral analysis.* Presented at the 2013 9th International Wireless Communications and Mobile Computing Conference (IWCMC), pp. 1833–1838. https://doi.org/10.1109/IWCMC.2013.6583835.

Barkema, H. W., Westrik, J. D., van Keulen, K. A. S., Schukken, Y. H. and Brand, A. 1994. The effects of lameness on reproductive performance, milk production and culling in Dutch dairy farms. *Prev. Vet. Med.* 20(4), 249–259. https://doi.org/10.1016/0167-5877(94)90058-2.

Barker, Z. E., Amory, J. R., Wright, J. L., Mason, S. A., Blowey, R. W. and Green, L. E. 2009. Risk factors for increased rates of sole ulcers, white line disease, and digital dermatitis in dairy cattle from twenty-seven farms in England and Wales. *J. Dairy Sci.* 92(5), 1971–1978. https://doi.org/10.3168/jds.2008-1590.

Barker, Z. E., Vázquez Diosdado, J. A., Codling, E. A., Bell, N. J., Hodges, H. R., Croft, D. P. and Amory, J. R. 2018. Use of novel sensors combining local positioning and acceleration to measure feeding behavior differences associated with lameness in dairy cattle. *J. Dairy Sci.* 101(7), 6310–6321. https://doi.org/10.3168/jds.2016-12172.

Barwick, J., Lamb, D. W., Dobos, R., Welch, M. and Trotter, M. 2018. Categorising sheep activity using a tri-axial accelerometer. *Comput. Electron. Agric.* 145, 289–297. https://doi.org/10.1016/j.compag.2018.01.007.

Beer, G., Alsaaod, M., Starke, A., Schuepbach-Regula, G., Müller, H., Kohler, P. and Steiner, A. 2016. Use of extended characteristics of locomotion and feeding behavior for automated identification of lame dairy cows. *PLoS ONE* 11(5), e0155796. https://doi.org/10.1371/journal.pone.0155796.

Bell, N. 2017. Preventing and managing lameness in dairy cows. In: Webster, J. (Ed.) *Achieving Sustainable Production of Milk* (Vol. 3). Dairy Herd Management and Welfare. Cambridge: Burleigh Dodds Science Publishing.

Bicalho, R. C., Vokey, F., Erb, H. N. and Guard, C. L. 2007a. Visual locomotion scoring in the first seventy days in milk: Impact on pregnancy and survival. *J. Dairy Sci.* 90(10), 4586–4591. https://doi.org/10.3168/jds.2007-0297.

Bicalho, R. C., Cheong, S. H., Cramer, G. and Guard, C. L. 2007b. Association between a visual and an automated locomotion score in lactating Holstein cows. *J. Dairy Sci.* 90(7), 3294–3300. https://doi.org/10.3168/jds.2007-0076.

Bicalho, R. C., Warnick, L. D. and Guard, C. L. 2008. Strategies to analyze milk losses caused by diseases with potential incidence throughout the lactation: A lameness example. *J. Dairy Sci.* 91(7), 2653–2661. https://doi.org/10.3168/jds.2007-0744.

Bikker, J. P., van Laar, H., Rump, P., Doorenbos, J., van Meurs, K., Griffioen, G. M. and Dijkstra, J. 2014. Technical note: Evaluation of an ear-attached movement sensor to record cow feeding behavior and activity. *J. Dairy Sci.* 97(5), 2974–2979. https://doi.org/10.3168/jds.2013-7560.

Blackie, N., Amory, J., Bleach, E. and Scaife, J. 2011. The effect of lameness on lying behaviour of zero grazed Holstein dairy cattle. *Appl. Anim. Behav. Sci.* 134(3–4), 85–91. https://doi.org/10.1016/j.applanim.2011.08.004.

Blackie, N. and Maclaurin, L. 2019. Influence of lameness on the lying behaviour of zero-grazed lactating Jersey dairy cattle housed in straw yards. *Animals (Basel)* 9(10), 829.

Bonde, M., Rousing, T., Badsberg, J. H. and Sørensen, J. T. 2004. Associations between lying-down behaviour problems and body condition, limb disorders and skin lesions of lactating sows housed in farrowing crates in commercial sow herds. *Livest. Prod. Sci.* 87, 179–187. https://doi.org/10.1016/j.livprodsci.2003.08.005.

Boyland, N. K., Mlynski, D. T., James, R., Brent, L. J. N. and Croft, D. P. 2016. The social network structure of a dynamic group of dairy cows: From individual to group level patterns. *Appl. Anim. Behav. Sci.* 174, 1–10. https://doi.org/10.1016/j.applanim.2015.11.016.

Briene, P., Szczodry, O., De Geest, P., Van Weyenberg, S., Van Nuffel, A., Vangeyte, J., Millet, S., Ampe, B., Tuyttens, F. A. M. and Maselyne, J. 2021. Testing the potential of the Sow Stance Information System (SowSIS) based on a force plate system built into an electronic sow feeder for on-farm automatic lameness detection in breeding sows. *Biosyst. Eng.* 204, 270–282. https://doi.org/10.1016/j.biosystemseng.2021.01.024.

Broom, D. M. 2006. Behaviour and welfare in relation to pathology. *Appl. Anim. Behav. Sci. Int. Soc. Appl. Ethol. (ISAE) Spec. Issue* 97, 73–83. https://doi.org/10.1016/j.applanim.2005.11.019.

Bruijnis, M., Beerda, B., Hogeveen, H. and Stassen, E. 2012. Foot disorders in dairy cattle: Impact on cow and dairy farmer. *Anim. Welf.* 21(1), 33–40. https://doi.org/10.7120/096272812X13345905673601.

Byabazaire, J., Olariu, C., Taneja, M. and Davy, A. 2019. *Lameness detection as a service: Application of machine learning to an Internet of cattle*. Presented at the 2019 16th IEEE Annual Conference on Communication Network Conference (CCNC), pp. 1–6. https://doi.org/10.1109/CCNC.2019.8651681.

Byrne, D. T., Esmonde, H., Berry, D. P., McGovern, F., Creighton, P. and McHugh, N. 2019. Sheep lameness detection from individual hoof load. *Comput. Electron. Agric.* 158, 241–248. https://doi.org/10.1016/j.compag.2019.01.048.

Calderon, D. F. and Cook, N. B. 2011. The effect of lameness on the resting behavior and metabolic status of dairy cattle during the transition period in a freestall-housed dairy herd. *J. Dairy Sci.* 94(6), 2883–2894. https://doi.org/10.3168/jds.2010-3855.

Chapinal, N., de Passillé, A. M., Weary, D. M., von Keyserlingk, M. A. G. and Rushen, J. 2009. Using gait score, walking speed, and lying behavior to detect hoof lesions in dairy cows. *J. Dairy Sci.* 92(9), 4365–4374. https://doi.org/10.3168/jds.2009-2115.

Chapinal, N., de Passillé, A. M., Rushen, J., and Wagner, S. 2010. Automated methods for detecting lameness and measuring analgesia in dairy cattle. *J. Dairy Sci.* 93(5), 2007–2013. https://doi.org/10.3168/jds.2009-2803.

Chopra, K., Hodges, H. R., Barker, Z. E., Vázquez Diosdado, J. A., Amory, J. R., Cameron, T. C., Croft, D. P., Bell, N. J. and Codling, E. A. 2020. Proximity interactions in a permanently housed dairy herd: Network structure, consistency, and individual differences. *Front. Vet. Sci.* 7, 583715. https://doi.org/10.3389/fvets.2020.583715.

Codling, E. A., Plank, M. J. and Benhamou, S. 2008. Random walk models in biology. *J. R. Soc. Interface* 5(25), 813–834. https://doi.org/10.1098/rsif.2008.0014.

Conte, S., Bergeron, R., Grégoire, J., Gète, M., D'Allaire, S., Meunier-Salaün, M. C. and Devillers, N. 2014. On-farm evaluation of methods to assess welfare of gestating sows. *Animal* 8(7), 1153–1161. https://doi.org/10.1017/S1751731114000949.

Cramer, G., Lissemore, K. D., Guard, C. L., Leslie, K. E. and Kelton, D. F. 2009. The association between foot lesions and culling risk in Ontario Holstein cows. *J. Dairy Sci.* 92(6), 2572–2579. https://doi.org/10.3168/jds.2008-1532.

Davis-Unger, J., Schwartzkopf-Genswein, K. S. G., Pajor, E. A., Hendrick, S., Marti, S., Dorin, C. and Orsel, K. 2019. Prevalence and lameness-associated risk factors in Alberta feedlot cattle. *Transl. Anim. Sci.* 3, 595–606. https://doi.org/10.1093/tas/txz008

D'Eath, R. 2012. Repeated locomotion scoring of a sow herd to measure lameness: Consistency over time, the effect of sow characteristics and inter-observer reliability. *Anim. Welf.* 21, 219–231. https://doi.org/10.7120/09627286.21.2.219.

de Freslon, I., Martínez-López, B., Belkhiria, J., Strappini, A. and Monti, G. 2019. Use of social network analysis to improve the understanding of social behaviour in dairy cattle and its impact on disease transmission. *Appl. Anim. Behav. Sci.* 213, 47–54. https://doi.org/10.1016/j.applanim.2019.01.006.

de Mol, R. M., André, G., Bleumer, E. J. B., van der Werf, J. T. N., de Haas, Y. and van Reenen, C. G. 2013. Applicability of day-to-day variation in behavior for the automated detection of lameness in dairy cows. *J. Dairy Sci.* 96(6), 3703–3712. https://doi.org/10.3168/jds.2012-6305.

De Vries, A. 2020. Symposium review: Why revisit dairy cattle productive lifespan? *J. Dairy Sci.* 103(4), 3838–3845. https://doi.org/10.3168/jds.2019-17361.

Dickins, A., Clark, C. C. A., Kaler, J., Ferguson, E., O'Kane, H. and Green, L. E. 2016. Factors associated with the presence and prevalence of contagious ovine digital dermatitis: A 2013 study of 1136 random English sheep flocks. *Prev. Vet. Med.* 130, 86-93. https://doi.org/10.1016/j.prevetmed.2016.06.009.

Duncan, S., Stewart, T. I., Oliver, M., Mavoa, S., MacRae, D., Badland, H. M. and Duncan, M. J. 2013. Portable global positioning system receivers: Static validity and environmental conditions. *Am. J. Prev. Med.* 44(2), e19-29. https://doi.org/10.1016/j.amepre.2012.10.013.

Dunthorn, J., Dyer, R. M., Neerchal, N. K., McHenry, J. S., Rajkondawar, P. G., Steingraber, G. and Tasch, U. 2015. Predictive models of lameness in dairy cows achieve high sensitivity and specificity with force measurements in three dimensions. *J. Dairy Res.* 82(4), 391-399. https://doi.org/10.1017/S002202991500028X.

Fabbri, G., Gianesella, M., Morgante, M., Armato, L., Bonato, O. and Fiore, E. 2020. Ultrasonographic alterations of bovine claws sole soft tissues associated with claw horn disruption lesions, body condition score and locomotion score in Holstein dairy cows. *Res. Vet. Sci.* 131, 146-152. https://doi.org/10.1016/j.rvsc.2020.04.016.

Fiore, E., Perillo, L., Marchesini, G., Piccione, G., Giudice, E., Zumbo, A., Armato, L., Fabbri, G. and Gianesella, M. 2019. Effect of parity on claw horn lesions in Holstein dairy cows: Clinical and radiological study. *Ann. Anim. Sci.* 19(1), 147-158. https://doi.org/10.2478/aoas-2018-0050.

Fjeldaas, T., Nafstad, O., Fredriksen, B., Ringdal, G. and Sogstad, Å. M. 2007. Claw and limb disorders in 12 Norwegian beef-cow herds. *Acta Vet. Scand.* 49, 24. https://doi.org/10.1186/1751-0147-49-24.

Flower, F. C., Sanderson, D. J. and Weary, D. M. 2005. Hoof pathologies influence kinematic measures of dairy cow gait. *J. Dairy Sci.* 88(9), 3166-3173. https://doi.org/10.3168/jds.S0022-0302(05)73000-9.

Flower, F. C. and Weary, D. M. 2006. Effect of hoof pathologies on subjective assessments of dairy cow gait. *J. Dairy Sci.* 89(1), 139-146. https://doi.org/10.3168/jds.S0022-0302(06)72077-X.

Floyd, R. E. 2015. RFID in animal-tracking applications. *IEEE Potentials* 34(5), 32-33. https://doi.org/10.1109/MPOT.2015.2410308.

Fogarty, E. S., Swain, D. L., Cronin, G. M., Moraes, L. E. and Trotter, M. 2020. Behaviour classification of extensively grazed sheep using machine learning. *Comput. Electron. Agric.* 169, 105175. https://doi.org/10.1016/j.compag.2019.105175.

Frankena, K., Somers, J. G. C. J., Schouten, W. G. P., van Stek, J. V., Metz, J. H. M., Stassen, E. N. and Graat, E. A. M. 2009. The effect of digital lesions and floor type on locomotion score in Dutch dairy cows. *Prev. Vet. Med.* 88(2), 150-157. https://doi.org/10.1016/j.prevetmed.2008.08.004.

Galindo, F. and Broom, D. M. 2002. The effects of lameness on social and individual behavior of dairy cows. *J. Appl. Anim. Welf. Sci.* 5(3), 193-201.

Garcia, E., Klaas, I., Amigo, J. M., Bro, R. and Enevoldsen, C. 2014. Lameness detection challenges in automated milking systems addressed with partial least squares discriminant analysis. *J. Dairy Sci.* 97(12), 7476-7486. https://doi.org/10.3168/jds.2014-7982.

Gillman, C. E., KilBride, A. L., Ossent, P. and Green, L. E. 2008. A cross-sectional study of the prevalence and associated risk factors for bursitis in weaner, grower and finisher pigs from 93 commercial farms in England. *Prev. Vet. Med.* 83(3–4), 308–322. https://doi.org/10.1016/j.prevetmed.2007.09.001.

Gillman, C. E., KilBride, A. L., Ossent, P. and Green, L. E. 2009. A cross-sectional study of the prevalence of foot lesions in post-weaning pigs and risks associated with floor type on commercial farms in England. *Prev. Vet. Med.* 91(2–4), 146–152. https://doi.org/10.1016/j.prevetmed.2009.05.023.

Goldhawk, C., Chapinal, N., Veira, D. M., Weary, D. M. and von Keyserlingk, M. A. G. 2009. Prepartum feeding behavior is an early indicator of subclinical ketosis. *J. Dairy Sci.* 92(10), 4971–4977. https://doi.org/10.3168/jds.2009-2242.

González, L. A., Tolkamp, B. J., Coffey, M. P., Ferret, A. and Kyriazakis, I. 2008. Changes in feeding behavior as possible indicators for the automatic monitoring of health disorders in dairy cows. *J. Dairy Sci.* 91(3), 1017–1028. https://doi.org/10.3168/jds.2007-0530.

Grégoire, J., Bergeron, R., D'Allaire, S., Meunier-Salaün, M.-C. and Devillers, N. 2013. Assessment of lameness in sows using gait, footprints, postural behaviour and foot lesion analysis. *Animal* 7, 1163–1173. https://doi.org/10.1017/S1751731113000098.

Griffiths, B. E., Grove White, D. and Oikonomou, G. 2018. A cross-sectional study into the prevalence of dairy cattle lameness and associated herd-level risk factors in England and Wales. *Front. Vet. Sci.* 5, 65. https://doi.org/10.3389/fvets.2018.00065.

Grimm, K., Haidn, B., Erhard, M., Tremblay, M. and Döpfer, D. 2019. New insights into the association between lameness, behavior, and performance in Simmental cows. *J. Dairy Sci.* 102(3), 2453–2468. https://doi.org/10.3168/jds.2018-15035.

Groenevelt, M., Main, D. C. J., Tisdall, D., Knowles, T. G. and Bell, N. J. 2014. Measuring the response to therapeutic foot trimming in dairy cows with fortnightly lameness scoring. *Vet. J.* 201(3), 283–288. https://doi.org/10.1016/j.tvjl.2014.05.017.

Guzhva, O., Ardö, H., Nilsson, M., Herlin, A. and Tufvesson, L. 2018. Now you see me: Convolutional neural network based tracker for dairy cows. *Front. Robot. AI* 5, 107. https://doi.org/10.3389/frobt.2018.00107.

Gygax, L., Neisen, G. and Bollhalder, H. 2007. Accuracy and validation of a radar-based automatic local position measurement system for tracking dairy cows in free-stall barns. *Comput. Electron. Agric.* 56(1), 23–33. https://doi.org/10.1016/j.compag.2006.12.004.

Haladjian, J., Haug, J., Nüske, S. and Bruegge, B. 2018. A wearable sensor system for lameness detection in dairy cattle. *Multimodal Technol. Interact.* 2(2), 27. https://doi.org/10.3390/mti2020027.

Harris-Bridge, G., Young, L., Handel, I., Farish, M., Mason, C., Mitchell, M. A. and Haskell, M. J. 2018. The use of infrared thermography for detecting digital dermatitis in dairy cattle: What is the best measure of temperature and foot location to use? *Vet. J.* 237, 26–33. https://doi.org/10.1016/j.tvjl.2018.05.008.

Higginson Cutler, J. H. H., Rushen, J., de Passillé, A. M., Gibbons, J., Orsel, K., Pajor, E., Barkema, H. W., Solano, L., Pellerin, D., Haley, D. and Vasseur, E. 2017. Producer estimates of prevalence and perceived importance of lameness in dairy herds with tiestalls, freestalls, and automated milking systems. *J. Dairy Sci.* 100(12), 9871–9880. https://doi.org/10.3168/jds.2017-13008.

Hirst, W. M., Murray, R. D., Ward, W. R. and French, N. P. 2002. A mixed-effects time-to-event analysis of the relationship between first-lactation lameness and subsequent lameness in dairy cows in the UK. *Prev. Vet. Med.* 54(3), 191–201. https://doi.org/10.1016/S0167-5877(02)00021-1.

Huhtala, A., Suhonen, K., Mäkelä, P., Hakojärvi, M. and Ahokas, J. 2007. Evaluation of instrumentation for cow positioning and tracking indoors. *Biosyst. Eng.* 96(3), 399-405. https://doi.org/10.1016/j.biosystemseng.2006.11.013.

Hultgren, J., Manske, T. and Bergsten, C. 2004. Associations of sole ulcer at claw trimming with reproductive performance, udder health, milk yield, and culling in Swedish dairy cattle. *Prev. Vet. Med.* 62(4), 233-251. https://doi.org/10.1016/j.prevetmed.2004.01.002.

Jarchi, D., Kaler, J. and Sanei, S. 2021. Lameness detection in cows using hierarchical deep learning and synchrosqueezed wavelet transform. *IEEE Sens. J.* 21(7), 9349-9358. https://doi.org/10.1109/JSEN.2021.3054718.

Jiang, B., Yin, X. and Song, H. 2020. Single-stream long-term optical flow convolution network for action recognition of lameness dairy cow. *Comput. Electron. Agric.* 175, 105536. https://doi.org/10.1016/j.compag.2020.105536.

Kaler, J., Wassink, G. J. and Green, L. E. 2009. The inter- and intra-observer reliability of a locomotion scoring scale for sheep. *Vet. J.* 180, 189-194. https://doi.org/10.1016/j.tvjl.2007.12.028.

Kaler, J., Mitsch, J., Vázquez-Diosdado, J. A., Bollard, N., Dottorini, T. and Ellis, K. A. 2020. Automated detection of lameness in sheep using machine learning approaches: Novel insights into behavioural differences among lame and non-lame sheep. *R. Soc. Open Sci.* 7(1), 190824. https://doi.org/10.1098/rsos.190824.

Kamphuis, C., Frank, E., Burke, J. K., Verkerk, G. A. and Jago, J. G. 2013. Applying additive logistic regression to data derived from sensors monitoring behavioral and physiological characteristics of dairy cows to detect lameness. *J. Dairy Sci.* 96(11), 7043-7053. https://doi.org/10.3168/jds.2013-6993.

Kays, R., Crofoot, M. C., Jetz, W. and Wikelski, M. 2015. Ecology: Terrestrial animal tracking as an eye on life and planet. *Science* 348(6240), aaa2478. https://doi.org/10.1126/science.aaa2478.

KilBride, A. L., Gillman, C. E., Ossent, P. and Green, L. E. 2008. A cross-sectional study of the prevalence and associated risk factors for capped hock and the associations with bursitis in weaner, grower and finisher pigs from 93 commercial farms in England. *Prev. Vet. Med.* 83(3-4), 272-284. https://doi.org/10.1016/j.prevetmed.2007.08.004.

Kofler, J., Fiedler, A., Charfeddine, N., Capion, N., Fjeldaas, T., Carmer, G., Bell, N. J., Müller, K. E., Christen, A., Thomas, G.-M., Heringstad, B., Stock, K. F., Holzhauer, M., Nieto, J. M., Egger-Danner, C. and Döpfer, D. 2020. Digital dermatitis stages (M-stages). *ICAR Claw Health Atlas* 40, Appendix 1.

Laister, S., Stockinger, B., Regner, A.-M., Zenger, K., Knierim, U. and Winckler, C. 2011. Social licking in dairy cattle–Effects on heart rate in performers and receivers. *Appl. Anim. Behav. Sci.* 130(3-4), 81-90. https://doi.org/10.1016/j.applanim.2010.12.003.

Langrock, R., King, R., Matthiopoulos, J., Thomas, L., Fortin, D. and Morales, J. M. 2012. Flexible and practical modeling of animal telemetry data: Hidden Markov models and extensions. *Ecology* 93(11), 2336-2342. https://doi.org/10.1890/11-2241.1.

Leach, K. A., Tisdall, D. A., Bell, N. J., Main, D. C. J. and Green, L. E. 2012. The effects of early treatment for hindlimb lameness in dairy cows on four commercial UK farms. *Vet. J.* 193(3), 626-632. https://doi.org/10.1016/j.tvjl.2012.06.043.

Leach, K. A., Whay, H. R., Maggs, C. M., Barker, Z. E., Paul, E. S., Bell, A. K. and Main, D. C. J. 2010. Working towards a reduction in cattle lameness: 1. Understanding barriers to lameness control on dairy farms. *Res. Vet. Sci.* 89(2), 311-317. https://doi.org/10.1016/j.rvsc.2010.02.014.

Lehrer, A. R., Lewis, G. S. and Aizinbud, E. 1992. Oestrus detection in cattle: Recent developments. *Anim. Reprod. Sci.* 28, 355–362. https://doi.org/10.1016/0378-4320(92)90121-S.

Lin, Y. C., Mullan, S. and Main, D. C. J. 2018. Optimising lameness detection in dairy cattle by using handheld infrared thermometers. *Vet. Med. Sci.* 29(4(3)), 218–226. https://doi.org/10.1002/vms3.104.

Liu, J., Dyer, R. M., Neerchal, N. K., Tasch, U. and Rajkondawar, P. G. 2011. Diversity in the magnitude of hind limb unloading occurs with similar forms of lameness in dairy cows. *J. Dairy Res.* 78, 168–177. https://doi.org/10.1017/S0022029911000057.

Machado, V. S., Caixeta, L. S., McArt, J. A. A. and Bicalho, R. C. 2010. The effect of claw horn disruption lesions and body condition score at dry-off on survivability, reproductive performance, and milk production in the subsequent lactation. *J. Dairy Sci.* 93(9), 4071–4078. https://doi.org/10.3168/jds.2010-3177.

Maertens, W., Vangeyte, J., Baert, J., Jantuan, A., Mertens, K. C., De Campeneere, S., Pluk, A., Opsomer, G., Van Weyenberg, S. and Van Nuffel, A. 2011. Development of a real time cow gait tracking and analysing tool to assess lameness using a pressure sensitive walkway: The GAITWISE system. *Biosyst. Eng.* 110(1), 29–39. https://doi.org/10.1016/j.biosystemseng.2011.06.003.

Main, D. C. J., Clegg, J., Spatz, A. and Green, L. E. 2000. Repeatability of a lameness scoring system for finishing pigs. *Vet. Rec.* 147, 574–576. https://doi.org/10.1136/vr.147.20.574.

Main, D. C. J., Barker, Z. E., Leach, K. A., Bell, N. J., Whay, H. R. and Browne, W. J. 2010. Sampling strategies for monitoring lameness in dairy cattle. *J. Dairy Sci.* 93(5), 1970–1978. https://doi.org/10.3168/jds.2009-2500.

Mandel, R., Harazy, H., Gygax, L., Nicol, C. J., Ben-David, A., Whay, H. R. and Klement, E. 2017. Short communication: Detection of lameness in dairy cows using a grooming device. *J. Dairy Sci.* 101(2), 1511–1517 https://doi.org/10.3168/jds.2017-13207.

Martiskainen, P., Järvinen, M., Skön, J.-P., Tiirikainen, J., Kolehmainen, M. and Mononen, J. 2009. Cow behaviour pattern recognition using a three-dimensional accelerometer and support vector machines. *Appl. Anim. Behav. Sci.* 119(1–2), 32–38. https://doi.org/10.1016/j.applanim.2009.03.005.

Maselyne, J., Saeys, W., Briene, P., Mertens, K., Vangeyte, J., De Ketelaere, B., Hessel, E. F., Sonck, B. and Van Nuffel, A. 2016. Methods to construct feeding visits from RFID registrations of growing-finishing pigs at the feed trough. *Comput. Electron. Agric.* 128, 9–19. https://doi.org/10.1016/j.compag.2016.08.010.

McConnel, C. S., Lombard, J. E., Wagner, B. A. and Garry, F. B. 2008. Evaluation of factors associated with increased dairy cow mortality on United States dairy operations. *J. Dairy Sci.* 91(4), 1423–1432. https://doi.org/10.3168/jds.2007-0440.

Miguel-Pacheco, G. G., Thomas, H. J., Huxley, J. N., Newsome, R. F. and Kaler, J. 2017. Effect of claw horn lesion type and severity at the time of treatment on outcome of lameness in dairy cows. *Vet. J.* 225, 16–22. https://doi.org/10.1016/j.tvjl.2017.04.015.

Miguel-Pacheco, G. G., Thomas, H. J., Kaler, J., Craigon, J. and Huxley, J. N. 2016. Effects of lameness treatment for claw horn lesions on lying behaviour in dairy cows. *Appl. Anim. Behav. Sci.* 179, 11–16. https://doi.org/10.1016/j.applanim.2016.04.001.

Mohling, C. M., Johnson, A. K., Coetzee, J. F., Karriker, L. A., Abell, C. E., Millman, S. T. and Stalder, K. J. 2014. Kinematics as objective tools to evaluate lameness phases in multiparous sows. *Livest. Sci.* 165, 120–128. https://doi.org/10.1016/j.livsci.2014.04.031.

Nagy, M., Vásárhelyi, G., Pettit, B., Roberts-Mariani, I., Vicsek, T. and Biro, D. 2013. Context-dependent hierarchies in pigeons. *Proc. Natl. Acad. Sci. U. S. A.* 110(32), 13049-13054. https://doi.org/10.1073/pnas.1305552110.

Nechanitzky, K., Starke, A., Vidondo, B., Müller, H., Reckardt, M., Friedli, K. and Steiner, A. 2016. Analysis of behavioral changes in dairy cows associated with claw horn lesions. *J. Dairy Sci.* 99(4), 2904–2914. https://doi.org/10.3168/jds.2015-10109.

Neveux, S., Weary, D. M., Rushen, J., von Keyserlingk, M. A. G. and de Passillé, A. M. 2006. Hoof discomfort changes how dairy cattle distribute their body weight. *J. Dairy Sci.* 89(7), 2503–2509. https://doi.org/10.3168/jds.S0022-0302(06)72325-6.

Newsome, R., Green, M. J., Bell, N. J., Chagunda, M. G. G., Mason, C. S., Rutland, C. S., Sturrock, C. J., Whay, H. R. and Huxley, J. N. 2016. Linking bone development on the caudal aspect of the distal phalanx with lameness during life. *J. Dairy Sci.* 99(6), 4512–4525. https://doi.org/10.3168/jds.2015-10202.

Newsome, R. F., Green, M. J., Bell, N. J., Bollard, N. J., Mason, C. S., Whay, H. R. and Huxley, J. N. 2017. A prospective cohort study of digital cushion and corium thickness. Part 2: Does thinning of the digital cushion and corium lead to lameness and claw horn disruption lesions? *J. Dairy Sci.* 100(6), 4759–4771. https://doi.org/10.3168/jds.2016-12013.

O'Leary, N. W., Byrne, D. T., O'Connor, A. H. and Shalloo, L. 2020. Invited review: Cattle lameness detection with accelerometers. *J. Dairy Sci.* 103(5), 3895–3911. https://doi.org/10.3168/jds.2019-17123.

Ouweltjes, W., Spoor, C. W., van Leeuwen, J. L. and Gussekloo, S. W. S. 2019. Spatial distribution of load induced soft-tissue strain in cattle claws. *Vet. J.* 248, 28–36. https://doi.org/10.1016/j.tvjl.2019.03.007.

Pairis-Garcia, M., Johnson, A., Stalder, K., Abell, C., Karriker, L., Coetzee, J. and Millman, S. 2015. Behavioural evaluation of analgesic efficacy for pain mitigation in lame sows. *Anim. Welf.* 24, 93–99. https://doi.org/10.7120/09627286.24.1.093.

Pandolfi, F., Kyriazakis, I., Stoddart, K., Wainwright, N. and Edwards, S. A. 2017. The "Real Welfare" scheme: Identification of risk and protective factors for welfare outcomes in commercial pig farms in the UK. *Prev. Vet. Med.* 146, 34–43. https://doi.org/10.1016/j.prevetmed.2017.07.008.

Pastell, M., Aisla, A. M., Hautala, M., Poikalainen, V., Praks, J., Veermäe, I. and Ahokas, J. 2006. Contactless measurement of cow behavior in a milking robot. *Behav. Res. Methods* 38(3), 479–486. https://doi.org/10.3758/BF03192802.

Pastell, M., Hänninen, L., de Passillé, A. M. and Rushen, J. 2010. Measures of weight distribution of dairy cows to detect lameness and the presence of hoof lesions. *J. Dairy Sci.* 93, 954–960. https://doi.org/10.3168/jds.2009-2385.

Pastell, M. and Frondelius, L. 2018. A hidden Markov model to estimate the time dairy cows spend in feeder based on indoor positioning data. *Comput. Electron. Agric.* 152, 182–185. https://doi.org/10.1016/j.compag.2018.07.005.

Pastell, M., Frondelius, L., Järvinen, M. and Backman, J. 2018. Filtering methods to improve the accuracy of indoor positioning data for dairy cows. *Biosyst. Eng.* 169, 22-31. https://doi.org/10.1016/j.biosystemseng.2018.01.008.

Pastell, M., Hautala, M., Poikalainen, V., Praks, J., Veermäe, I., Kujala, M. and Ahokas, J. 2008a. Automatic observation of cow leg health using load sensors. *Comput. Electron. Agric. Precis. Livest. Farm. (PLF)* 62(1), 48–53. https://doi.org/10.1016/j.compag.2007.09.003.

Pastell, M., Kujala, M., Aisla, A.-M., Hautala, M., Poikalainen, V., Praks, J., Veermäe, I. and Ahokas, J. 2008b. Detecting cow's lameness using force sensors. *Comput. Electron.*

Agric. Smart Sens. Precis. Livest. Farm. 64(1), 34–38. https://doi.org/10.1016/j.compag.2008.05.007.

Pastell, M. and Madsen, H. 2008. Application of CUSUM charts to detect lameness in a milking robot. *Expert Syst. Appl.* 35(4), 2032–2040. https://doi.org/10.1016/j.eswa.2007.08.087.

Pastell, M. E. and Kujala, M. 2007. A probabilistic neural network model for lameness detection. *J. Dairy Sci.* 90(5), 2283–2292. https://doi.org/10.3168/jds.2006-267.

Patterson, T. A., Thomas, L., Wilcox, C., Ovaskainen, O. and Matthiopoulos, J. 2008. State-space models of individual animal movement. *Trends Ecol. Evol.* 23(2), 87–94. https://doi.org/10.1016/j.tree.2007.10.009.

Perisho, S. and Hajnal, A. 2021. Characterization of swine behavior and production using measurements collected via indoor positioning system. *Appl. Anim. Behav. Sci.* 234, 105164. https://doi.org/10.1016/j.applanim.2020.105164.

Piette, D., Norton, T., Exadaktylos, V. and Berckmans, D. 2020. Individualised automated lameness detection in dairy cows and the impact of historical window length on algorithm performance. *Animal* 14(2), 409–417. https://doi.org/10.1017/S1751731119001642.

Pluk, A., Bahr, C., Leroy, T., Poursaberi, A., Song, X., Vranken, E., Maertens, W. and Berkmans, D. 2010. Evaluation of step overlap as an automatic measure in dairy cow locomotion. *Trans. ASABE* 53(4), 1305–1312.

Pluk, A., Bahr, C., Poursaberi, A., Maertens, W., van Nuffel, A. and Berckmans, D. 2012. Automatic measurement of touch and release angles of the fetlock joint for lameness detection in dairy cattle using vision techniques. *J. Dairy Sci.* 95(4), 1738–1748. https://doi.org/10.3168/jds.2011-4547.

Pluym, L. M., Maes, D., Vangeyte, J., Mertens, K., Baert, J., Van Weyenberg, S., Millet, S. and Van Nuffel, A. 2013. Development of a system for automatic measurements of force and visual stance variables for objective lameness detection in sows: SowSIS. *Biosyst. Eng.* 116(1), 64–74. https://doi.org/10.1016/j.biosystemseng.2013.06.009.

Porto, S. M. C., Arcidiacono, C., Giummarra, A., Anguzza, U. and Cascone, G. 2014. Localisation and identification performances of a real-time location system based on ultra wide band technology for monitoring and tracking dairy cow behaviour in a semi-open free-stall barn. *Comput. Electron. Agric.* 108, 221–229. https://doi.org/10.1016/j.compag.2014.08.001.

Poursaberi, A., Bahr, C., Pluk, A., Van Nuffel, A. and Berckmans, D. 2010. Real-time automatic lameness detection based on back posture extraction in dairy cattle: Shape analysis of cow with image processing techniques. *Comput. Electron. Agric.* 74(1), 110–119. https://doi.org/10.1016/j.compag.2010.07.004.

Prosser, N. S., Purdy, K. J. and Green, L. E. 2019. Increase in the flock prevalence of lameness in ewes is associated with a reduction in farmers using evidence-based management of prompt treatment: A longitudinal observational study of 154 English sheep flocks 2013–2015. *Prev. Vet. Med.* 173, 104801. https://doi.org/10.1016/j.prevetmed.2019.104801

Proudfoot, K. L., Jensen, M. B., Weary, D. M. and von Keyserlingk, M. A. G. 2014. Dairy cows seek isolation at calving and when ill. *J. Dairy Sci.* 97:2731–2739. https://doi.org/10.3168/jds.2013-7274.

Quinn, A. J., Boyle, L. A., KilBride, A. L. and Green, L. E. 2015. A cross-sectional study on the prevalence and risk factors for foot and limb lesions in piglets on commercial farms in Ireland. *Prev. Vet. Med.* 119(3–4), 162–171. https://doi.org/10.1016/j.prevetmed.2015.02.016.

Rajkondawar, P. G., Tasch, U., Lefcourt, A. M., Erez, B., Dyer, R. M. and Varner, M. A. 2002. A system for identifying lameness in dairy cattle. *Appl. Eng. Agric.* 18(1), 87-69. https://doi.org/10.13031/2013.7707.

Randall, L. V., Green, M. J., Chagunda, M. G. G., Mason, C., Green, L. E. and Huxley, J. N. 2016. Lameness in dairy heifers; impacts of hoof lesions present around first calving on future lameness, milk yield and culling risk. *Prev. Vet. Med.* 133, 52-63. https://doi.org/10.1016/j.prevetmed.2016.09.006.

Randall, L. V., Green, M. J. and Huxley, J. N. 2018. Use of statistical modelling to investigate the pathogenesis of claw horn disruption lesions in dairy cattle. *Vet. J.* 238, 41-48. https://doi.org/10.1016/j.tvjl.2018.07.002.

Rice, E. M., Ferrell, J., Vanzant, E., Jackson, J. and Costa, J. 2020. Real-time localization system for livestock dairy cattle: Validation of static positioning in a commercial facility. 2020 ASABE Annual International Virtual Meeting, Paper number 2000797. https://doi.org/10.13031/aim.202000797.

Rocha, L. E. C., Terenius, O., Veissier, I., Meunier, B. and Nielsen, P. P. 2020. Persistence of sociality in group dynamics of dairy cattle. *Appl. Anim. Behav. Sci.* 223, 104921. https://doi.org/10.1016/j.applanim.2019.104921.

Roland, L., Schweinzer, V., Kanz, P., Sattlecker, G., Kickinger, F., Lidauer, L., Berger, A., Auer, W., Mayer, J., Sturm, V., Efrosinin, D., Breitenberger, S., Drillich, M. and Iwersen, M. 2018. Technical note: Evaluation of a triaxial accelerometer for monitoring selected behaviors in dairy calves. *J. Dairy Sci.* 101(11), 10421-10427. https://doi.org/10.3168/jds.2018-14720.

Rørvang, M. V., Herskin, M. S. and Jensen, M. B. 2018. The motivation-based calving facility: Social and cognitive factors influence isolation seeking behaviour of Holstein dairy cows at calving. *PLoS ONE* 13(1), e0191128. https://doi.org/10.1371/journal.pone.0191128.

Salau, J. and Krieter, J. 2020. Analysing the space-usage-pattern of a cow herd using video surveillance and automated motion detection. *Biosyst. Eng.* 197, 122-134. https://doi.org/10.1016/j.biosystemseng.2020.06.015.

Scheel, C., Traulsen, I., Auer, W., Müller, K., Stamer, E. and Krieter, J. 2017. Detecting lameness in sows from ear tag-sampled acceleration data using wavelets. *Animal* 11, 2076-2083. https://doi.org/10.1017/S1751731117000726.

Schieltz, J. M., Okanga, S., Allan, B. F. and Rubenstein, D. I. 2017. GPS tracking cattle as a monitoring tool for conservation and management. *Afr. J. Range Forage Sci.* 34, 173-177. https://doi.org/10.2989/10220119.2017.1387175.

Scott, G. B. 1989. Changes in limb loading with lameness for a number of Friesian cattle. *Br. Vet. J.* 145(1), 28-38. https://doi.org/10.1016/0007-1935(89)90005-5.

Sepúlveda-Varas, P., Lomb, J., von Keyserlingk, M. A. G., Held, R., Bustamante, H. and Tadich, N. 2018. Claw horn lesions in mid-lactation primiparous dairy cows under pasture-based systems: Association with behavioral and metabolic changes around calving. *J. Dairy Sci.* 101(10), 9439-9450. https://doi.org/10.3168/jds.2018-14674.

Sepúlveda-Varas, P., Proudfoot, K. L., Weary, D. M. and von Keyserlingk, M. A. G. 2016. Changes in behaviour of dairy cows with clinical mastitis. *Appl. Anim. Behav. Sci. Behav. Indic. Health* 175, 8-13. https://doi.org/10.1016/j.applanim.2014.09.022.

Sepúlveda-Varas, P., Weary, D. M. and von Keyserlingk, M. A. G. 2014. Lying behavior and postpartum health status in grazing dairy cows. *J. Dairy Sci.* 97(10), 6334-6343. https://doi.org/10.3168/jds.2014-8357.

Sogstad, A. M., Østerås, O. and Fjeldaas, T. 2006. Bovine claw and limb disorders related to reproductive performance and production diseases. *J. Dairy Sci.* 89(7), 2519-2528. https://doi.org/10.3168/jds.S0022-0302(06)72327-X.

Solano, L., Barkema, H. W., Mason, S., Pajor, E. A., LeBlanc, S. J. and Orsel, K. 2016. Prevalence and distribution of foot lesions in dairy cattle in Alberta, Canada. *J. Dairy Sci.* 99(8), 6828-6841. https://doi.org/10.3168/jds.2016-10941.

Song, X., Leroy, T., Vranken, E., Maertens, W., Sonck, B. and Berckmans, D. 2008. Automatic detection of lameness in dairy cattle–Vision-based trackway analysis in cow's locomotion. *Comput. Electron. Agric. Smart Sens. Precis. Livest. Farm.* 64(1), 39-44. https://doi.org/10.1016/j.compag.2008.05.016.

Sprecher, D. J., Hostetler, D. E. and Kaneene, J. B. 1997. A lameness scoring system that uses posture and gait to predict dairy cattle reproductive performance. *Theriogenology* 47(6), 1179-1187. https://doi.org/10.1016/S0093-691X(97)00098-8.

Stavrakakis, S., Guy, J. H., Syranidis, I., Johnson, G. R. and Edwards, S. A. 2015. Pre-clinical and clinical walking kinematics in female breeding pigs with lameness: A nested case-control cohort study. *Vet. J.* 205(1), 38-43. https://doi.org/10.1016/j.tvjl.2015.04.022.

Stokes, J. E., Leach, K. A., Main, D. C. J. and Whay, H. R. 2012. An investigation into the use of infrared thermography (IRT) as a rapid diagnostic tool for foot lesions in dairy cattle. *Vet. J.* 193, 674-678. https://doi.org/10.1016/j.tvjl.2012.06.052.

Szyszka, O., Tolkamp, B. J., Edwards, S. A. and Kyriazakis, I. 2013. Do the changes in the behaviours of cattle during parasitism with Ostertagia ostertagi have a potential diagnostic value? *Vet. Parasitol.* 193(1-3), 214-222. https://doi.org/10.1016/j.vetpar.2012.10.023.

Tadich, N., Flor, E. and Green, L. 2010. Associations between hoof lesions and locomotion score in 1098 unsound dairy cows. *Vet. J.* 184(1), 60-65. https://doi.org/10.1016/j.tvjl.2009.01.005.

Tadich, N., Tejeda, C., Bastias, S., Rosenfeld, C. and Green, L. E. 2013. Nociceptive threshold, blood constituents and physiological values in 213 cows with locomotion scores ranging from normal to severely lame. *Vet. J.* 197(2), 401-405. https://doi.org/10.1016/j.tvjl.2013.01.029.

Ter-Sarkisov, A., Ross, R. and Kelleher, J. 2017. *Bootstrapping labelled dataset construction for cow tracking and behavior analysis*. Presented at the 2017 14th Conference on Computer and Robot Vision (CRV), pp. 277-284. https://doi.org/10.1109/CRV.2017.25.

Thompson, A. J., Weary, D. M., Bran, J. A., Daros, R. R., Hötzel, M. J. and von Keyserlingk, M. A. G. 2019. Lameness and lying behavior in grazing dairy cows. *J. Dairy Sci.* 102(7), 6373-6382. https://doi.org/10.3168/jds.2018-15717.

Thorup, V. M., do Nascimento, O. F., Skjøth, F., Voigt, M., Rasmussen, M. D., Bennedsgaard, T. W. and Ingvartsen, K. L. 2014. Short communication: Changes in gait symmetry in healthy and lame dairy cows based on 3-dimensional ground reaction force curves following claw trimming. *J. Dairy Sci.* 97(12), 7679-7684. https://doi.org/10.3168/jds.2014-8410.

Traulsen, I., Breitenberger, S., Auer, W., Stamer, E., Müller, K. and Krieter, J. 2016. Automatic detection of lameness in gestating group-housed sows using positioning and acceleration measurements. *Animal* 10:970-977. https://doi.org/10.1017/S175173111500302X.

Tullo, E., Fontana, I., Gottardo, D., Sloth, K. H. and Guarino, M. 2016. Technical note: Validation of a commercial system for the continuous and automated monitoring of dairy cow activity. *J. Dairy Sci.* 99(9), 7489–7494. https://doi.org/10.3168/jds.2016-11014.

Tunstall, J., Mueller, K., Grove White, D., Oultram, J. W. H. and Higgins, H. M. 2019. Lameness in beef cattle: UK farmers' perceptions, knowledge, barriers, and approaches to treatment and control. *Front. Vet. Sci.* 6, 94. https://doi.org/10.3389/fvets.2019.00094.

Vacek, M., Stádník, L. and Stipkova, M. 2007. Relationships between the incidence of health disorders and the reproduction traits of Holstein cows in the Czech Republic. *Czech J. Anim. Sci.* 52(8), 227–235.

Van De Gucht, T., Saeys, W., Van Weyenberg, S., Lauwers, L., Mertens, K., Vandaele, L., Vangeyte, J. and Van Nuffel, A. 2017. Automatic cow lameness detection with a pressure mat: Effects of mat length and sensor resolution. *Comput. Electron. Agric.* 134, 172–180. https://doi.org/10.1016/j.compag.2017.01.011.

van der Tol, P. P. J., Metz, J. H. M., Noordhuizen-Stassen, E. N., Back, W., Braam, C. R. and Weijs, W. A. 2002. The pressure distribution under the bovine claw during square standing on a flat substrate. *J. Dairy Sci.* 85(6), 1476–1481. https://doi.org/10.3168/jds.S0022-0302(02)74216-1.

van der Tol, P. P. J., Metz, J. H. M., Noordhuizen-Stassen, E. N., Back, W., Braam, C. R. and Weijs, W. A. 2003. The vertical ground reaction force and the pressure distribution on the claws of dairy cows while walking on a flat substrate. *J. Dairy Sci.* 86(9), 2875–2883. https://doi.org/10.3168/jds.S0022-0302(03)73884-3.

van der Tol, P. P. J., Metz, J. H. M., Noordhuizen-Stassen, E. N., Back, W., Braam, C. R. and Weijs, W. A. 2005. Frictional forces required for unrestrained locomotion in dairy cattle. *J. Dairy Sci.* 88(2), 615–624. https://doi.org/10.3168/jds.S0022-0302(05)72725-9.

Van Hertem, T., Maltz, E., Antler, A., Romanini, C. E. B., Viazzi, S., Bahr, C., Schlageter-Tello, A., Lokhorst, C., Berckmans, D. and Halachmi, I. 2013. Lameness detection based on multivariate continuous sensing of milk yield, rumination, and neck activity. *J. Dairy Sci.* 96(7), 4286–4298. https://doi.org/10.3168/jds.2012-6188.

Van Hertem, T., Schlageter Tello, A., Viazzi, S., Steensels, M., Bahr, C., Romanini, C. E. B., Lokhorst, K., Maltz, E., Halachmi, I. and Berckmans, D. 2018. Implementation of an automatic 3D vision monitor for dairy cow locomotion in a commercial farm. *Biosyst. Eng.* 173, 166–175. https://doi.org/10.1016/j.biosystemseng.2017.08.011.

Van Nuffel, A., Saeys, W., Sonck, B., Vangeyte, J., Mertens, K. C., De Ketelaere, B. and Van Weyenberg, S. 2015. Variables of gait inconsistency outperform basic gait variables in detecting mildly lame cows. *Livest. Sci.* 177, 125–131. https://doi.org/10.1016/j.livsci.2015.04.008.

Van Nuffel, A., Sprenger, M., Tuyttens, F. and Maertens, W. 2009. Cow gait scores and kinematic gait data: Can people see gait irregularities? *Anim. Welf.* 18, 433–439.

Van Nuffel, A., Vangeyte, J., Mertens, K. C., Pluym, L., De Campeneere, S., Saeys, W., Opsomer, G. and Van Weyenberg, S. 2013. Exploration of measurement variation of gait variables for early lameness detection in cattle using the GAITWISE. *Livest. Sci.* 156(1–3), 88–95. https://doi.org/10.1016/j.livsci.2013.06.013.

Vazquez Diosdado, J. A., Barker, Z. E., Hodges, H. R., Amory, J. R., Croft, D. P., Bell, N. J. and Codling, E. A. 2015. Classification of behaviour in housed dairy cows using an accelerometer-based activity monitoring system. *Anim. Biotelemetry* 3, 15. https://doi.org/10.1186/s40317-015-0045-8.

Vázquez Diosdado, J. A., Barker, Z. E., Hodges, H. R., Amory, J. R., Croft, D. P., Bell, N. J. and Codling, E. A. 2018. Space-use patterns highlight behavioural differences linked to lameness, parity, and days in milk in barn-housed dairy cows. *PLoS ONE* 13(12), e0208424. https://doi.org/10.1371/journal.pone.0208424.

Veissier, I., Mialon, M. M. and Sloth, K. H. 2017. Short communication: Early modification of the circadian organization of cow activity in relation to disease or estrus. *J. Dairy Sci.* 100(5), 3969–3974. https://doi.org/10.3168/jds.2016-11853.

Viazzi, S., Bahr, C., Schlageter-Tello, A., Van Hertem, T., Romanini, C. E. B., Pluk, A., Halachmi, I., Lokhorst, C. and Berckmans, D. 2013. Analysis of individual classification of lameness using automatic measurement of back posture in dairy cattle. *J. Dairy Sci.* 96(1), 257–266. https://doi.org/10.3168/jds.2012-5806.

Viazzi, S., Bahr, C., Van Hertem, T., Schlageter-Tello, A., Romanini, C. E. B., Halachmi, I., Lokhorst, C. and Berckmans, D. 2014. Comparison of a three-dimensional and two-dimensional camera system for automated measurement of back posture in dairy cows. *Comput. Electron. Agric.* 100, 139–147. https://doi.org/10.1016/j.compag.2013.11.005.

Volkmann, N., Kulig, B., Hoppe, S., Stracke, J., Hensel, O. and Kemper, N. 2021. On-farm detection of claw lesions in dairy cows based on acoustic analyses and machine learning. *J. Dairy Sci.* 104(5), 5921–5931. https://doi.org/10.3168/jds.2020-19206.

von Keyserlingk, M. A. G., Barrientos, A., Ito, K., Galo, E. and Weary, D. M. 2012. Benchmarking cow comfort on North American freestall dairies: Lameness, leg injuries, lying time, facility design, and management for high-producing Holstein dairy cows. *J. Dairy Sci.* 95, 7399–7408. https://doi.org/10.3168/jds.2012-5807.

Voulodimos, A. S., Patrikakis, C. Z., Sideridis, A. B., Ntafis, V. A. and Xylouri, E. M. 2010. A complete farm management system based on animal identification using RFID technology. *Comput. Electron. Agric.* 70, 380–388. https://doi.org/10.1016/j.compag.2009.07.009.

Walker, A. M., Pfau, T., Channon, A. and Wilson, A. 2010. Assessment of dairy cow locomotion in a commercial farm setting: The effects of walking speed on ground reaction forces and temporal and linear stride characteristics. *Res. Vet. Sci.* 88(1), 179–187. https://doi.org/10.1016/j.rvsc.2009.05.016.

Weigele, H. C., Gygax, L., Steiner, A., Wechsler, B. and Burla, J. B. 2018. Moderate lameness leads to marked behavioral changes in dairy cows. *J. Dairy Sci.* 101(3), 2370–2382. https://doi.org/10.3168/jds.2017-13120.

Westin, R., Vaughan, A., de Passillé, A. M., DeVries, T. J., Pajor, E. A., Pellerin, D., Siegford, J. M., Vasseur, E. and Rushen, J. 2016. Lying times of lactating cows on dairy farms with automatic milking systems and the relation to lameness, leg lesions, and body condition score. *J. Dairy Sci.* 99(1), 551–561. https://doi.org/10.3168/jds.2015-9737.

Whay, H. R., Waterman, A. E., Webster, A. J. and O'Brien, J. K. 1998. The influence of lesion type on the duration of hyperalgesia associated with hindlimb lameness in dairy cattle. *Vet. J. Lond. Engl.* 156(1), 23–29.

Willgert, K. J. E., Brewster, V., Wright, A. J. and Nevel, A. 2014. Risk factors of lameness in sows in England. *Prev. Vet. Med.* 113, 268–272. https://doi.org/10.1016/j.prevetmed.2013.10.004

Williams, M. L., Parthaláin, N. M., Brewer, P., James, W. P. J. and Rose, M. T. 2016. A novel behavioral model of the pasture-based dairy cow from GPS data using data mining and machine learning techniques. *J. Dairy Sci.* 99(3), 2063–2075. https://doi.org/10.3168/jds.2015-10254.

Willshire, J. A. and Bell, N. J. 2009. An economic review of cattle lameness. *Cattle Pract.* 17, 136–141.

Winter, J. R., Kaler, J., Ferguson, E., KilBride, A. L. and Green, L. E. 2015. Changes in prevalence of, and risk factors for, lameness in random samples of English sheep flocks: 2004–2013. *Prev. Vet. Med.* 122(1–2), 121–128. https://doi.org/10.1016/j.prevetmed.2015.09.014.

Wolfger, B., Jones, B. W., Orsel, K. and Bewley, J. M. 2017. Technical note: Evaluation of an ear-attached real-time location monitoring system. *J. Dairy Sci.* 100(3), 2219–2224. https://doi.org/10.3168/jds.2016-11527.

Yunta, C., Guasch, I. and Bach, A. 2012. Short communication: Lying behavior of lactating dairy cows is influenced by lameness especially around feeding time. *J. Dairy Sci.* 95(11), 6546–6549. https://doi.org/10.3168/jds.2012-5670.

Zhuang, S., Maselyne, J., Van Nuffel, A., Vangeyte, J. and Sonck, B. 2020. Tracking group housed sows with an ultra-wideband indoor positioning system: A feasibility study. *Biosyst. Eng.* 200, 176–187. https://doi.org/10.1016/j.biosystemseng.2020.09.011.

Zillner, J. C., Tücking, N., Plattes, S., Heggemann, T. and Büscher, W. 2018. Using walking speed for lameness detection in lactating dairy cows. *Livest. Sci.* 218, 119–123. https://doi.org/10.1016/j.livsci.2018.10.005.

Chapter 11

Developments in automated monitoring of livestock fertility/pregnancy

Michael Iwersen and Marc Drillich, University of Veterinary Medicine Vienna, Austria

1 Introduction

Livestock production has been characterized by the intensification and specialization of production leading to larger farms with increased productivity. While a substantial increase in average milk production per cow has been achieved worldwide (Fig. 1), reproductive performance has been compromised over the same period of time.

For US Holstein-Friesian (HF) herds, for example, the first service conception rate decreased by 0.45% per year between 1975 and 1997 (Butler and Smith, 1989; Beam and Butler, 1999). A similar trend was found by Royal et al. (2000), who reported a decrease in first service pregnancy rate of 1% on average per year between 1975 and 1998 to approximately 40% for HF cows in England. In the Netherlands, first insemination success decreased from 55.5% to 45.5% over 10 years (Jorritsma and Jorritsma, 2000). This development is of significance because reproductive performance is one of the most important factors influencing the economic success of a dairy farm. In addition to reducing the amount of milk produced on a farm, reproductive inefficiency reduces the number of offsprings and further increases the involuntary culling rate (Gröhn and Rajala-Schultz, 2000). Today's average time of a dairy cow staying in the herd is only 4.5–5.5 years or 2.5–3.5 lactations (Wathes et al., 2008; Knaus, 2009). The biologically achievable age of a dairy cow of approximately 20 years (De Vries and Marcondes, 2020) is thus far from exhausted, and hence the breeding

http://dx.doi.org/10.19103/AS.2021.0090.12

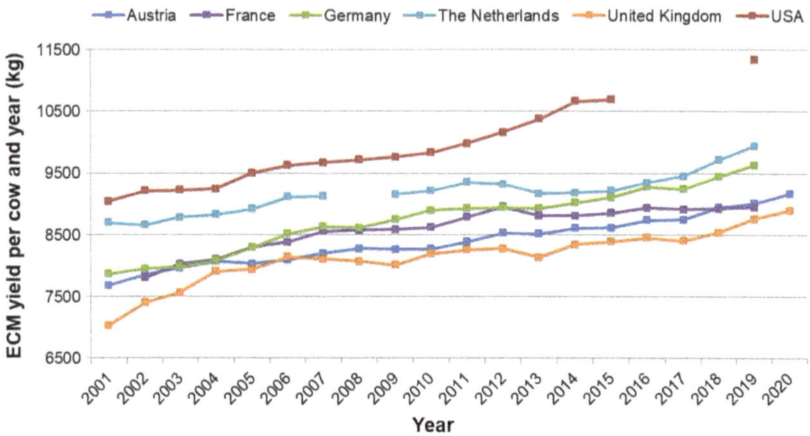

Figure 1 Average energy-corrected milk yield (ECM; based on 4.0% butterfat and 3.4% protein) for Holstein dairy cows per year in selected countries. Data provided by the International Committee for Animal Recording (ICAR, 2021).

potential of the animals, as well as the genetic progress of a herd, is not being realized on most farms. In this context, detecting cows in oestrus and getting cows pregnant are still two major challenges on the farm (Denis-Robichaud et al., 2018).

Many resources have been spent on research to elucidate the causation of the potential antagonism between increased milk yield and decreased reproductive performance, and controversial results have been reported in this regard. However, declines in reproductive performance are rarely monocausal, but rather multifactorial in nature (Leroy et al., 2006). If any effect of increased milk production on fertility performance can be shown at all, it is considered to be of minor importance compared to other factors (Lucy, 2001). Diseases exert a far greater influence on fertility performance than, for example, environmental or other animal-specific factors. In addition to diseases of the uterus and ovaries, whose direct influence on fertility is obvious, mastitis and metabolic diseases such as ketosis (Walsh et al., 2007; Chapinal et al., 2012b) are also reported to have a detrimental effect on reproductive performance. However, it is worth mentioning that diseases affect only part of the herd, while environmental factors affect the entire herd (Loeffler et al., 1999). This indicates the potential for optimizing management and housing conditions to meet the requirements of high-producing dairy cows and, hence, to ensure good reproductive performance, animal health, and well-being.

In modern dairy farming, the reproduction of dairy cows is performed mainly by artificial insemination (AI). Compared to the previous use of bulls, individual mating can be achieved, for example, to exploit the genetic potential of the herd. Furthermore, the risk of transmission of venereal diseases by natural services and the risk of accidents when handling bulls can be minimized. However, when AI is used, the optimal insemination time must be detected by farmers (or technologies); this task is otherwise performed continuously and reliably by the bull.

Over the past decades, the percentage of animals in oestrus showing a 'standing-to-be-mounted' (StbM) behaviour has declined from 80% to 50%. In the same period, the duration of StbM has declined from 15 h to 5 h and first-service pregnancy rate has dropped from 70% to 40% (Dobson et al., 2008). Furthermore, a large proportion of dairy cows show oestrus during the evening and night hours. This further complicates the visual oestrus detection by farmers and/or their personnel and may result in overlooked events. A well-managed dairy farm should achieve a heat detection rate of at least 70% and a pregnancy rate of 50–60% (Esslemont and Kossaibati, 2000). The average oestrus detection rate in high-yielding dairy herds, however, is often around 50% (Esslemont, 1992; Drillich, 1999; Mee et al., 2002; Gaude et al., 2021), or in other words, every other heat is missed. Due to the aforementioned challenges, many farmers use hormone programs, for example, for oestrus induction, followed by timed AI (i.e. OvSynch). These programs, however, are associated with lower conception rates, increased costs per AI, and in contrast to consumers' perceptions of dairy farming (Barkema et al., 2015; Pieper et al., 2016). These data and facts support the need for the use of reliable and non-pharmaceutical methods to assist farmers in heat detection, for example, by use of sensor-technologies.

2 The oestrous cycle in dairy cows

The success of AI in a dairy herd is largely dependent on an accurate and efficient detection of oestrus, which is also referred to as 'standing heat' or 'heat.'

This chapter will explain the physiological basics of the reproductive cycle in dairy cows. Special emphasis is paid to oestrus-associated changes in animal behavioural traits and physiological parameters. For detailed information on the (patho)physiology of animal reproduction, please refer to the relevant literature.

After puberty, heifers enter the phase of reproductive cyclicity, which is associated with the alternation between readiness to be mated and the rejection of the sexual partner. The sexual cycle generally persists into the older age of the animals, and a biological lifespan of a dairy cow of approximately 20 years is reported (De Vries and Marcondes, 2020). Domesticated cattle breeds show constant cyclicity throughout the year, that is, they are aseasonally polyestrous. The interval between the onset of two periods of sexual receptivity (oestrus) is defined as the oestrous cycle. The duration of a dairy cow's oestrous cycle is described with a mean of 21 days and a range of 18–24 days (Grunert, 1999; Hartigan, 2004; Forde et al., 2011; Greenham et al., 2019). In a retrospective data analysis of approximately 40 000 dairy cows in the UK, 59% of inter-service intervals fell within this range, while 90% fell within an even broader range of 18–28 days (Remnant et al., 2015).

The oestrous cycle can be divided into two main phases (Fig. 2). The 'luteal phase' lasts approximately 14–18 days and is characterized by the formation and maintenance of a corpus luteum (CL). The 'follicular phase' lasts approximately 4–6 days and is characterized by luteolysis (regression of the CL), follicular growth, and ovulation of the dominant follicle (Forde et al., 2011).

During the oestrous cycle, well-orchestrated and transient changes in the concentration of hormones in the blood can be observed, which enable communication among various tissues and lead to specific behaviours. Based on the hormonal dynamics illustrated in Fig. 2, the oestrous cycle can be divided into proestrus, oestrus, metestrus, and diestrus (Rathbone et al., 2001). The detailed description of the fine-tuned and complex hormonal regulation of the oestrous cycle is beyond the scope of this chapter and is presented in detail elsewhere.

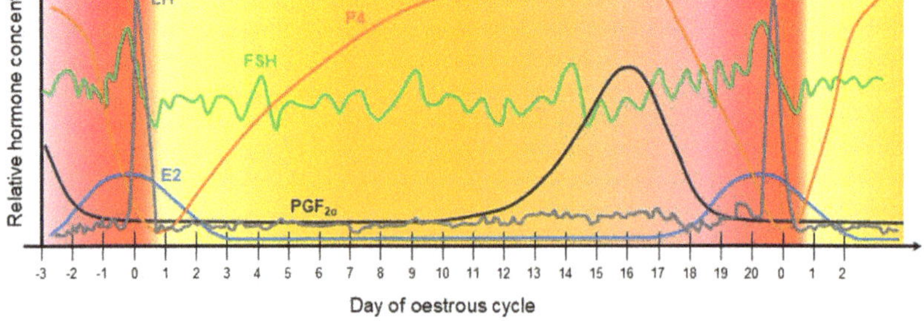

Figure 2 Schematic illustration of the hormonal dynamic during the oestrous cycle and associated structures on the surface of the ovaries: (a) maturing 'Graafian' follicle; (b) ovulation with the release of the ovum; (c) formation of the corpus luteum from remaining granulosa and theca internal cells; (d) functional corpus luteum; (e) initiation of luteal regression; (f) maturation of a dominant follicle which further develops into (g) a new Graafian follicle. P4, progesterone; FSH, follicle-stimulating hormone; E2, estradiol; LH, Luteinizing Hormone; $PGF_{2\alpha}$, prostaglandin $F_{2\alpha}$.

2.1 Proestrus

Proestrus is the period from luteolyses to the onset of oestrus and lasts approximately 2–3 days. After regression of the CL, the dominant follicle continues to develop into the pre-ovulatory follicle. External signs of proestrus include the onset of vulvar swelling, hyperemic and increasingly moist vaginal mucosa, and an increasing accumulation of mucus in the vulva. The quality of the mucus changes from moderately viscous to transparent and stringy during proestrus. With the progress of proestrus, signs of oestrus increase. In particular, the restlessness of the animals as well as contact and mounting attempts of herd mates can be observed.

2.2 Oestrus

Referring to Senger (2012), the term 'oestrus' originates from a Greek word meaning 'gadfly, sting or frenzy' and was used to describe *Oestridae*, a family of parasitic biting insects. Oestrus-relates behaviours were deemed similar to those observed during insect attacks (i.e. restlessness, tail flailing), and hence the term oestrus was applied to the period of sexual receptivity. Standing-to-be-mounted behaviour, that is, the cow in oestrus stands immobile for a few seconds when mounted by herdmates or a bull, is the primary and most reliable observable behaviour, indicating that a cow is in oestrus and ready to be mated or inseminated. Compared to other oestrus-associated behaviours, StbM is rarely seen at other times of the oestrous cycle (Van Eerdenburg et al., 1996; Roelofs et al., 2005a). Mounted cows are more likely to be in oestrus than mounting cows (Hurnik et al., 1975). Unfortunately, pregnant cows also sometimes show oestrus behaviour (Roelofs et al., 2010; Senger, 2012), for example, StbM. Inaccurate AI, that is, artificial insemination of a cow not in oestrus, has impacts on farm profitability due to semen wastage and iatrogenic pregnancy loss. Literature data on the prevalence of inaccurate AI show considerable variation and were reported to range between 3% and 31% in a recent publication by Kelly et al. (2021).

Other behaviours associated with the occurrence of oestrus include increasing acts of aggression (butting), investigatory behaviours (licking and sniffing at the anogenital region of other animals, chin resting on the back, orientation behind an animal with or without mounting), disordered mounting attempts (without standing and/or disoriented), and, to a lesser extent, Flehmen, that is, curling back of the upper lip by exposing the front teeth to transfer pheromones and other scents into the vomeronasal organ (Esslemont et al., 1980; Roelofs et al., 2005a). Other classic (external) signs of oestrus triggered by estrogens include increased edema and tissue tone, as well as increased secretory activity of the cervical glands, often observed as a discharge of more viscous and clear mucus from the external genital tract (Roelofs et al., 2010). Internal signs of oestrus that can be palpated or observed, for example by vaginoscopy, include maximal contraction of uterus, the presence of a large (approximately 12–16 mm) bulging follicle, the straw-to-pencil-sized opening of the cervical channel, hyperemia of the vaginal mucosa, and increased accumulation of mucus with resultant wetness.

2.3 Metestrus

The transition between oestrus and metestrus, which is also referred to as postestrus, is smooth. Metestrus begins with the end of the animal's readiness to mate and the disappearance of external and internal signs of oestrus and lasts 1–4 days. At the beginning of this period, the concentrations of luteinizing hormone (LH), follicle stimulating hormone (FSH), and estrogens in the blood decrease rapidly (Senger, 2012). Ovulation of the mature follicle occurs on the first day of postestrus. As an external sign of postestrus, the discharge of a bloody mucus from the vagina and/or blood stains on the tail can sometimes be observed (Diskin and Sreenan, 2000). The blood admixtures originate from the estrogen-induced hyperemic uterine mucosa and enter the lumen of the uterus via capillary hemorrhage (Ohtani et al., 1993). The appearance of bloody mucus merely indicates that ovulation has occurred but does not indicate whether or not fertilization has taken place (Grunert, 1999).

2.4 Diestrus

The diestrus, which is also referred to as interestrus, starts approximately between the third and fifth day of the cycle and is the longest period of the sexual cycle, lasting about 15 days. This period, characterized by sexual quiescence, is dominated by the action of the progesterone (P4) producing CL and lasts until the onset of the subsequent proestrus.

3 Oestrus detection, pregnancy diagnosis and reproductive performance

3.1 Economic aspects of oestrus detection

Worldwide 'reproductive failures,' that is, the failure to conceive, are among the most common causes for involuntary culling of dairy cows (Schuster et al., 2020). For example, the percentage of culling due to infertility reported in 2019 by farmers to dairy herd improvement associations was 24.2% in Austria (ZuchtData, 2020) and 19.4% in Germany (VIT, 2020). These proportions are in agreement with the reported 20% in the US (Bascom and Young, 1998).

Besides increased replacement costs due to involuntary culling, additional farm losses are caused by reduced annual milk yield, increased use of semen and veterinary services, lowered calf output, and hence, non-exhaustion of genetic progress, as well as by increased maintenance costs for keeping the dry cow for longer (Esslemont and Kossaibati, 2000). The costs for extended 'days open' (DO, i.e. the number of days from calving to conception) are estimated at US$3–5 per day by dairy professionals. However, these costs may vary in expenses and revenues over time and between farms. A missed oestrus event and/or a failure to conceive requires an additional 21 days until the next chance to detect the cow in oestrus (Fetrow et al., 2007). Senger (1994) estimated annual losses for the US dairy industry due to poor oestrus detection to be at least US$300 million. Diskin and Sreenan (2000) reported that approximately 10% of non-detected oestruses are due to issues of the cow, while 90% are caused by poor management and/or housing conditions. Examples of poor management procedures related to poor oestrus detection include too few observations per day to check for oestrus activity, too little time taken to observe cows, and observing cows at the wrong time or place, such as at feeding times or in the holding area at milking times (Reith and Hoy, 2018).

To sum up, establishing effective oestrus detection is a key factor in reducing involuntary culling and thus increasing the economic success of a dairy farm. It is one of the most important tasks in daily herd management.

3.2 Factors affecting oestrus expression and oestrus detection

In addition to regular cyclic activity, animals must show signs of oestrus, which in turn must be detected by the farmer or the technologies that are used for this purpose. Although definitions and methods of determining the duration of oestrus vary between studies, there is evidence that the length of time cows exhibit oestrous behaviour has steadily shortened in recent decades (Table 1). In a review by Sheldon et al. (2006), a decrease in oestrus duration from 14.9 to 7.1 h in the period from 1976 to 1998, accompanied by a decrease in the average number of mounts per cow from 56.3 to 8.5, was reported.

The expression and duration of oestrous behaviours are influenced by technological factors, such as housing type, pen size, stocking density, and floor (Becker et al., 2005) and biological factors, including climatic conditions, animal nutrition, stage of lactation, milk yield, and hormonal treatments, as well as various diseases such as lameness (Gwazdauskas et al., 1983). These and other factors, such as endocrine disruptors, for example, from the chemical industry [reviewed by Petro et al. (2012)] can disturb the finely tuned hormonal cascade at different levels. For example, stressors leading to elevated levels of corticosteroids have been associated with a delayed or blocked pre-ovulatory peak of the LH affecting oestrous expression, as reviewed by Stevenson (2001).

Table 1 Reported oestrus duration

Reference	Number of cows or events,* breed	Average duration of oestrus (h, mean ± SD)
Esslemont and Bryant (1976)	60 HF	14.9 ± 4.7
Schams et al. (1977)	61 BS, HF	16.9 ± 4.9
Walker et al. (1996)	51 HF	9.5 ± 6.9
Xu et al. (1998)	98 (48 HF, 50 JE)	8.5 ± 0.5
Dransfield et al. (1998)	2661 AI, 17 herds	7.1 ± 5.4
At-Taras and Spahr (2001)	71, breed n.a.	5.8 ± 0.8
Lopez et al. (2002)	23 HF	3.6 ± 0.8
Roelofs et al. (2005b)	67 HF	11.8 ± 4.4
Sveberg et al. (2011)	20 HF	12.9 ± 1.8

*HF, Holstein Friesian; JE, Jersey; BS, Brown Swiss; n.a., not available.

3.2.1 Milk yield

Various epidemiological studies have been conducted to evaluate the association between milk yield and reproductive performance in dairy cows (Dematawewa and Berger, 1998; Fleming et al., 2019; Dominguez-Castaño et al., 2020). Valid data sets show that a reliable association between milk yield and reproductive performance is rare and 'temporal associations do not imply causation' (Leblanc, 2010). Furthermore, the management and housing conditions (e.g. ration balancing, feeding, use of hired workers) in dairy farming have changed over time and have to be considered in the analyses. If any effect of increased milk production on fertility performance can be shown at all, it is considered to be of minor importance compared to other factors (Lucy, 2001). Furthermore, the influence of milk quantity on oestrus intensity and duration is frequently discussed. Based on average milk production during 10 days before the day of oestrus, Lopez et al. (2004) reported shorter oestrus durations (6.2 ± 0.5 h vs. 10.9 ± 0.7 h), standing events (6.3 ± 0.4 h vs. 8.8 ± 0.6 h), and standing time (21.7 ± 1.3 h vs. 28.2 ± 1.9 h) for high-producing cows (≥39.5 kg/day) compared to lower producers (<39.5 kg/day). The authors assumed that an increased metabolism in higher-performing animals is associated with a more rapid degradation and clearance of steroid hormones due to elevated feed consumption and increased blood flow in the liver. In a meta-analysis by Bedere et al. (2018) the interval from calving to the first observed oestrus was used as an indicator for oestrus expression. For each additional 1 kg of milk yield, both, at peak and during the first 14 weeks of lactation, the meta-analysis revealed a delay in this interval by 1.1 days. The type of calving system (i.e. seasonal vs. the year round), genetics, level of concentrate feeding, and parity were identified as factors affecting the association between peak milk yield and the interval between calving and first observed oestrus.

3.2.2 Herd size

Traditionally, identification of cows in oestrus is often performed by visual observation. To achieve a sufficiently high oestrus detection rate, at least 3 observation periods of 20 min each per manageable group of animals and day are recommended (Van Eerdenburg et al., 1996). With increasing herd size, the time required by farmers and employees for oestrus monitoring increases and can often not or only insufficiently be achieved during their day-to-day business.

The appearance of oestrus-associated social and agonistic behaviours is influenced by space availability and stocking density, among other factors. If the stocking density is too high, the expression of oestrus-related behaviours is reduced, making oestrus detection more challenging (Diskin and Sreenan, 2000). On the other hand, increasing herd size facilitates oestrus detection because the possibility to form 'sexually active groups' increases (Roelofs et al., 2010). The intensity of heat expression, and thus the possibility of oestrus detection, increases with the number of cows in oestrus at the same time (Hurnik et al., 1975; Helmer and Britt, 1985; Van Vliet and Van Eerdenburg, 1996; Diskin and Sreenan, 2000; Roelofs et al., 2005b; Cutullic et al., 2009; Sveberg et al., 2011).

Cows that are in the luteal phase of the cycle and pregnant cows show less interest in mounting a cow in oestrus. As the number of pregnant cows increases the likelihood of detecting cows in oestrus at the same time decreases, thus making oestrus detection more difficult (Diskin and Sreenan, 2000) in smaller herds or groups. Other factors influencing group dynamics (e.g. frequency of regrouping, hierarchy, age structure) are also relevant.

3.2.3 Ambient temperature

Heat-stress has several negative effects related to fertility in cows, including decreased oestrus detection rates [reviewed by De Rensis and Scaramuzzi (2003)]. Gwazdauskas et al. (1983) reported that oestrous activity can be maximized by keeping the environmental temperature within the animals' comfort zone; cold temperatures and temperatures above 30°C inhibit oestrus expression. This aspect is particularly important in the context of global warming and represents an additional challenge for heat detection, conception, and the maintenance of pregnancy.

3.2.4 Lameness

Along with fertility problems and mastitis, farmers perceive the occurrence of lameness as one of the top three health problems in their herds (Denis-Robichaud et al., 2018), with prevalence ranges from 10% to 55% (von Keyserlingk et al., 2012; Adams et al., 2017). Lameness is a painful disease that influences the standing and laying behaviour of the animals. Affected cows show greater lying times and less time walking or standing (Roelofs et al., 2010). Furthermore, lame cows are mounted less frequently and show less intense oestrus behaviour (Walker et al., 2010).

3.2.5 Negative energy balance

Poor animal nutrition and a decrease in body condition due to negative energy balance (NEB) are reported to have negative effects on oestrous intensity (Walsh et al., 2011). Rutherford et al. (2016) reported a lower peak activity and shorter duration of activity clusters in cows with subclinical ketosis [ß-hydroxybutyrate (BHB) concentration in blood of 1.2–2.9 mmol/L] compared to healthy cows. Moreover, calving to first oestrus, calving to first insemination, and calving to pregnancy intervals were prolonged in cows with hyperketonemia. In this context, Lucy (2000) reported that an increased NEB is associated with decreased pulsatile secretion of LH and Insulin-like growth factor 1 (IGF-1) concentrations (Diskin et al., 2003) and hence compromised follicular competence for E2 synthesis, which result in poorer oestrus expression.

The association between increased ketone body concentration (as a result of excessive NEB) and reproductive performance is controversial in the literature. Walsh et al. (2007) reported that the probability of pregnancy was reduced by 20% in cows suffering from NEB and subclinical ketosis (BHB concentration in blood of ≥1.4 mmol/L) in the first or second week of lactation. Cows suffering from subclinical ketosis in both weeks were 50% less likely to conceive after the first AI. In contrast, other authors could not demonstrate any association between increased ketone body concentration and reproductive performance (Chapinal et al., 2012a; McArt et al., 2012).

3.3 Pregnancy testing

Consistent performed and early detection of pregnancy is a cornerstone of reproductive management and a safety measure to improve reproductive performance and profitability on commercial farms. It can be assumed that approximately every fourth cow or heifer remains non-pregnant after the first AI, because either the insemination time was chosen wrongly or because of embryonic death during pregnancy. Early detection of non-pregnancy is a prerequisite for bringing open cows and heifers back into oestrus (e.g. by use of hormones) immediately after detection, so that they can be inseminated or mated again. By reducing the interval between successive AI services, reproductive performance and pregnancy rates can be increased. Due to the non-negligible early embryonic death, which was estimated to be 11% in the period between 28 and 42 days after breeding (Vasconcelos et al., 1997), it should be noted that cows diagnosed pregnant in early gestation should undergo at least one subsequent pregnancy test.

Close animal observation between 28 and 32 days after AI or mating for signs of oestrus is considered as easiest and last costly method by farmers. Not only because of the previously mentioned low visual oestrus detection rates, animal observation is rather uncertain for pregnancy detection and only gives an indication of pregnancy. Transrectal palpation of the uterus and its contents is the oldest and most widely used method for pregnancy diagnosis in cattle and considered as an industry standard for evaluating new direct and indirect methods for pregnancy diagnosis in dairy cows (Fricke et al., 2016). Other methods commonly used in practice for pregnancy testing are transrectal ultrasound examination (Müller et al., 1999; Heuwieser et al., 2000), determination of progesterone concentration in milk and blood (actually, this test can only confirm a non-pregnancy), and detection of bovine pregnancy-associated glycoproteins (PAGs) (Green and Hennessy, 2018). Evaluating the suitability of other biomarkers (e.g. interferon-tau and interferon-tau-stimulated genes; circulating microRNAs; lactoferrin, lactotransferrin, and alpha1G in milk) to detect pregnancy is currently under research (Szenci, 2021). Cost-effective and automated detection of these biomarkers using sensor technologies is also being worked on.

3.4 Detection performance

The aim of oestrus detection as well as pregnancy testing is, on the one hand, high efficiency (how many cows are detected in oestrus/as pregnant) and, on the other hand, high accuracy, that is, of those animals detected in oestrus/as pregnant, what proportion is really in oestrus/pregnant (Fetrow et al., 2007). With regard to oestrus detection, high efficiency means that as many animals as possible eligible for insemination are detected in oestrus and inseminated per time unit (for example in 21 or 42 days).

For method validation purposes, the terminology 'true positive' (TP), 'false positive' (FP), 'true negative' (TN), and 'false negative' (FN) refers to the comparison of the predicted result with the true condition, as presented in Fig. 3 for oestrus detection. This and the following also apply for, for example, pregnancy tests.

With regard to animal reproduction, correct outcomes in oestrus detection include the cases where an animal is either in oestrus with a positive oestrus test (TP) or not in oestrus with a negative test result (TN). Unrecognized oestrus events (FN) are problematic for the economic success of a dairy enterprise because the opportunity of pregnancy and hence 'maximizing the profits' are not realized. On the other hand, oestrus detection methods that lead to a high rate of false positive events can also have a detrimental effect on parameters of reproductive performance as well as on the farm economics.

Predicted oestrus	*Reproduction cycle status:* ***oestrus***	*Reproduction cycle status:* ***not in oestrus***	
positive	True positive (TP)	False positive (FP)	**Positive predictive value =** True positive (TP)/ Predicted condition positive (TP+FP)
negative	False negative (FN)	True negative (TN)	**Negative predictive value =** True negative (TN)/ Predicted condition negative (FN+TN)
	Sensitivity= True positive (TP)/ True condition positive (TP+FN)	**Specificity=** True negative (TN)/ True condition negative (FP+TN)	

Figure 3 Contingency table for the evaluation of oestrus detection tools.

4 Methods for oestrus detection in cows

Due to the various oestrus-associated behavioural changes, numerous analog and digital technologies have been used for several decades to support visual oestrus monitoring by farmers and/or to provide fully automated oestrus detection. An optimal oestrus detection method ideally ensures compliance with the following criteria: reliable identification of the animals, continuous monitoring (24/7), minimal or no labor requirement, low investment and maintenance costs, and high detection rate combined with high accuracy (Senger, 1994; Diskin and Sreenan, 2000). So far, none of the oestrus detection methods used on farms completely fulfils the before-mentioned criteria. Hence, to achieve maximum oestrus detection rates, many farmers rely on a combination of different methods, some based on traditional heat detection methods and some on sensor-based heat detection, which are described below.

4.1 Traditional and non-automated methods

Complementary to visual animal observation, numerous non-automated tools and methods can be used for heat detection. Some of them are widely used in practice, while others are interesting approaches rarely implemented in practice.

4.1.1 Visual observation

Visual observation of cows and heifers (with or without the use of additional tools) remains the most widely used oestrus detection method worldwide.

In a study by Van Eerdenburg et al. (1996) cows eligible for insemination were observed for 30 min every 2 h (i.e. 12 intervals per day) and various signs of oestrus were recorded (Table 2). Each symptom was given a score and scores were summed over 24 h. If the sum of the scores was >100 within 24 h, an animal was considered as being in oestrus. Using this procedure, 100% of oestruses (confirmed by P4 concentrations) were detected, with StbM events observed in only 37% of oestruses. When reducing the observation periods to three per day, which is more in line with practice, StbM events were recorded in less than 25% of oestruses (Van Eerdenburg et al., 1996; Gaude et al., 2021).

To achieve a heat detection rate of >80%, at least three observation periods of 20 min each per observable group of animals are recommended throughout the day (Van Eerdenburg et al., 1996). This 'ideal situation' is not feasible on many farms due to economic pressure, increasing herd sizes, and often prevailing personnel shortages.

4.1.2 Colour-based oestrus detection aids

To assist farmers in visual oestrus detection, a variety of color change-based aids are used in practice. For example, chalk, livestock paint, or markers are used to mark the tail-head of the animals. Abraised or smeared color marks give an indication that the animal has been mounted by other herdmates. Other tools consist of scratchcards and pressure-sensitive color cartridge systems, which are self-adhesive or glued to the tail-head. Examples are the Chin-Ball Mating Device® (Paviour Ltd., New Zealand) and the KaMaR Heat Mount Detector® (Kamar Inc., USA). A special feature is the Bovine Beacon System® (OmniGlow

Table 2 Number of observed signs of oestrus during a 6-week observation in 2 Dutch dairy herds performed by Van Eerdenburg et al. (1996) together with suggested scorings

	Observations (n) during		
Behaviour	Oestrus	Diestrus	Score
Mucous vaginal discharge	13	44	3
Flehmen	34	69	3
Restlessness	73	21	5
Being mounted w/o standing	28	24	10
Sniffing at anogenital region of other cows	246	113	10
Resting with chin on other cows	228	54	15
Mounting attempts to other cows	218	5	35
Mounting head side of other cows	11	0	45
Standing while mounted	34	0	100

LLC, USA): the fluorescent dye used here is highly visible even in low-light conditions.

For effective use of the above-mentioned systems, timely application, regular checks, interpretation, and, if necessary, a re-staining are required (Gordon, 2011). As an example, in a study by Cavalieri et al. (2003) in which tail paint and a color cartridge system (Kamar Heatmount detector, Agri-Gene, Wangaratta) were used with sensitivities and PPVs of 91% and 92%, and of 86% and 93%, respectively. However, depending on the housing conditions, high error rates are reported for colour-based oestrus detection aids, for example, when using cow brushes or animals are in close contact with the barn equipment (Becker et al., 2005).

4.1.3 Rectal palpation

Rectal palpation is used by many practitioners as an additional source of information in oestrus detection or confirmation. For example, rectal palpation of the uterus and ovaries can provide an assessment of the status of the cow's oestrus cycle. Characteristics of the rectal gynecological examination of an animal in oestrus are a contractile uterus, a CL that has regressed to a size of <10 mm, and a prominent, fluctuating follicle.

At this point, it should be mentioned that routine rectal gynecological examination of cows aims primarily at detecting animals with problems (e.g. endometritis, ovarian cysts, inactive ovaries, silent heat) that can be treated early and, subsequently, inseminated in a timely manner. A general examination of animals presumed to be in oestrus may be counterproductive and is not recommended, since manual palpation of the ovaries increases the risk of prematurely triggered ovulation.

4.1.4 Teaser animals

The use of teaser animals, with or without marking devices, is common on some farms. For this purpose, fertilization and/or mating with incapable bulls or androgenized cows (Mortimer et al., 1990) are used to detect cows in oestrus via olfactory stimuli obtained at the vulva or from urine, for instance. These animals vary in their sexual aggressiveness and have the potential to generally increase the sexual activity in the herd (Morgan and Dawson, 2008). As described in Section 3.2, the more animals are sexually active at the same time, the more frequently each animal is mounted in oestrus. To mark animals in oestrus, teaser animals can be equipped with a harness for color marking. Mortimer et al. (1990) reported an increased number of mounts (10.3 vs. 9.4 per observation period, $P<0.05$) and attempts to mount (16.2 vs. 14.4 per observation period, $P<0.05$) cows in oestrus by adding a freemartin heifer with testosterone propionate and estradiol benzoate implant, kept in a lot of dairy cows. The authors concluded that the addition of an androgenized heifer with a 30-minute observation period improved the likelihood of detecting cows in oestrus. In contrast, Gwazdauskas et al. (1990) reported a relatively high incidence of ineffectiveness of androgenized females when used as the sole method for oestrus detection. In addition to animal welfare concerns associated with animal infertility procedures in bulls (Becker et al., 2005), there is a risk of transmission of venereal diseases with the use of animals capable of mating. Besides general costs of maintaining a bull, the risk of injury to farm personnel when bulls are used in the herd should not be underestimated. Every year, farmers and other personnel are killed accidentally.

4.2 Sensor technologies

Since the 1980s, sensors have been developed to record animal-related parameters in livestock farming. For dairy cows, the development started with identification of individual animals and was followed by sensors for mastitis monitoring and pedometers to record the animal's activity for oestrus detection.

While general herd management procedures in the past focused on larger groups or pens of animals, modern sensor technologies have the potential to pay more attention to an individual cow within the herd. For managers of larger farms, sensors offer the opportunity to remove subjectivity from decision-making processes and reduce the need for skilled co-workers experienced in animal handling and management (Borchers, 2015), in particular, in oestrus detection.

In general, sensors used in dairy cows can be divided into attached and nonattached systems. Attached sensors can either be placed inside the cow's body (in-cow), for example, as bolus, or fixed on the cow's body (on-cow), for example, with collars, wristbands, and on the ears. Nonattached (off-cow) sensors were developed for measurements while cows pass by, through or over the device. Furthermore, nonattached sensors can be classified as in-line and on-line sensors. While in-line sensors were designed for continuous measurement of a fluid (e.g. milk) flowing through a pipe (e.g. in the milk line), on-line sensors take a sample for analyses (Rutten et al., 2013).

Due to the various changes in physiology as well as behaviours associated with oestrus, various sensor technologies have been developed for oestrus detection (Fig. 4). While some technologies remain in the experimental stage or need further development, other technologies are successfully used in practice for (automated) oestrus detection.

While oestrus detection by earlier electronic systems was often based on exceeding fixed thresholds, newer and more sophisticated developments use artificial intelligence methods to determine animal-specific thresholds. This

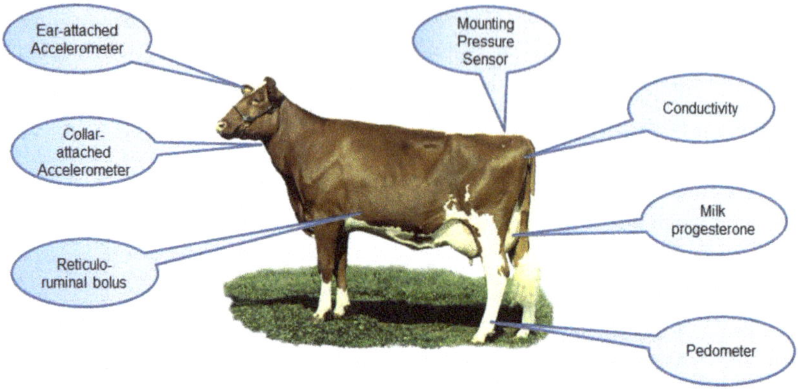

Figure 4 Examples of available technologies for heat detection attached to or in the animal.

allows the detection of even subtle changes in animal behaviour, which have been reported for several cows in oestrus (Yániz et al., 2003). In addition to the technologies used on and in the animal, technologies are increasingly being developed that are located in the environment of the animals and are based, for example, on image, video, or sound recordings. By monitoring groups of animals, these developments can help to reduce costs. Since these technologies do not come into direct contact with the animal, they can be considered non-invasive and allow the animals to show undisturbed behaviours.

4.2.1 Heat mount detectors

Pressure-sensitive devices activated by mounting events are fixed to the tail-head of the cows by glue or the use of a strap. An example of a stand-alone system is the DEC device (IMV Technologies, France). The number, length, and interval between each mounting are analyzed by a device-internal microprocessor. If a specific threshold is exceeded, an oestrus alarm is signaled by a flashing LED. The flashing frequency is proportional to the time elapsed since the first mounting. The onset of oestrus is indicated by the number of flashes within a 10-second period (Diskin and Sreenan, 2000; Saumande, 2002). An example of the technological advancement is the Heat Watch (DDX Inc., Denver, CO, USA) system. The device records the time and length of mountings and sends the data via radio-telemetry to receivers for further processing and storage on a computer. Software algorithms examine animal-specific mounting profiles, and information is relayed back to each cow's transmitter. The processed data are available in the form of lists, tables, and graphs. Based on the number of detected StbM events within a period of 4 h, animals are classified as being 'in oestrus' (≥3 mountings) or 'suspected of being in oestrus' (1–2 mountings).

Cows with no detected StbM event within 25 days of being equipped with the system are considered as 'inactive,' and those detected with more than one oestrus event in a period of 13 days are listed on a 'short-cycler' list (At-Taras and Spahr, 2001).

The use of tail-fixed pressure sensors can cause problems in terms of equipment loss or unreliable data when cow brushes are used in the pens. Investment costs of approximately US$80 for a single sensor and about US$5000–10 000 for an entire system have been reported by Becker et al. (2005).

4.2.2 Conductivity of vaginal discharge

A swollen vulva is one of the classic, visually observable symptoms of a cow in oestrus. The swelling is the result of increasing hydration of the tissues associated with changes in tissue and secretion composition as well as in their electrical resistance. Research into the previously described changes to be used for oestrus detection had its peak in the 1970s and was reviewed by Senger (1994). The electrical resistance of vaginal fluid is high during the luteal phase, decreases during the follicular phase and is lowest during oestrus (Leidl and Stolla, 1976; Foote et al., 1979). Repeated measurements are needed to detect physiological changes over time. This requires repeated (manual) sampling of the cows by the farmer, which is labor-intensive, or the use of sensor technology surgically implanted or inserted into the vagina (Feldmann et al., 1978; Lewis et al., 1989). Frequent intervention in the vagina or insertion of sensors can lead to irritation of the vaginal mucosa and increases the risk of inflammation, which can distort the measurements. In a review of various oestrus detection methods conducted by Becker et al. (2005), the cost of vaginal probes was reported to range from US$400 to US$2000 per animal.

Due to the previously mentioned disadvantages, costs, and insufficient test characteristics, this method for oestrus detection has not gained acceptance in practice. Nevertheless, this method is at times being worked on again in research (Hockey et al., 2010; Liu and He, 2019).

4.2.3 eNoses

Olfactory technologies have been used for identifying changes in bovine body fluids, for example, occurring during oestrus. For this purpose, 'electronic noses' (eNoses) have been developed for the identification of volatile organic constituents emitted from the anogenital region, from feces and also from breath sampling (Lane and Wathes, 1998; Wiegerinck et al., 2011; Manzoli et al., 2019). To analyze these chemical compounds in the gaseous phase, an array of different chemical sensors is used, and pattern recognition models are used for classification (Wiegerinck et al., 2011).

In a pilot study on 52 cows, breath-samples exhaled into a handheld bucket-like sampling device were analyzed by the use of an electronic nose, consisting of three different metal-oxide sensors. Using a dataset of 36 oestrus and 35 non-oestrus events, a Se of 83% and Sp of 86% were achieved (Sanderink et al., 2017).

Although some technologies have been described as suitable for oestrus detection (Sanderink et al., 2017; Manzoli et al., 2019), this technology has not yet been established on farms, yet. A disadvantage of current technologies is the repeated manual sampling for which cows must be fixed.

4.2.4 Temperature measurement

Research activities on oestrus- and pregnancy-associated changes in body temperature have been conducted since at least the 1950s (Wrenn et al., 1958). In recent decades, sensors and implants have increasingly been used to measure temperature directly in various areas in or on the animal. Furthermore, milk temperature has been used as an indirect measure of changes in body temperature for oestrus detection. As an example, temperature measurements on the body surface, and in the deep body, the vagina and milk should be mentioned here, as reviewed by Lehrer et al. (1992), among others.

Wrenn et al. (1958) reported a decrease in vaginal temperature towards oestrus, reaching a minimum 2 days before its onset. The vaginal temperature increased sharply on the day of oestrus and decreased again one day later (i.e. after ovulation). An increase in the intravaginal temperature of 0.3–1.1°C around oestrus and between 21 h and 12 h before ovulation has been reported (Clapper et al., 1990; Mosher et al., 1990; Wang et al., 2020). However, ambient temperature accounts for more of the variance than the oestrus event itself. Burfeind et al. (2010) found that the measurement of rectal temperature can vary by up to 0.5°C depending on the person performing the measurement as well as on the model and the penetration depth of the thermometer. However, repeated manual measurement is labor-intensive and leads to an increased risk of pathogen introduction into the vagina and subsequent infections. Therefore, the method is considered unsuitable for routine oestrus monitoring.

As an indirect measure of body temperature, the increase in milk temperature has also been used to predict oestrus events. Based on a 2-year experiment, Schlünsen et al. (1987) reported that in only 42% of the cows, the optimal time of oestrus was detected by measuring milk temperature. When milk temperature is measured only twice - in the morning and the evening - the authors suggested that the short oestrus-induced temperature peak might not be detected.

For continuous measurement of the core body temperature, rumen-reticular boluses, for example, the SmaXtec (Smaxtec, Graz, Austria) and

TempTrack (DVM Systems, Greeley, CO, USA) devices, have been developed. The evaluation of the TempTrack bolus by Burnett et al. (2020) with 190 true oestrus events from 97 HF cows, however, revealed a wide variation in alerts in predicting oestrus and ovulation events and thus the system was not recommended by the authors.

All of the above-mentioned methods for temperature monitoring require continuous measurements. Confounding factors, such as the ambient temperature, or water intake in case of reticulo-rumen boluses, must be considered. Implanting and inserting the sensors into the animal allows for continuous measurement under constant conditions, but compared to rectal or vaginal temperature measurement by a standard thermometer, it is more invasive and costly. Furthermore, the implantation of a sensor is questionable in terms of permanent energy supply and also in terms of preservation of the carcass.

4.2.5 Variability in milk yield and milk composition

During oestrus, milk yield decreases between 2% and 6% with a rebound to the next milking (King, 1977; Horrell et al., 1984; Schofield et al., 1991; Lopez et al., 2004; Toledo-Alvarado et al., 2018). The decline in milk yield is considered to be due to a reduction in milk ejection rather than in milk secretion (Horrell et al., 1985; Schofield et al., 1991). Many farms are now equipped with milk yield recording devices in their milking parlors, so data on the variation in milk yield are often available. However, there are only a few and often older studies that have been conducted into the variations in milk quantity and composition for heat detection. The results of these studies are often contradictory. Schofield et al. (1991) reported that this decrease in milk yield was not evident in all animals and that only one-third of cows in heat could be identified based on this parameter.

Recent research by Toledo-Alvarado et al. (2018) used the components and physical properties of approximately fifty-seven thousand milk samples determined by Fourier-transform infrared (FTIR) spectroscopy to investigate their association with the oestrous cycle. The composition of milk, in particular fat, protein, and lactose, showed significant differences during the oestrous cycle. Fat concentration increased by 0.14% from diestrus to oestrus, whereas protein decreased by 0.03% at the same time. The lactose concentration appeared to remain relatively constant during diestrus, increasing 1 day before oestrus and then gradually decreasing. Furthermore, fatty acid profiles analyzed by the use of gas chromatography showed significant differences during the oestrous cycle. The concentrations of acetic, valeric, capropic, and myristoleic acids were greater on the day of oestrus compared to 2 weeks later (Zebari et al., 2019).

Considering the development of cheaper and mobile FTIR devices and the increase in the use of automated milking systems on farms where more cows are milked per unit relative to conventional milking systems, further development of this technology appears promising, as other metabolites, for example, ketones, can also be measured in milk by FTIR. Milk profiles including milk composition could be useful in automated management systems to identify cows in oestrus or to predict pregnancy, but further research on prediction accuracy is needed.

4.2.6 Variability in progesterone concentration

In-line systems for use on farms, for example, in automated milking systems or parlors, are available for supporting herd health management. An example of an automated milk analysis system that has been commercially available for some time and is able to quantify P4 is the Herd Navigator (HN, DeLaval, Tumba, Sweden) system. The system takes representative milk samples from specific cows during milking for use in, for example, reproduction, ketosis, and mastitis management. Prior to milking, an algorithm selects the individual cows and specific parameters to be tested. The fully automated device relies on dry stick analyses, using colorimetric reactions for the determination of lactate dehydrogenase, urea, and BHB, and an immuno-assay for measuring progesterone (Mazeris, 2010).

To detect spontaneous ovarian activity, the in-line system starts analyzing milk samples at approximately 3 weeks after calving and adjusts the sampling frequency according to the stage of the oestrous cycle (Friggens and Chagunda, 2005). Supporting reproductive management, key features of the system include monitoring the resumption of luteal activity after parturition and subsequent luteal phases, detecting ovarian dysfunctions, estimating the pregnancy status and declaring impending oestrus based on the end of a luteal phase, and early estimation of oestrus and pregnancy status based on P4 profiles (Saint-Dizier and Chastant-Maillard, 2012; Bruinjé and Ambrose, 2019; Bruinjé et al., 2019; Antanaitis et al., 2020).

4.2.7 Activity meters and pedometers

Increasing restlessness is among the prominent external signs of an animal in heat. The use of activity meters and pedometers, starting in the early 1980s, is based on the significant increase in animal activity during oestrus.

Whereas pedometers are fixed to the leg of a cow, activity meters can be attached to the side or below the neck with a collar or integrated into an ear-tag (Fig. 5) or in reticulo-rumen boluses. Activity is measured, for example, by a mercury switch (Firk et al., 2002) or by the use of a magnetic ball placed between two copper coils (Dietrich, 2012). Animal activity activates the switch or leads to the movement of the ball, which generates a measurable current in the copper coils. Further development of integrated circuits and accelerometers led to the introduction of lower-cost tri-axial accelerometer systems in dairy farming in the 1990s (Mottram, 2016). Compared to animals in sexual silence, cows in oestrus show two- to fourfold greater locomotion activities (Schofield et al., 1991). Oestrus detection can be based on different activity thresholds. These include, for example, the increase in activity above the mean by 1 or 2 standard deviations, the exceeding of a scaled value, or the use of confidence intervals (Firk et al., 2002). The detection rates of some exemplary studies are shown in Table 3.

Some authors reported that pedometers provide more reliable oestrus detection compared to collar-mounted activity meters (Eradus et al., 1992; Sakaguchi et al., 2007). Similar activity patterns and insignificant differences in oestrus detection between the front and rear legs were found by Liu and Spahr (1993).

By detecting specific patterns in the acceleration data, the use of accelerometer systems is not limited to oestrus detection, but they are also used for monitoring rumination (Schirmann et al., 2009; Borchers et al., 2016; Reiter et al., 2018), parturition prediction (Titler et al., 2015; Krieger et al., 2019), and disease detection (Stangaferro et al., 2016a,b,c; Gusterer et al., 2020), among others.

Fig. 6 shows an example of the animal activity and ruminating time presented in the herd management software of an ear-tag-based accelerometer system, as well as the probability of oestrus derived from it.

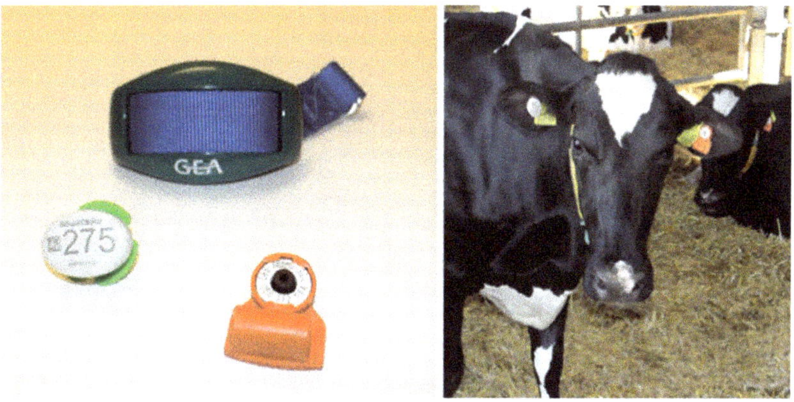

Figure 5 Examples of commercially available ear and neck fixed accelerometers for oestrus detection and cow equipped with various sensors for study purposes.

Table 3 Test characteristics of selected oestrus detection systems tested in indoor-housed dairy cows

Reference	Parameter/type and position of the device	Number of cows	Reference	Se	Sp
Holman et al. (2011)[a]	Activity/collar-mounted accelerometer	67	P4 milk	59	94
Holman et al. (2011)[a]	Steps/leg-mounted pedometer	67	P4 milk	63	74
Chanvallon et al. (2014)[a]	Activity, steps/pedometer	63	P4 milk	64-77[b]	n.a.
Chanvallon et al. (2014)[a]	Activity, steps/pedometer	63	P4 milk	55-69[b]	n.a.
Chanvallon et al. (2014)[a]	Activity, steps/pedometer	63	P4 milk	54-68[b]	n.a.
Dolecheck et al. (2015)[a]	Activity, resting, rumination/accelerometer	18	P4 blood	100[c]	99–100[c]
Dolecheck et al. (2015)[a]	Activity, rumination/accelerometer	18	P4 blood	60–100[c]	96–99[c]
Dolecheck et al. (2015)[a]	Steps, lying bouts, and time/accelerometer	17	P4 blood	80–100[c]	98–100[c]
Dolecheck et al. (2015)[a]	Activity, lying time/accelerometer	14	P4 blood	100[c]	91–97[c]
Higaki et al. (2019)[a]	Vaginal temperature/wearable vaginal sensor	17	Ultrasound	71–94[d]	n.a.
Higaki et al. (2019)[a]	Vaginal conductivity/wearable vaginal sensor	17	Ultrasound	24–41[d]	n.a.
Mayo et al. (2019)[a]	Activity, lying bouts, and time, steps, rumination, eating time/accelerometer	109[f]	P4 blood	57-90	83-100
Schweinzer et al. (2019)	Activity, rumination/ear-attached accelerometer	116	Confirmed pregnancy	97	98

Reference	Parameter/type and position of the device	Number of cows	Reference	Se	Sp
Benaissa et al. (2020)[a]	Rumination, feeding, resting/neck-attached accelerometer	12	Visual observation	42–76[e]	93–95[e]
Benaissa et al. (2020)[a]	Lying bouts and times, steps/leg-attached accelerometer	12	Visual observation	51–76[e]	92–95[e]
Benaissa et al. (2020)[a]	Localization/neck-attached localization	12	Visual observation	56–78[e]	93–94[e]
Schilkowsky et al. (2021)	Activity, rumination/ear-attached accelerometer	199	Ultrasound, pressure sensor	93	92

[a]Different devices were tested.
[b]Data derived from the operating period of all three devices which were tested in the study.
[c]Depending on different machine learning techniques (random forest, linear discriminant analysis, neural networks) used for oestrus detection.
[d]Depending on different machine learning techniques (decision tree, support vector machine, artificial neural network) used for oestrus detection.
[e]Depending on different time intervals (2, 3, 8, 12, 24 h) before AI.
[f]109 animals were enrolled, but only functional devices (65–109) were reflected in the final analyses.

4.2.8 Computer vision

Computer vision, also known as machine vision, processes and analyzes camera-captured images and videos by various methods to gain high-level understanding of content and/or to recognize geometric structures. Through the automated recognition of structures and an understanding of their properties, which corresponds to the human visual system, process automation is possible. Computer vision is an interdisciplinary field of research, and rapid developments in digital imaging and computer technology have made it a standard tool for scientists and engineers in various disciplines. Past and present applications in industry include autonomous navigation, robotic assembly, and inspections. Detecting specific events or organizing information for indexing databases of images and image sequences are among the newer techniques also used in non-industrial areas such as traffic planning, sports analysis, and medicine. In order to recognize an object in the context of machine vision and to distinguish it from others, it must first be described mathematically. For this, various methods have been developed.

Video-based heat detection systems identify general animal activity and/or typical oestrous behaviours such as chin resting on the back or StbM events (Bruyère et al., 2012; Guo et al., 2019). Depending on the size of the barn, several cameras may be required for this purpose, which zoom in on the respective image section when oestrous behaviour is detected. The corresponding video frequency is then stored with a few seconds of lead and lag time. In a recent study, Hirata et al. (2019) proposed a method that focused on the direction of vectors by optical flow by combining frame differences

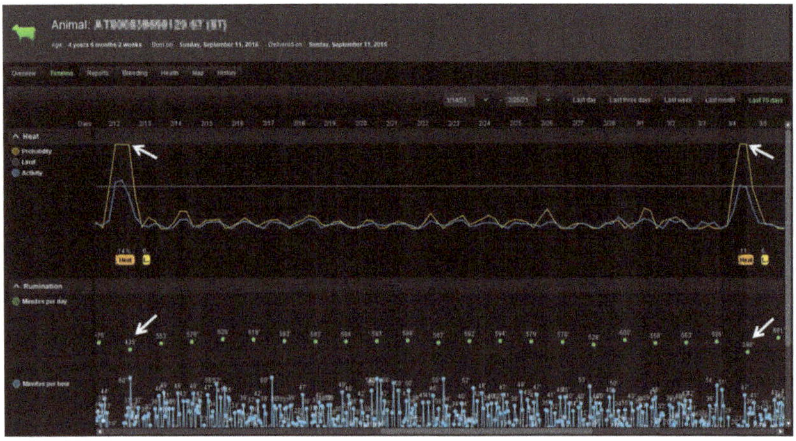

Figure 6 Graphical presentation of animal activity (blue line) and oestrus probability (orange line) showing two peaks 21 days apart (white arrows). The bottom of the figure shows the daily rumination time (green dots), which is decreased on the day of oestrus.

with 'Motion History Image' for detecting mounting behaviour in cattle. The methodological approach was confirmed in an animal experiment with high heat detection rates, but the number of animals used was low. Hence, further research on various farms with differing environmental conditions is needed.

Because for most systems animal identification must be done retrospectively by the user, a distinct and unique animal identification is essential, especially in larger herds. In addition, poor lighting conditions in the barn and dirty cameras can make animal identification difficult. These are possibly the reasons why video surveillance for heat detection has not yet gained acceptance in practice.

4.2.9 Animal vocalization

Evidence exists that in some mammals the call of females provides information about their physiological and psychological state (Semple and McComb, 2000). In cattle, the vocalizations of animals are reported to contain information about age, sex, and dominance and reproductive status (Watts and Stookey, 2000).

Schön et al. (2007) developed a method for the continuous and automated recording of cattle vocalization. In their pilot study of 10 tethered heifers, they attached a wireless microphone to a neck harness in close proximity to the head. The sounds were digitized and processed by an algorithm, generating serial 250-ms signal windows. Two different structures could be detected in the calls: (1) 'moo' calls or call sections were characterized by a harmonic sound with regular frequency bands, while (2) 'bellow' calls or call sections were nonharmonic and more noisy. A sharp increase in vocalization rate began 1 day before until oestrus, which was confirmed by visual observation, in 70–171 calls. A comparable trend was found for 'moo' and 'bellow' calls. Two days before oestrus, the number of 'bellow' sounds increased compared to 'moo' sounds, but animal-specific reactions were different; this was confirmed by large standard errors in the analyses.

Despite the possible advantages of the method, which is based on non-invasive and quantitative measurement of various dimensions, the method has not yet been established for oestrus detection under practical farm conditions. One possible reason for this is the complex technical implementation under modern husbandry conditions, which the authors describe as challenging.

4.2.10 Combination of parameters

None of the above-mentioned methods provides perfect oestrus detection rates. Furthermore, when using only one trait for oestrus detection, high numbers of false positive oestrous alerts occur (Roth et al., 1987; Schlünsen et al., 1987). The additional use of second reliable oestrous traits may contribute to an overall improvement in oestrus detection. For this, traits should be used that are easily and routinely recorded (Firk et al., 2002).

The bundling of various sensor-based information and management functions into an overarching herd management software has been recommended (Williams et al., 1981; Roth et al., 1987), to lead to reductions in costs and labor compared to single systems. When combining different oestrus detection traits monitored by one or more systems and applying suitable statistical methods, more valid oestrus detection rates are obtained (Firk et al., 2002). For example, the detection rate increased from 84% with visual observation plus assembly detectors to a high of 93% with visual observation plus activity measurements (Williams et al., 1981). Combining activity and vaginal temperature measurements resulted in a 90% detection rate and a 28% error rate compared to visual oestrus detection of 66% (Redden et al., 1993).

Ideally, different parameters can be detected reliably by single-sensor technologies. Senger (1994) described the integration of a pedometer, a pressure sensor to detect mountings, and an impedance sensor into one implantable device as an ideal system.

But, of course, it should also be possible to combine parameters from different sensor technologies; this requires suitable interfaces and software solutions for evaluation of the data.

4.3 Economic aspects

Today's agriculture is characterized by increasing mechanization and automation of production processes, which, among others, aim to reduce physical labor, increase labor efficiency, and thus reduce production costs. The fact that in many countries of the world the cost of the capital used for this purpose is lower than the cost of labor has accelerated this trend in recent years (De Koning, 2010; Rutten et al., 2013). In this context, farmers have to make increasingly complex and long-term decisions considering various investment options. Whereas in the past decisions were often based on a gut feeling, experience or the recommendation of other farmers and/or consultants, today cost:benefit assessments are increasingly used. However, there is often a lack of reliable data to support farmers in their decision-making, which must take into account biological and economic factors but also socioeconomic aspects, for example work–life balance and consumer concerns. In addition to the economic benefits, technical support, training on the system, and user-friendliness are increasingly influencing farmers' investment decisions (Banhazi et al., 2012).

Considering economic cost models, it is often the farmer's attitude towards sensor technologies and their ability to use the information that determines the economic success of these technologies. The personal benefits of using these technologies, for example, on-job satisfaction or work–life balance, are difficult to quantify in economic terms.

In the field of reproductive management, some economic evaluations of different sensor technologies for oestrus detection in cattle have already been performed (Dolecheck et al., 2016; Adenuga et al., 2020; Pfeiffer et al., 2020). The economic value depends on numerous factors, which vary from farm to farm and situation to situation, for instance, on general reproductive management, milk yield, milk price, and labor costs. Furthermore, the economic evaluation also depends on the number of features offered by one sensor system. With an increasing number of frequently used features, the economic benefit of a system will increase. Considering different scenarios, Pfeiffer et al. (2020) reported the probability of a positive net return of investments in activity meters of 85-99% for HF cows. In these calculations, the potential additional benefit of using the system in heifers or for health monitoring was not included.

The economic impact of primarily pedometer and accelerometer systems used for oestrus detection was recently reviewed by Adenuga et al. (2020). For this purpose, articles published in peer-reviewed journals from 1970 to 2019 were selected and various parameters that can be used to measure the viability of investments in sensor technologies were analyzed. For most studies, a positive 'net present value' was reported, which implies that the net revenues during the lifespan of the sensor technology exceed the initial investment costs. Depending on the technology, the number of years needed to pay back the initial investment costs (i.e. the payback period) was 3.5 to 8 years. Furthermore, a profit per cow and year between €8.50 and US$94.30 was reported.

In addition to numerous simulation models for comparing different reproductive management strategies (Galvão et al., 2013; Ricci et al., 2020), on-farm tools have also been developed that take into account the complex conditions of the individual farm. With these tools, some of which are available online, farmers and consultants can make farm-specific adjustments to general farm management, reproduction management, and the proposed oestrus detection system, among other things. Based on these entries, the estimated days open, cull rate due to failure to conceive, net present value, and payback period, for example, are calculated for the investment decision (Bewley, 2010; Dolecheck et al., 2016; Thomas et al., 2019). Ideally, decision-making tools should be kept as simple as possible, be easy to use, and be as accurate as possible, taking into account the key influencing factors. Information that is already available from other sources should be included in the decision model without having to re-enter that data. This requires appropriate interfaces.

5 Conclusion and future trends in research

The successful and continuous implementation of decades of research into agricultural practice, for example, from the areas of genetics, feeding, and animal husbandry, as well as farm and herd (health) management, is well documented by the milk yield achieved by farms today. However, this progress has not been matched in all areas of dairy farming. While a significant increase in milk yield has been realized worldwide, achieving good reproductive performance is still challenging for farmers and veterinarians.

Modern dairy herd management aims to monitor continuously the related areas, to identify weaknesses and, if necessary, to introduce countermeasures for further optimization. For continuous monitoring of the environment and/or the animals, computer and sensor technologies are increasingly used.

It has been estimated that approximately 10% of the reasons for non-detection of oestrus are due to issues related to the cow, while 90% are due to poor management and/or housing conditions. As a result, many strategies and technologies for oestrus detection have been developed. Automatic oestrus detection systems can be classified into (1) mount detection, (2) activity monitoring, (3) image analysis, and (4) milk progesterone measuring systems. Various sensors and devices are commercially available for monitoring mounting and animal activity (e.g. pedometers, activity meters, 3D-accelerometers) and belong to the class of 'on-cow' sensors, which are attached to the cow's body (e.g. the ear, neck, leg, or sacrum) or 'in-cow' sensors, for example, when a bolus in the cow's reticulo-rumen is used. In addition, other oestrus-associated changes in the cows' physiology and behaviour can be observed and measured, such as body temperature, vocalization, rumination, and feeding time, for which additional sensors may be used on the farm. Recent developments have focused on the use of automated image and video analysis for oestrus detection, but so far only a few prototypes are in use. Whereas in the past the exceeding of fixed, parameter- and herd-specific thresholds was mainly used for oestrus detection, more sophisticated systems rely on animal-specific thresholds detected by various artificial intelligence methods. When using such systems, sensitivities and specificities in oestrus detection of more than 90% each can be achieved.

An ideal method for animal (health) monitoring incorporates the reliable identification of animals, continuous monitoring (24/7), minimal or no labor requirement, low investment and maintenance costs, and valid recording of multiple parameters. At the same time, the system used should be as little- or non-invasive as possible and should not impair the natural behaviour of the animals.

Considering the surveillance and sensor costs per animal and the sometimes invasive attachment of sensors to the animal, the development of reliable 'off-cow' and 'in-line' technologies is becoming more important. The availability of cheaper, smaller, and more robust sensors and analysis techniques (e.g. image and sound sensors, FTIR spectroscopy devices) as well as the increasing use of automated milking systems favor these developments.

In this context, intensive research activities are focusing on the identification of additional biomarkers (e.g. interferon-tau, and lactoferrin, lactotransferrin, alpha1G in milk) associated with animal reproduction, for example, those used for the reliable assessment of pregnancy and ovarian cysts. Furthermore, with these non-invasive sensor technologies, it is possible to monitor various animals and different parameters (e.g. mounting, individual animal distances, animal localization) with one device. On the other hand, the use of 'off-cow' sensors leads to new challenges, for example, when using cameras with regard to animal identification in changing light conditions or with dirty lenses.

Before introducing new technologies into practice, the question arises as to the economic return on the investment. One advantage of algorithm-based systems for oestrus detection is that the introduction of new features is usually not associated with a replacement of the entire sensor system, but, for example, 'only' with modifications in the software. The optimization of algorithms during operation can also lead to an increase in economic efficiency, but further research on the cost:benefit ratio by use of sensor technologies in livestock farming is needed. In this context, socio-economic aspects, for example, work–life balance, consumer concerns, and animal welfare issues have to be considered; these are factors that are difficult to quantify in economic terms.

New approaches for the simple and reliable presentation and recording of, for example, examination data directly on the animal, for example, by using voice control or optical aids, are also under development. These data collected by the farmer or veterinarian (e.g. insemination data or pregnancy confirmation) can provide additional information to a sensor system and further improve the accuracy of, for example, oestrus detection by adjusting threshold values, for instance.

In the past, factors influencing reproductive performance at the animal or herd level, such as animal nutrition, husbandry, climate, and milk yield, were often analyzed independently without consideration of possible interactions. Recent research has focused on integrating heterogeneous farm and animal data obtained from multiple sources. In addition to facilitating the exchange of data between systems, the integrated analysis of (big) data using, for example, artificial intelligence, enables new insights into the development of fertility disorders in dairy cows as well as into the interactions between reproductive performance and milk yield, and also environmental factors. In this more 'holistic' approach, concepts of animal science, behavioural research, biostatistics, computer science, economics, and product development need to interact across disciplines to gain additional information. For this, a close and trusting cooperation between, for example, animal science, veterinary medicine, engineering, economics, and social sciences is needed. In this context, special attention should be paid to a transparent presentation of the developed technologies to the public to ensure an adequate picture of increasingly automated agriculture. Automated systems already widely used in agriculture today, such as milking robots, are associated by many consumers with negative associations, such as 'animal suffering,' 'imprisoned,' and 'not animal-appropriate' (Pfeiffer et al., 2021).

Despite all the optimism regarding the use of sensor technologies and the application of complex mathematical methods to identify animals in oestrus, there is still a need to maintain and sharpen our 'common sense' in animal husbandry. The use of sensor technologies for oestrus detection should not lead to exclusive reliance on these technologies while overlooking obvious deficiencies in, for example, housing conditions (e.g. slippery floors), animal nutrition, and animal health. The use of sensor technologies is the continuation of the general technical development in the dairy industry, which is intended to support farmers and his or her personnel as well as veterinarians in their daily work and decision-making but not to replace their expertise.

6 Where to look for further information

A comprehensive introduction to the principles of reproductive physiology is found in 'Pathways to Pregnancy and Parturition' presented by Senger (2012).

- An example for a user-friendly decision support tool for investment analysis of oestrus detection tools is presented by Dolecheck et al. (2016).
- Decision tools, for example, for investment analysis of heat detection technologies and calculating farm-specific costs for additional 'days open' are found at the 'Dairy Extension Resources' (https://afs.ca.uky.edu/dairy/decision-tools) of the University of Kentucky College of Agriculture, Food and Environment.
- The homepage of Data Driven Decision For Farmers (4D4F; https://4d4f.eu) offers a good overview of different sensor systems and their functionalities.

Journals and regular conferences providing recent research, include:

- The *Journal of Dairy Science* is a renowned peer-reviewed dairy research journal.
- The European Association on Precision Livestock Farming (EA-PLF) holds regular meetings on PLF.
- The European Conference on Precision Livestock Farming (ECPLF) presents innovations in PLF every 2 years.
- The International Conference on Precision Dairy Farming, held alternately in Europe and the Americas, presents innovations in precision dairy farming every 3 years.

- The World Buiatrics Congress (WBC) promotes knowledge transfer within the international veterinary community.

There are a number of current research projects on the use of modern management tools to support farmers in oestrus and pregnancy monitoring, including:

- ClearFarm aims to exploit new technologies to monitor and improve animal welfare in pigs and cattle throughout the production chain, thus contributing to improved production (www.clearfarm.eu).
- The 'HealthyLivestock' project is working on concepts for improved animal health and welfare to reduce the use of antibiotics in pigs and poultry (www.healthylivestock.net).

7 References

Adams, A. E., Lombard, J. E., Fossler, C. P., Román-Muñiz, I. N. and Kopral, C. A. (2017). Associations between housing and management practices and the prevalence of lameness, hock lesions, and thin cows on US dairy operations. *Journal of Dairy Science* 100(3): 2119–2136.

Adenuga, A. H., Jack, C., Olagunju, K. O. and Ashfield, A. (2020). Economic viability of adoption of automated oestrus detection technologies on dairy farms: a review. *Animals* 10(7).

Antanaitis, R., Malašauskienė, D., Televičius, M., Juozaitienė, V., Žilinskas, H. and Baumgartner, W. (2020). Dynamic changes in progesterone concentration in cows' milk determined by the at-line milk analysis system herd navigatortm. *Sensors* 20(18): 1-11.

At-Taras, E. E. and Spahr, S. L. (2001). Detection and characterization of estrus in dairy cattle with an electronic heatmount detector and an electronic activity tag. *Journal of Dairy Science* 84(4): 792–798.

Banhazi, T. M., Lehr, H., Black, J. L., Crabtree, H., Schofield, P., Tscharke, M. and Berckmans, D. (2012). Precision Livestock Farming: an international review of scientific and commercial aspects. *International Journal of Agricultural and Biological Engineering* 5: 1-9.

Barkema, H. W., von Keyserlingk, M. A. G., Kastelic, J. P., Lam, T. J. G. M., Luby, C., Roy, J. P., LeBlanc, S. J., Keefe, G. P. and Kelton, D. F. (2015). Invited review: changes in the dairy industry affecting dairy cattle health and welfare. *Journal of Dairy Science* 98(11): 7426–7445.

Bascom, S. S. and Young, A. J. (1998). A summary of the reasons why farmers cull cows. *Journal of Dairy Science* 81(8): 2299–2305.

Beam, S. W. and Butler, W. R. (1999). Effects of energy balance on follicular development and first ovulation in postpartum dairy cows. *Journal of Reproduction and Fertility. Supplement* 54: 411–424.

Becker, F., Kanitz, W. and Heuwieser, W. (2005). Advantages and disadvantages of different methods of heat detection in cattle. *Zuchtungskunde* 77: 140–150.

Bedere, N., Cutullic, E., Delaby, L., Garcia-Launay, F. and Disenhaus, C. (2018). Meta-analysis of the relationships between reproduction, milk yield and body condition score in dairy cows. *Livestock Science* 210: 73–84.

Benaissa, S., Tuyttens, F. A. M., Plets, D., Trogh, J., Martens, L., Vandaele, L., Joseph, W. and Sonck, B. (2020). Calving and estrus detection in dairy cattle using a combination of indoor localization and accelerometer sensors. *Computers and Electronics in Agriculture* 168: 1–10.

Bewley, J. (2010). Precision dairy farming: advanced analysis solutions for future profitability. The First North American Conference on Precision Dairy Management. Toronto, Canada.

Borchers, M. R. (2015). An evaluation of precision dairy farming technology adoption, perception, effectiveness, and use. *Theses and Dissertations. Animal and Food Sciences.* 45. University of Kentucky. http://uknowledge.uky.edu/animalsci_etds/45.

Borchers, M. R., Chang, Y. M., Tsai, I. C., Wadsworth, B. A. and Bewley, J. M. (2016). A validation of technologies monitoring dairy cow feeding, ruminating, and lying behaviors. *Journal of Dairy Science* 99(9): 7458–7466.

Bruinjé, T. C. and Ambrose, D. J. (2019). Technical note: validation of an automated in-line milk progesterone analysis system to diagnose pregnancy in dairy cattle. *Journal of Dairy Science* 102(4): 3615–3621.

Bruinjé, T. C., Colazo, M. G., Ribeiro, E. S., Gobikrushanth, M. and Ambrose, D. J. (2019). Using in-line milk progesterone data to characterize parameters of luteal activity and their association with fertility in Holstein cows. *Journal of Dairy Science* 102(1): 780–798.

Bruyère, P., Hétreau, T., Ponsart, C., Gatien, J., Buff, S., Disenhaus, C., Giroud, O. and Guérin, P. (2012). Can video cameras replace visual estrus detection in dairy cows? *Theriogenology* 77(3): 525–530.

Burfeind, O., von Keyserlingk, M. A. G., Weary, D. M., Veira, D. M. and Heuwieser, W. (2010). Short communication: repeatability of measures of rectal temperature in dairy cows. *Journal of Dairy Science* 93(2): 624–627.

Burnett, T. A., Kaur, M., Polsky, L. and Cerri, R. L. A. (2020). Rumen-reticular temperature during estrus and ovulation using automated activity monitors in dairy cows. *Frontiers in Veterinary Science* 7: 597512.

Butler, W. R. and Smith, R. D. (1989). Interrelationships between energy balance and postpartum reproductive function in dairy cattle. *Journal of Dairy Science* 72(3): 767–783.

Cavalieri, J., Eagles, V. E., Ryan, M. and Macmillan, K. L. (2003). Comparison of four methods for detection of oestrus in dairy cows with resynchronised oestrous cycles. *Australian Veterinary Journal* 81(7): 422–425.

Chanvallon, A., Coyral-Castel, S., Gatien, J., Lamy, J. M., Ribaud, D., Allain, C., Clément, P. and Salvetti, P. (2014). Comparison of three devices for the automated detection of estrus in dairy cows. *Theriogenology* 82(5): 734–741.

Chapinal, N., Carson, M. E., LeBlanc, S. J., Leslie, K. E., Godden, S., Capel, M., Santos, J. E. P., Overton, M. W. and Duffield, T. F. (2012a). The association of serum metabolites in the transition period with milk production and early-lactation reproductive performance. *Journal of Dairy Science* 95(3): 1301–1309.

Chapinal, N., LeBlanc, S. J., Carson, M. E., Leslie, K. E., Godden, S., Capel, M., Santos, J. E. P., Overton, M. W. and Duffield, T. F. (2012b). Herd-level association of serum metabolites in the transition period with disease, milk production, and early lactation reproductive performance. *Journal of Dairy Science* 95(10): 5676–5682.

Clapper, J. A., Ottobre, J. S., Ottobre, A. C. and Zartman, D. L. (1990). Estrual rise in body temperature in the bovine I. Temporal relationships with serum patterns of reproductive hormones. *Animal Reproduction Science* 23(2): 89–98.

Cutullic, E., Delaby, L., Causeur, D., Michel, G. and Disenhaus, C. (2009). Hierarchy of factors affecting behavioural signs used for oestrus detection of Holstein and Normande dairy cows in a seasonal calving system. *Animal Reproduction Science* 113(1-4): 22–37.

De Koning, C. (2010). Automatic milking - Common practice on dairy farms. Proceedings of the 2nd North American Conference on Robotic Milking. Precision Dairy Operators, Toronto, Canada, V59–V63.

De Rensis, F. and Scaramuzzi, R. J. (2003). Heat stress and seasonal effects on reproduction in the dairy cow - a review. *Theriogenology* 60(6): 1139–1151.

De Vries, A. and Marcondes, M. I. (2020). Review: overview of factors affecting productive lifespan of dairy cows. *Animal* 14(S1): s155–s164.

Dematawewa, C. M. B. and Berger, P. J. (1998). Genetic and phenotypic parameters for 305-day yield, fertility, and survival in Holsteins. *Journal of Dairy Science* 81(10): 2700–2709.

Denis-Robichaud, J., Cerri, R. L. A., Jones-Bitton, A. and LeBlanc, S. J. (2018). Dairy producers' attitudes toward reproductive management and performance on Canadian dairy farms. *Journal of Dairy Science* 101(1): 850–860.

Dietrich, O. (2012). *Etablierung einer neuen Methode zur automatisierten Brunsterkennung beim Rind*. Munich: Ludwig-Maximilians-Universität München.

Diskin, M. G., Mackey, D. R., Roche, J. F. and Sreenan, J. M. (2003). Effects of nutrition and metabolic status on circulating hormones and ovarian follicle development in cattle. *Animal Reproduction Science* 78(3–4): 345–370.

Diskin, M. G. and Sreenan, J. M. (2000). Expression and detection of oestrus in cattle. *Reproduction, Nutrition, Development* 40(5): 481–491.

Dobson, H., Walker, S. L., Morris, M. J., Routly, J. E. and Smith, R. F. (2008). Why is it getting more difficult to successfully artificially inseminate dairy cows? *Animal* 2(8): 1104–1111.

Dolecheck, K. A., Heersche Jr., G. and Bewley, J. M. (2016). Retention payoff-based cost per day open regression equations: application in a user-friendly decision support tool for investment analysis of automated estrus detection technologies. *Journal of Dairy Science* 99(12): 10182–10193.

Dolecheck, K. A., Silvia, W. J., Heersche, G., Jr., Chang, Y. M., Ray, D. L., Stone, A. E., Wadsworth, B. A. and Bewley, J. M. (2015). Behavioral and physiological changes around estrus events identified using multiple automated monitoring technologies. *Journal of Dairy Science* 98(12): 8723–8731.

Dominguez-Castaño, P., Vargas de Oliveira, M. H., El Faro, L. and de Vasconcelos Silva, J. A. I. (2020). Relationship between reproductive and productive traits in Holstein cattle using multivariate analysis. *Reproduction in Domestic Animals* 55(7): 770–776.

Dransfield, M. B. G., Nebel, R. L., Pearson, R. E. and Warnick, L. D. (1998). Timing of insemination for dairy cows identified in estrus by a radiotelemetric estrus detection system. *Journal of Dairy Science* 81(7): 1874–1882.

Drillich, M. (1999). Vergleich des strategischen Einsatzes von Prostaglandin F2alpha mit konventionellen Methoden des Fruchtbarkeitsmanagements in zwei Milchviehbetrieben. In: *Tierklinik für Fortpflanzung*. Berlin: Freie Universität Berlin, 135.

Eradus, W. J., Rossing, W., Hogewerf, P. H. and Benders, E. (1992). Signal processing of activity data for oestrus detection in dairy cattle. In: *Proceedings of the International Symposium on Prospects for Automatic Milking* Ipema, A. H., Lippus, A. C., Metz, J. H., Rossing, W. (Eds). Wageningen, The Netherlands: Pudoc Scientific, 360-369.

Esslemont, R. J. (1992). Measuring dairy herd fertility. *Veterinary Record* 131(10): 209-212.

Esslemont, R. J. and Bryant, M. J. (1976). Oestrous behaviour in a herd of dairy cows. *Veterinary Record* 99(24): 472-475.

Esslemont, R. J., Glencross, R. G., Bryant, M. J. and Pope, G. S. (1980). A quantitative study of pre-ovulatory behaviour in cattle (British Friesian heifers). *Applied Animal Ethology* 6(1): 1-17.

Esslemont, R. J. and Kossaibati, M. A. (2000). The use of databases to manage fertility. *Animal Reproduction Science* 60-61: 725-741.

Feldmann, F., Aizinbud, E., Schindler, H. and Broda, H. (1978). The electrical conductivity inside the bovine vaginal wall. *Animal Science* 26(1): 61-65.

Fetrow, J., Stewart, S., Eicker, S. and Rapnicki, P. (2007). Reproductive health programs for dairy herds: analysis of assessment of reproductive performance. In: *Current Therapy in Large Animal Theriogenology* Youngquist, R. (Ed). St. Louis, PA: W.B. Saunders, 473-489.

Firk, R., Stamer, E., Junge, W. and Krieter, J. (2002). Automation of oestrus detection in dairy cows: a review. *Livestock Production Science* 75(3): 219-232.

Fleming, A., Baes, C. F., Martin, A. A. A., Chud, T. C. S., Malchiodi, F., Brito, L. F. and Miglior, F. (2019). Symposium review: the choice and collection of new relevant phenotypes for fertility selection. *Journal of Dairy Science* 102: 3722-3734.

Foote, R. H., Oltenacu, E. A. B., Mellinger, J., Scott, N. R. and Marshall, R. A. (1979). Pregnancy rate in dairy cows inseminated on the basis of electronic probe measurements. *Journal of Dairy Science* 62(1): 69-73.

Forde, N., Beltman, M. E., Lonergan, P., Diskin, M., Roche, J. F. and Crowe, M. A. (2011). Oestrous cycles in Bos taurus cattle. *Animal Reproduction Science* 124(3-4): 163-169.

Fricke, P. M., Ricci, A., Giordano, J. O. and Carvalho, P. D. (2016). Methods for and implementation of pregnancy diagnosis in dairy cows. *Veterinary Clinics of North America. Food Animal Practice* 32(1): 165-180.

Friggens, N. C. and Chagunda, M. G. G. (2005). Prediction of the reproductive status of cattle on the basis of milk progesterone measures: model description. *Theriogenology* 64(1): 155-190.

Galvão, K. N., Federico, P., De Vries, A. and Schuenemann, G. M. (2013). Economic comparison of reproductive programs for dairy herds using estrus detection, timed artificial insemination, or a combination. *Journal of Dairy Science* 96(4): 2681-2693.

Gaude, I., Kempf, A., Strüve, K. D. and Hoedemaker, M. (2021). Estrus signs in Holstein Friesian dairy cows and their reliability for ovulation detection in the context of visual estrus detection. *Livestock Science* 245: 1-6.

Gordon, P. (2011). Oestrus detection in dairy cattle. *In Practice* 33(10): 542-546.

Green, J. A. and Hennessy, M. E. (2018). Pregnancy associated glycoproteins. *Encyclopedia of Reproduction* 2: 508-513.

Greenham, T., Oikonomou, G. and Grove-White, D. H. (2019). A description of interestrus and interservice intervals and associated fertility in 16 United Kingdom dairy herds. *Journal of Dairy Science* 102(1): 824–832.

Gröhn, Y. T. and Rajala-Schultz, P. J. (2000). Epidemiology of reproductive performance in dairy cows. *Animal Reproduction Science* 60–61: 605–614.

Grunert, E. (1999). Sexualzyklus. In: *Fertilitätsstörungen beim weiblichen Rind* (3rd edn.) Grunert, E. and de Kruif, A. (Eds). Berlin: Parey, 3–26.

Guo, Y., Zhang, Z., He, D., Niu, J. and Tan, Y. (2019). Detection of cow mounting behavior using region geometry and optical flow characteristics. *Computers and Electronics in Agriculture* 163: 1–11.

Gusterer, E., Kanz, P., Krieger, S., Schweinzer, V., Süss, D., Lidauer, L., Kickinger, F., Öhlschuster, M., Auer, W., Drillich, M. and Iwersen, M. (2020). Sensor technology to support herd health monitoring: using rumination duration and activity measures as unspecific variables for the early detection of dairy cows with health deviations. *Theriogenology* 157: 61–69.

Gwazdauskas, F. C., Lineweaver, J. A. and McGilliard, M. L. (1983). Environmental and management factors affecting estrous activity in dairy cattle. *Journal of Dairy Science* 66(7): 1510–1514.

Gwazdauskas, F. C., Nebel, R. L., Sprecher, D. J., Whittier, W. D. and McGilliard, M. L. (1990). Effectiveness of rump-mounted devices and androgenized females for detection of estrus in dairy cattle. *Journal of Dairy Science* 73(10): 2965–2970.

Hartigan, P. J. (2004). Reproductive physiology in cattle. In: *Bovine Medicine: Diseases and Husbandry of Cattle* Andrews, A. H., Blowley, R. W., Boyd, H., Eddy, R. G. (Eds). Oxford, UK: Blackwell Publishing Science Ltd., 471–505.

Helmer, S. D. and Britt, J. H. (1985). Mounting behavior as affected by stage of estrous cycle in Holstein heifers. *Journal of Dairy Science* 68(5): 1290–1296.

Heuwieser, W., Müller, K. and Ahlers, D. (2000). Pregnancy diagnosis in cattle: Part 2 - Consequences and liability in sonographic pregnancy examination on a herd health basis. *Praktische Tierarzt* 81: 587–589.

Higaki, S., Miura, R., Suda, T., Andersson, L. M., Okada, H., Zhang, Y., Itoh, T., Miwakeichi, F. and Yoshioka, K. (2019). Estrous detection by continuous measurements of vaginal temperature and conductivity with supervised machine learning in cattle. *Theriogenology* 123: 90–99.

Hirata, T., Zin, T. T., Kobayashi, I. and Hama, H. (2019). A study on estrus detection of cattle combining video image and sensor information. *Advances in Intelligent Systems and Computing* 744: 267–273.

Hockey, C. D., Morton, J. M., Norman, S. T. and McGowan, M. R. (2010). Improved prediction of ovulation time may increase pregnancy rates to artificial insemination in lactating dairy cattle. *Reproduction in Domestic Animals* 45(6): e239–e248.

Holman, A., Thompson, J., Routly, J. E., Cameron, J., Jones, D. N., Grove-White, D., Smith, R. F. and Dobson, H. (2011). Comparison of oestrus detection methods in dairy cattle. *Veterinary Record* 169(2): 47.

Horrell, R. I., Kilgour, R., MacMillan, K. L. and Bremner, K. (1984). Evaluation of fluctuations in milk yield and parlour behaviour as indicators of oestrus in dairy cows. *Veterinary Record* 114(2): 36–39.

Horrell, R. I., Macmillan, K. L., Kilgour, R. and Bremner, K. (1985). Changes in milk yield and composition at oestrus in dairy cows. *Journal of Dairy Research* 52(1): 9–16.

Hurnik, J. F., King, G. J. and Robertson, H. A. (1975). Estrous and related behaviour in postpartum Holstein cows. *Applied Animal Ethology* 2(1): 55-68.

ICAR ICfAR. (2021). Milk recording surveys on cow, sheep, and goats. International Committee for Animal Recording. Available from: https://www.icar.org/index.php/publications-technical-materials/milk-yearly-surveys-cow-sheep-goats/ (accessed 1 May 2022).

Jorritsma, H. and Jorritsma, R. (2000). Overview of fertility and milk production on 15 dairy farms in south-east Friesland. *Tijdschrift voor Diergeneeskunde* 125(6): 180-184.

Kelly, E. T., McAloon, C. G., Crowe, M. A. and Beltman, M. E. (2021). Estimation of the true prevalence of inaccurate artificial inseminations in Irish milk recording dairy cows using a Bayesian latent class analysis. *Preventive Veterinary Medicine* 197: 105502.

King, J. O. (1977). The effect of oestrus on milk production in cows. *Veterinary Record* 101(6): 107-108.

Knaus, W. (2009). Dairy cows trapped between performance demands and adaptability. *Journal of the Science of Food and Agriculture* 89(7): 1107-1114.

Krieger, S., Oczak, M., Lidauer, L., Berger, A., Kickinger, F., Öhlschuster, M., Auer, W., Drillich, M. and Iwersen, M. (2019). An ear-attached accelerometer as an on-farm device to predict the onset of calving in dairy cows. *Biosystems Engineering* 184: 190-199.

Lane, A. J. P. and Wathes, D. C. (1998). An electronic nose to detect changes in perineal odors associated with estrus in the cow. *Journal of Dairy Science* 81(8): 2145-2150.

Leblanc, S. (2010). Assessing the association of the level of milk production with reproductive performance in dairy cattle. *Journal of Reproduction and Development* 56(Suppl.): S1-S7.

Lehrer, A. R., Lewis, G. S. and Aizinbud, E. (1992). Oestrus detection in cattle: recent developments. *Animal Reproduction Science* 28(1-4): 355-362.

Leidl, W. and Stolla, R. (1976). Measurement of electric resistance of the vaginal mucus as an aid for heat detection. *Theriogenology* 6(2-3): 237-249.

Leroy, J. L., Vanholder, T., Opsomer, G., Van Soom, A. and de Kruif, A. (2006). The in vitro development of bovine oocytes after maturation in glucose and beta-hydroxybutyrate concentrations associated with negative energy balance in dairy cows. *Reproduction in Domestic Animals* 41(2): 119-123.

Lewis, G. S., Aizinbud, E. and Lehrer, A. R. (1989). Changes in electrical resistance of vulvar tissue in Holstein cows during ovarian cycles and after treatment with prostaglandin F2α. *Animal Reproduction Science* 18(1-3): 183-197.

Liu, X. and Spahr, S. L. (1993). Automated electronic activity measurement for detection of estrus in dairy cattle. *Journal of Dairy Science* 76(10): 2906-2912.

Liu, Z. and He, D. (2019). Research of implantable sensor and wireless monitoring system for cow's vaginal resistance. *Nongye Jixie Xuebao/Transactions of the Chinese Society for Agricultural Machinery* 50: 175-185.

Loeffler, S. H., De Vries, M. J. and Schukken, Y. H. (1999). The effects of time of disease occurrence, milk yield, and body condition on fertility of dairy cows. *Journal of Dairy Science* 82(12): 2589-2604.

Lopez, H., Bunch, T. D. and Shipka, M. P. (2002). Estrogen concentrations in milk at estrus and ovulation in dairy cows. *Animal Reproduction Science* 72(1-2): 37-46.

Lopez, H., Satter, L. D. and Wiltbank, M. C. (2004). Relationship between level of milk production and estrous behavior of lactating dairy cows. *Animal Reproduction Science* 81(3-4): 209-223.

Lucy, M. C. (2000). Regulation of ovarian follicular growth by somatotropin and insulin-like growth factors in cattle. *Journal of Dairy Science* 83(7): 1635-1647.
Lucy, M. C. (2001). Reproductive loss in high-producing dairy cattle: where will it end? *Journal of Dairy Science* 84(6): 1277-1293.
Manzoli, A., Steffens, C., Paschoalin, R. T., Graboski, A. M., De Mello Brandão, H., de Carvalho, B. C., Bellini, J. L. and de Paula Herrmann, P. S., Jr. (2019). Volatile compounds monitoring as indicative of female cattle fertile period using electronic nose. *Sensors and Actuators. Part B* 282: 609-616.
Mayo, L. M., Silvia, W. J., Ray, D. L., Jones, B. W., Stone, A. E., Tsai, I. C., Clark, J. D., Bewley, J. M. and Heersche, G., Jr. (2019). Automated estrous detection using multiple commercial precision dairy monitoring technologies in synchronized dairy cows. *Journal of Dairy Science* 102(3): 2645-2656.
Mazeris, F. (2010). DeLaval Herd Navigator: proactive herd management. Proceedings of the First North American Conference on Precision Dairy Management. Toronto, Canada, 26-27.
McArt, J. A. A., Nydam, D. V. and Oetzel, G. R. (2012). Epidemiology of subclinical ketosis in early lactation dairy cattle. *Journal of Dairy Science* 95(9): 5056-5066.
Mee, J. F., Moyes, T., Gleeson, D. and O'Brien, B. (2002). A questionnaire survey of fertility management on dairy farms in the Republic of Ireland. *Irish Veterinary Journal* 55: 122-128.
Morgan, G. L. and Dawson, L. J. (2008). Development of teaser bulls under field conditions. *Veterinary Clinics of North America. Food Animal Practice* 24(3): 443-53, v.
Mortimer, R. G., Salman, M. D., Gutierrez, M. and Olson, J. D. (1990). Effects of androgenizing dairy heifers with ear implants containing testosterone and estrogen on detection of estrus. *Journal of Dairy Science* 73(7): 1773-1778.
Mosher, M. D., Ottobre, J. S., Haibel, G. K. and Zartman, D. L. (1990). Estrual rise in body temperature in the bovine II. The temporal relationship with ovulation. *Animal Reproduction Science* 23(2): 99-107.
Mottram, T. (2016). Animal board invited review: precision livestock farming for dairy cows with a focus on oestrus detection. *Animal* 10(10): 1575-1584.
Müller, K., Drillich, M., Tenhagen, B. A. and Heuwieser, W. (1999). Sensitivity and specificity of pregnancy examinations by means of ultrasound between day 20 and 34 after insemination in dairy cattle. *Praktische Tierarzt* 80: 1097-1104.
Ohtani, S., Okuda, K., Nishimura, K. and Mohri, S. (1993). Histological changes in bovine endometrium during the estrous cycle. *Theriogenology* 39(5): 1033-1042.
Petro, E. M. L., Leroy, J. L. M. R., Van Cruchten, S. J. M., Covaci, A., Jorssen, E. P. A. and Bols, P. E. J. (2012). Endocrine disruptors and female fertility: focus on (bovine) ovarian follicular physiology. *Theriogenology* 78(9): 1887-1900.
Pfeiffer, J., Gabriel, A. and Gandorfer, M. (2021). Understanding the public attitudinal acceptance of digital farming technologies: a nationwide survey in Germany. *Agriculture and Human Values* 38(1): 107-128.
Pfeiffer, J., Gandorfer, M. and Ettema, J. F. (2020). Evaluation of activity meters for estrus detection: a stochastic bioeconomic modeling approach. *Journal of Dairy Science* 103(1): 492-506.
Pieper, L., Doherr, M. G. and Heuwieser, W. (2016). Consumers' attitudes about milk quality and fertilization methods in dairy cows in Germany. *Journal of Dairy Science* 99(4): 3162-3170.

Rathbone, M. J., Kinder, J. E., Fike, K., Kojima, F., Clopton, D., Ogle, C. R. and Bunt, C. R. (2001). Recent advances in bovine reproductive endocrinology and physiology and their impact on drug delivery system design for the control of the estrous cycle in cattle. *Advanced Drug Delivery Reviews* 50(3): 277–320.

Redden, K. D., Kennedy, A. D., Ingalls, J. R. and Gilson, T. L. (1993). Detection of estrus by radiotelemetric monitoring of vaginal and ear skin temperature and pedometer measurements of activity. *Journal of Dairy Science* 76(3): 713–721.

Reiter, S., Sattlecker, G., Lidauer, L., Kickinger, F., Öhlschuster, M., Auer, W., Schweinzer, V., Klein-Jöbstl, D., Drillich, M. and Iwersen, M. (2018). Evaluation of an ear-tag-based accelerometer for monitoring rumination in dairy cows. *Journal of Dairy Science* 101(4): 3398–3411.

Reith, S. and Hoy, S. (2018). Review: behavioral signs of estrus and the potential of fully automated systems for detection of estrus in dairy cattle. *Animal* 12(2): 398–407.

Remnant, J. G., Green, M. J., Huxley, J. N. and Hudson, C. D. (2015). Variation in the interservice intervals of dairy cows in the United Kingdom. *Journal of Dairy Science* 98(2): 889–897.

Ricci, A., Li, M., Fricke, P. M. and Cabrera, V. E. (2020). Short communication: economic impact among 7 reproductive programs for lactating dairy cows, including a sensitivity analysis of the cost of hormonal treatments. *Journal of Dairy Science* 103(6): 5654–5661.

Roelofs, J., López-Gatius, F., Hunter, R. H. F., van Eerdenburg, F. J. C. M. and Hanzen, Ch (2010). When is a cow in estrus? Clinical and practical aspects. *Theriogenology* 74(3): 327–344.

Roelofs, J. B., Van Eerdenburg, F. J. C. M., Soede, N. M. and Kemp, B. (2005a). Pedometer readings for estrous detection and as predictor for time of ovulation in dairy cattle. *Theriogenology* 64(8): 1690–1703.

Roelofs, J. B., Van Eerdenburg, F. J. C. M., Soede, N. M. and Kemp, B. (2005b). Various behavioral signs of estrous and their relationship with time of ovulation in dairy cattle. *Theriogenology* 63(5): 1366–1377.

Roth, H., Schlünsen, D. and Schön, H. (1987). Computer-aided oestrus detection in dairy cattle. *ASAE* 1987/4018: 1–11.

Royal, M. D., Darwash, A. O., Flint, A. P. F., Webb, R., Woolliams, J. A. and Lamming, G. E. (2000). Declining fertility in dairy cattle: changes in traditional and endocrine parameters of fertility. *Animal Science* 70(3): 487–501.

Rutherford, A. J., Oikonomou, G. and Smith, R. F. (2016). The effect of subclinical ketosis on activity at estrus and reproductive performance in dairy cattle. *Journal of Dairy Science* 99(6): 4808–4815.

Rutten, C. J., Velthuis, A. G. J., Steeneveld, W. and Hogeveen, H. (2013). Invited review: sensors to support health management on dairy farms. *Journal of Dairy Science* 96(4): 1928–1952.

Saint-Dizier, M. and Chastant-Maillard, S. (2012). Towards an automated detection of oestrus in dairy cattle. *Reproduction in Domestic Animals* 47(6): 1056–1061.

Sakaguchi, M., Fujiki, R., Yabuuchi, K., Takahashi, Y. and Aoki, M. (2007). Reliability of estrous detection in Holstein heifers using a radiotelemetric pedometer located on the neck or legs under different rearing conditions. *Journal of Reproduction and Development* 53(4): 819–828.

Sanderink, F. E. P., Gerritsen, J. W., Groot Koerkamp, P. W. G. and van Mourik, S. (2017). Automatic detection of oestrus cows via breath sampling with an electronic nose: a pilot study. *Biosystems Engineering* 156: 1-6.

Saumande, J. (2002). Electronic detection of oestrus in postpartum dairy cows: efficiency and accuracy of the DEC® (showheat) system. *Livestock Production Science* 77(2-3): 265-271.

Schams, D., Schallenberger, E., Hoffmann, B. and Karg, H. (1977). The oestrous cycle of the cow: hormonal parameters and time relationships concerning oestrus, ovulation, and electrical resistance of the vaginal mucus. *Acta Endocrinologica* 86(1): 180-192.

Schilkowsky, E. M., Granados, G. E., Sitko, E. M., Masello, M., Perez, M. M. and Giordano, J. O. (2021). Evaluation and characterization of estrus alerts and behavioral parameters generated by an ear-attached accelerometer-based system for automated detection of estrus. *Journal of Dairy Science* 104(5): 6222-6237.

Schirmann, K., von Keyserlingk, M. A. G., Weary, D. M., Veira, D. M. and Heuwieser, W. (2009). Technical note: validation of a system for monitoring rumination in dairy cows. *Journal of Dairy Science* 92(12): 6052-6055.

Schlünsen, D., Roth, H., Schön, H., Paul, W. and Speckmann, H. (1987). Automatic health and oestrus control in dairy husbandry through computer aided systems. *Journal of Agricultural Engineering Research* 38(4): 263-279.

Schofield, S. A., Phillips, C. J. C. and Owens, A. R. (1991). Variation in the milk production, activity rate and electrical impedance of cervical mucus over the oestrous period of dairy cows. *Animal Reproduction Science* 24(3-4): 231-248.

Schön, P. C., Hämel, K., Puppe, B., Tuchscherer, A., Kanitz, W. and Manteuffel, G. (2007). Altered vocalization rate during the estrous cycle in dairy cattle. *Journal of Dairy Science* 90(1): 202-206.

Schuster, J. C., Barkema, H. W., De Vries, A., Kelton, D. F. and Orsel, K. (2020). Invited review: academic and applied approach to evaluating longevity in dairy cows. *Journal of Dairy Science* 103(12): 11008-11024.

Schweinzer, V., Gusterer, E., Kanz, P., Krieger, S., Süss, D., Lidauer, L., Berger, A., Kickinger, F., Öhlschuster, M., Auer, W., Drillich, M. and Iwersen, M. (2019). Evaluation of an ear-attached accelerometer for detecting estrus events in indoor housed dairy cows. *Theriogenology* 130: 19-25.

Semple, S. and McComb, K. (2000). Perception of female reproductive state from vocal cues in a mammal species. *Proceedings. Biological Sciences* 267(1444): 707-712.

Senger, P. L. (1994). The estrus detection problem: new concepts, technologies, and possibilities. *Journal of Dairy Science* 77(9): 2745-2753.

Senger, P. L. (2012). Pathways to pregnancy & parturition. *Current Conceptions*, 3rd edn, 381 p.

Sheldon, I. M., Wathes, D. C. and Dobson, H. (2006). The management of bovine reproduction in elite herds. *Veterinary Journal* 171(1): 70-78.

Stangaferro, M. L., Wijma, R., Caixeta, L. S., Al-Abri, M. A. and Giordano, J. O. (2016a). Use of rumination and activity monitoring for the identification of dairy cows with health disorders: part I. Metabolic and digestive disorders. *Journal of Dairy Science* 99(9): 7395-7410.

Stangaferro, M. L., Wijma, R., Caixeta, L. S., Al-Abri, M. A. and Giordano, J. O. (2016b). Use of rumination and activity monitoring for the identification of dairy cows with health disorders: part II. Mastitis. *Journal of Dairy Science* 99(9): 7411-7421.

Stangaferro, M. L., Wijma, R., Caixeta, L. S., Al-Abri, M. A. and Giordano, J. O. (2016c). Use of rumination and activity monitoring for the identification of dairy cows with health disorders: part III. Metritis. *Journal of Dairy Science* 99(9): 7422–7433.
Stevenson, J. S. (2001). A review of oestrous behaviour and detection in dairy cows. *BSAP Occasional Publication*. BSAP Occasional Publication 26(1): 43–62.
Sveberg, G., Refsdal, A. O., Erhard, H. W., Kommisrud, E., Aldrin, M., Tvete, I. F., Buckley, F., Waldmann, A. and Ropstad, E. (2011). Behavior of lactating Holstein-Friesian cows during spontaneous cycles of estrus. *Journal of Dairy Science* 94(3): 1289–1301.
Szenci, O. (2021). Recent possibilities for the diagnosis of early pregnancy and embryonic mortality in dairy cows. *Animals* 11(6).
Thomas, E. B., Dolecheck, K. A., Mark, T. B., Eastwood, C. R., Dela Rue, B. T. and Bewley, J. M. (2019). A decision-support tool for investment analysis of automated oestrus detection technologies in a seasonal dairy production system. *Animal Production Science* 59(12): 2280–2287.
Titler, M., Maquivar, M. G., Bas, S., Rajala-Schultz, P. J., Gordon, E., McCullough, K., Federico, P. and Schuenemann, G. M. (2015). Prediction of parturition in Holstein dairy cattle using electronic data loggers. *Journal of Dairy Science* 98(8): 5304–5312.
Toledo-Alvarado, H., Vazquez, A. I., de los Campos, G., Tempelman, R. J., Gabai, G., Cecchinato, A. and Bittante, G. (2018). Changes in milk characteristics and fatty acid profile during the estrous cycle in dairy cows. *Journal of Dairy Science* 101(10): 9135–9153.
Van Eerdenburg, F. J. C. M., Loeffler, H. S. H. and van Vliet, J. H. (1996). Detection of oestrus in dairy cows: a new approach to an old problem. *Veterinary Quarterly* 18(2): 52–54.
Van Vliet, J. H. and Van Eerdenburg, F. J. C. M. (1996). Sexual activities and oestrus detection in lactating Holstein cows. *Applied Animal Behaviour Science* 50(1): 57–69.
Vasconcelos, J., Silcox, R., Lacerda, J., Pursley, J. and Wiltbank, M. (1997). Pregnancy rate, pregnancy loss, and response to heat stress after AI at 2 different times from ovulation in dairy cows. *Biology of Reproduction*: 230–230.
VIT VITwV. (2020). *Jahresbericht 2019: Trends, Fakten, Zahlen*. Vereinigte Informationssysteme Tierhaltung w.V. (vit). Available from: https://www.vit.de/wir-sind-vit/veroeffentlichungen (accessed 1 May 2022).
von Keyserlingk, M. A., Barrientos, A., Ito, K., Galo, E. and Weary, D. M. (2012). Benchmarking cow comfort on North American freestall dairies: lameness, leg injuries, lying time, facility design, and management for high-producing Holstein dairy cows. *Journal of Dairy Science* 95(12): 7399–7408.
Walker, S. L., Smith, R. F., Jones, D. N., Routly, J. E., Morris, M. J. and Dobson, H. (2010). The effect of a chronic stressor, lameness, on detailed sexual behaviour and hormonal profiles in milk and plasma of dairy cattle. *Reproduction in Domestic Animals* 45(1): 109–117.
Walker, W. L., Nebel, R. L. and McGilliard, M. L. (1996). Time of ovulation relative to mounting activity in dairy cattle. *Journal of Dairy Science* 79(9): 1555–1561.
Walsh, R. B., Walton, J. S., Kelton, D. F., LeBlanc, S. J., Leslie, K. E. and Duffield, T. F. (2007). The effect of subclinical ketosis in early lactation on reproductive performance of postpartum dairy cows. *Journal of Dairy Science* 90(6): 2788–2796.
Walsh, S. W., Williams, E. J. and Evans, A. C. O. (2011). A review of the causes of poor fertility in high milk producing dairy cows. *Animal Reproduction Science* 123(3–4): 127–138.

Wang, S., Zhang, H., Tian, H., Chen, X., Li, S., Lu, Y., Li, L. and Wang, D. (2020). Alterations in vaginal temperature during the estrous cycle in dairy cows detected by a new intravaginal device–a pilot study. *Tropical Animal Health and Production* 52(5): 2265-2271.

Wathes, D. C., Brickell, J. S., Bourne, N. E., Swali, A. and Cheng, Z. (2008). Factors influencing heifer survival and fertility on commercial dairy farms. *Animal* 2(8): 1135-1143.

Watts, J. M. and Stookey, J. M. (2000). Vocal behaviour in cattle: the animal's commentary on its biological processes and welfare. *Applied Animal Behaviour Science* 67(1-2): 15-33.

Wiegerinck, W., Setkus, A., Buda, V., Borg-Karlson, A. -K., Mozuraitis, R. and De Gee, A. (2011). BOVINOSE: pheromone-based sensor system for detecting estrus in dairy cows. *Procedia Computer Science* 7: 340-342.

Williams, W. F., Yver, D. R. and Gross, T. S. (1981). Comparison of estrus detection techniques in dairy heifers. *Journal of Dairy Science* 64(8): 1738-1741.

Wrenn, T. R., Bitman, J. and Sykes, J. F. (1958). Body temperature variations in dairy cattle during the estrous cycle and pregnancy. *Journal of Dairy Science* 41(8): 1071-1076.

Xu, Z. Z., McKnight, D. J., Vishwanath, R., Pitt, C. J. and Burton, L. J. (1998). Estrus detection using radiotelemetry or visual observation and tail painting for dairy cows on pasture. *Journal of Dairy Science* 81(11): 2890-2896.

Yániz, J., Santolaria, P. and López-Gatius, F. (2003). Relationship between fertility and the walking activity of cows at oestrus. *Veterinary Record* 152(8): 239-240.

Zebari, H. M., Rutter, S. M. and Bleach, E. C. L. (2019). Fatty acid profile of milk for determining reproductive status in lactating Holstein Friesian cows. *Animal Reproduction Science* 202: 26-34.

ZuchtData ZE-DG. (2020). *ZuchtData Jahresbericht 2019*. ZuchtData EDV-Dienstleistungen GmbH. Availble from: https://www.rinderzucht.at/downloads/jahresberichte.html (accessed 1 May 2022).

Chapter 12

Advances in automatic milking systems

Bernadette O'Brien and Deirdre Hennessy, Teagasc, Ireland

1 Introduction

Automatic milking (AM) represents a very significant technology entering dairy farming in recent years. This process became commercially available in the Netherlands in 1992 and worldwide adoption has been recently estimated at 50 000 units approximately. Automatic milking continues to represent a popular option in the Netherlands; almost half of all new milking systems installed in 2017 were automatic and more than 22% of Dutch farmers now operate such a system. The adoption of an AM system is highly innovative, as it undertakes the physical task of milking normally performed by the operator and considerably

http://dx.doi.org/10.19103/AS.2021.0090.13

reduces the manual work traditionally associated with milking a herd of cows, thus, it can mean a new order and method of management on dairy farms. The defining features of AM include (i) cows volunteer themselves for milking; (ii) milking is distributed over a 24-h period; and (iii) milking is conducted without human presence. The basic components of the system include the milking equipment (vacuum pump), milking crate, a teat position sensor (laser), a robotic arm for automatic teat-cup application and removal, and a gate system for controlling cow traffic. When the cow voluntarily enters the crate, she is identified via a tag on the leg, neck or ear; the software determines if the cow is due to be milked based on the current daily milk yield and the time since her last milking. If not due for milking, front gates open, and the cow is released. If due for milking, teats are cleaned and the teat cups are individually attached with the assistance of a scanning laser which determines the location of each teat. Teat cups are removed based on flow from individual quarters and the milk enters the bulk tank. The date and time of milking, the duration of milking and milk flowrate are all recorded. During milking, sensors check the volume and composition of the milk, quality of the milk and the health of the cow. This information is recorded and is available to the farmer to use in the daily management of his/her herd (de Koning, 2011). Sensors that monitor cow health and milk quality by measuring the temperature of the cow, colour and conductivity of the milk can allow subsequent separation of milk based on that information; milk can be diverted to separate tanks using the software settings, if requested by the operator. Because of the hands-off nature of this technology, remote monitoring is an essential component, for example it can send alerts to a mobile phone when there is a technical or cow problem. The system may also have the facility to manage cow movements remotely, for example separation of cows for later inspection/treatment and drafting gates that direct cows to different areas on the farm. Most current developmental work is associated with refining sensors, data analysis and software. Potential also exists for the automated separation of different milk flows, which could be used for a diversified and added value dairy product range. A current challenge is to have sensors developed that can analyse the milk faster than previously, select and separate it based on, for example protein levels, omega acids, milk for formula or another specific combination.

Different configurations of AM units exist which are generally aligned with different manufacturing or supply companies (Fig. 1). The most common AM unit is the single box/unit configuration. This formation allows one cow in the milking unit at a time and the automatic arm performs tasks such as cup attachment and removal, and teat disinfection post-milking. These units can accommodate approximately 70 cows milking twice daily. A further type of AM configuration may also be employed, that is an automatic milking rotary (AMR) involving a carousel that (a) has 4 milking arms and 1 spray arm and can milk about 90 cows per hour or (b) has its own arm at each cow position and can

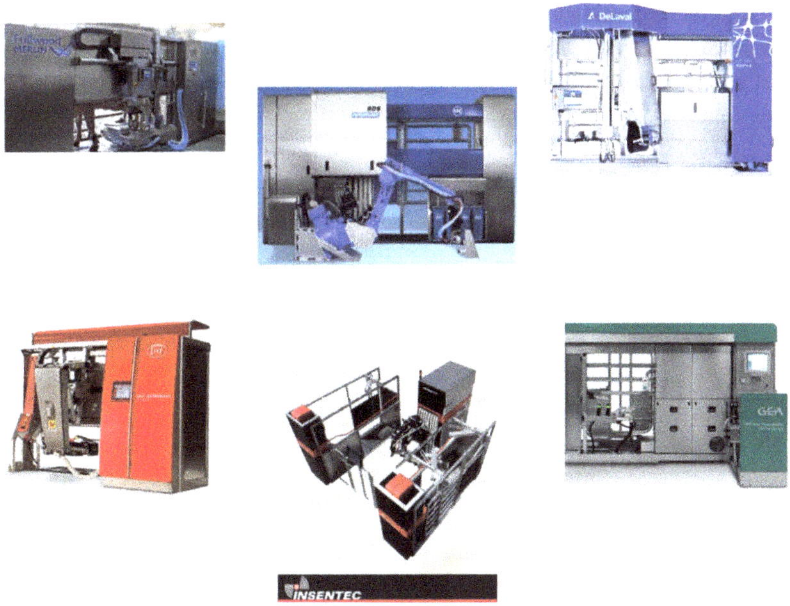

Figure 1 Examples of different AM units from various manufacturers/suppliers.

milk 140–170 cows per hour on a platform with 28 positions. More recently, approximately 30–40% of applications for new AM units concern leasing. In order to maintain a viable and profitable international dairy industry, the average productivity per operator and per hectare must increase. The future may offer two potential options (a) high capacity milking parlours with a high throughput in milk/person hours or (b) automatic systems where manual labour is replaced by an automatic milking system (AMS). Increased labour costs and decreasing milk prices could mean more widespread adoption of AM systems. Greater uptake and adoption rates of AM have occurred on farms that are smaller in size, often family owned and operated, where suitably skilled labour is difficult to source and is expensive.

Until relatively recently the majority of AM systems have been installed in combination with indoor/restricted grazing production systems, and indoor feeding systems have been well adapted to AM. Nevertheless, grazing has many advantages for the economy, the environment, animal welfare and product quality. Thus, from a sustainability perspective, it is important that dairy systems with new precision grazing technologies are developed, in which grazing and AM can be easily and profitably integrated.

Success with AM is dependent on a high level of management and realistic expectations. Milk harvested per unit is directly affected by the total number of milkings per unit and the milk production per milking. However, many other variables interact and influence (positively or negatively) this performance. The challenge for the farmer is to manage the variables as best as possible in order to optimise the operation and output. The challenge for research is to study and develop such management systems and the tools to apply them, particularly with regard to the efficient use of data for decision-making by future AM farmers. Enhanced milk production has been observed with AM in the past, and as with conventional milking (CM) systems, there are a number of factors that influence this, such as breed, stage of lactation, feeding level and so on. However, some management factors may be more critical for AM, for example milking frequency (MF) and milking interval (MI), feeding level and the importance of attention to management detail on AM farms have been highlighted (Svennersten-Sjaunja and Pettersson, 2008). In recent times, the study of Cogato et al. (2021) has indicated a high frequency (18.5% of 585 studies examined) of studies as having been associated with milk production in AMS. Cogato et al. (2021) and Hogenboom et al. (2019) both focused on milk quality as a very important aspect of AM from a research perspective and identified gaps relating to cleaning operations. Cogato et al. (2021) also indicated that industrial research is currently focusing on attempting to perfect the cleaning systems of AM systems. Benefits of AM can include reduced labour, better social life for families, increased milk yield due to more frequent milking and an improved and more precise management system. Additionally, if operated correctly AM offers the ultimate means of assuming milk quality and safety, through extensive sensor analysis and potential segregation. Potential risks of the system include a reduction in herd fertility or an increase in somatic cell count (SCC), but both are dependent on management.

The direction and duration of AM development have been quite different for the farmer, the broader production industry, the research community and the consumer. But it has mainly been about four main areas: process, animal, components and technology, as identified by Cogato et al. (2021). They showed that most of the scientific studies on AMS focused on topics related to animals and processes, while patents were mainly associated with technology and components. This chapter will also concentrate largely on process and animal, with a specific focus on optimisation of AM system performance, efficiency and utilisation, together with operator and animal well-being, profitability and quality of product output (milk). In doing so, it contributes knowledge on different animal and milking management aspects, including operation in pasture-based systems. Further research and extension work will likely enhance the attractiveness of the technology and future adoption.

The term AM is often used interchangeably with robotic milking; however, for consistency, AM will be used throughout this chapter. It should be noted that the term AMS refers to the overall dairy management system within which

AM operates; this can be quite different from a CM system in aspects of milk quality, cow feeding, cow traffic, grazing and animal behaviour management.

2 Cow breed

Holstein-Friesian (HF) genetics have been widely used to increase milk production potential (Harris and Kolver, 2001; Evans et al., 2006). Additionally, a multi-component payment system applied by many milk processors has sometimes influenced dairy farmers to focus on maximising milk solids production per hectare (Shalloo et al., 2007). These factors have initiated interest in crossbreeding HF cows with Jersey (J) sires, and many international studies have examined the benefits of the subsequent J × HF (JEX) offspring. In an Irish context, Prendiville et al. (2009) concluded that the JEX was an animal highly suited to grazing systems due to their high intake capacity at pasture, combined with the added benefit of improved production and feed efficiency. On a global basis, the JEX was shown to have improved production (Penno, 1998), fertility (Auldist et al., 2007; Heins et al., 2008), intake capacity (Goddard and Grainger, 2004) and subsequently improved profitability (Lopez-Villalobos et al., 2000). Furthermore, Clark et al. (2006) showed that under a once-daily milking regime, both milk yield and yield of milk solids of Jersey cows were maintained to a greater extent than those of HF cows. Subsequently, Walsh et al. (2007) established the Norwegian Red (NR) breed as a highly suitable breed for crossing with the HF in pasture-based systems. Begley (2008) showed that NR × HF (NRX) cows had a similar production potential and significantly improved reproductive performance and udder health compared with the HF.

However, little is known about the performance of these breeds in pasture-based AM systems and how breed affects cow traffic and ultimately how this could affect AM unit utilisation and efficiency. As voluntary cow traffic is the foundation on which successful AM operates, it is essential to develop an understanding of the suitability of different breeds for AM pasture-based systems. Shortall et al. (2018a) studied the performance of three breeds in a pasture-based AM system with respect to milk production, cow traffic and milking characteristics. The breeds examined were HF, JEX and NRX. The experiment was conducted in mid-lactation and variables measured included MF, MI, milk yield/milking and milk yield/day, wait time/visit and wait time/day, return time/visit and the daily distribution of milking events. The breed groups grazed as one herd of 80 cows (50 trial cows and 30 non-trial cows), within a 3-way (A, B, C) grazing system. Numerical differences existed between the breeds for milk yield/day and milk yield/milking, with HF cows yielding more milk/day and per milking, respectively, than the JEX cows. While the overall differences in milk yield between the breeds were not significant, the difference between HF and JEX approached significance. The performance of the JEX cows in the study of Shortall et al (2018a) was similar to those observed in previous studies (Auldist et al., 2007; Prendiville et al., 2011; Vance et al., 2013), with JEX cows having somewhat less milk volume than HF. Milking frequency did not differ significantly between breeds. However, the JEX cows did record a significantly lower and higher proportion of successful and failed milking events, respectively, compared to the other breeds. Further examination of the raw data showed that the JEX cows had a far greater variation in MI than the HF and NRX breed; they had a greater proportion of milkings with intervals both <12 h and >20 h. This variation in MI creates the potential for large changes in udder shape between milking, resulting in greater difficulty achieving successful cup attachment (Jago et al., 2007). While the HF breed was intermediate between the JEX and NRX, they also achieved significantly lower successful and higher failed milking compared to the NRX cows. However, the differences recorded (2%) were not as pronounced as those between the JEX and NRX (5%); thus, HF performance was more comparable to NRX than JEX.

Regarding cow traffic, significant differences existed, with the NRX breed returning from pasture to the milking yard more quickly (42 minutes sooner) than the JEX cows. The NRX cows had the shortest return time, followed by the HF, while the JEX cows had the longest. Jago et al. (2003) observed that when cows have negative experiences on roadways or in milking waiting yards, including encounters with cows of higher social rank within the herd, this can reduce the motivation of those cows to move from pasture to the AM unit. The NRX cows had a significantly shorter wait time/visit and wait time/day in the pre-milking yard of 0.9 h and 1.3 h, respectively, compared with the HF cows of 1.3 h, 1.7h and JEX cows of 1.5 h and 1.9 h, respectively.

A significant difference was also observed between breeds for daily activity measurements; JEX cows had significantly greater activity levels than HF cows, with the former recording 84 minutes more activity daily. The JEX also recorded a significantly greater proportion of milkings than the NRX and HF cows during the early morning hours of 04:00 h–06:00 h. The distribution of milking events showed that the HF and NRX cows were visiting the yard in anticipation of the allocation of fresh pasture. These data indicated that the NRX cows were trafficked most efficiently through the system and were a more dominant breed of cow. Ketelaar-de Lauwere et al. (1996) reported that lower social ranking cows visited the AM unit more frequently during the midnight to 6:00 am period in an indoor system, while Jago et al. (2003) found that more milking of lower ranking cows occurred during the late evening period in a pasture system. Based on the findings of those studies, and the distribution of milking events among the breeds in the current experiment, it could be hypothesised that HF and NRX breeds were among the more highly ranked cows in the study, while the JEX cows represented some of the more lowly ranked cows.

There was no significant effect of breed on milking duration/milking and milking duration/day and milk flow rate. Despite having a numerically lower milk yield, the JEX had the longest milking duration/milking and milking duration/day, with the HF having intermediate levels and NRX having the shortest milking duration. This may be explained by the slower milking speed of 1.5 kg/min for JEX compared with 1.7 kg/min for HF and NRX.

It is important to consider AM in the context of a whole-farm system, where key performance indicators (KPIs) such as milk solids production, fertility, survivability and efficient conversion of feed to high-value milk are vital aspects of a profitable milk production system. For optimisation of the AM system, it is important to have an even distribution of milking throughout the day; this may be best achieved by a mixed breed herd rather than a single breed herd.

3 Milk yield and quality

The main factors influencing milk production level in either CM or AM systems are genetic potential, nutritional status and lactation number of the cow. There is an additional factor associated with AM milking, that is MF. The positive effects of increased MF on milk yield were initially shown on CM farms when twice and three times daily milking regimes (at regular intervals) were compared (Friggens and Rasmussen, 2001). Smith et al. (2002) reported that the milk yield was increased by up to 14% by increasing fixed MFs (in a CM system) from 2 to 3 × per day while McNamara et al. (2008) showed that milk yield was reduced by 24% when MF was decreased to once per day. Subsequent studies have shown that AM increases MF; the average number of daily milkings per cow in indoor (as opposed to pasture-based) AM systems is generally considered to be in the range of 2.5–3.0, and increases in milk yield of 5–10% compared with fixed-frequency daily milking in CM systems have been observed. Increased milk yield recorded with increased fixed-frequency milking in CM systems may be associated with alteration of secretory cells and mammary functions. However, the effects of higher MF on milk yield in AM systems are more complicated, as MIs become variable from cow to cow and even for the same cow on a daily basis. Erdman and Varner (1995) compiled data from several studies on MF and concluded that increasing MF had a positive effect on milk production (3.5 kg/d throughout the lactation) independent of milk production level. Bach et al. (2007) found that offering concentrate to cows in an AM system did increase MF of cows that returned to the AM unit on a voluntary basis but did not increase MF of those that normally needed to be fetched. The effect of increased MF was found to be more important for multiparous cows than for heifers (Pettersson et al., 2011). This was again shown in a later study by Bogucki et al., 2017), which examined the association between MFs (≤2.40, 2.41–2.80 and >2.80 times/day) (in AM systems) in primiparous and multiparous cows and their daily milk yield. They found that regardless of the daily production level, most primiparous cows were milked 2.41–2.80 times/day and most multiparous cows>2.80 times/day. Higher daily milk yield and a positive effect of increasing MF on milk yield were observed in multiparous cows, with an increase from 29.7 kg (MF ≤ 2.40 times/day) to 31.1 kg milk (MF > 2.80 times/day). In primiparous cows, this increase was smaller (0.8 kg milk; 29.2–30.0 kg). The responsible mechanism underlying the increase in milk production response to MF has not been elucidated, but it has been suggested that it may be due to increased mammary epithelial cells, increased cell activity and frequent removal of the feedback inhibitor of lactation from the mammary glands (Vijayakumar et al., 2017). While this study (Vijayakumar et al., 2017) showed that the highest milk yield was recorded during a cow's third lactation, in early stage lactation and with a MF of four times per day, this may not always be practical. Monitoring of lactation is considered very important as it allows the control of milk yield dynamics and AMS performance (Masía et al., 2020). This study (Masía et al., 2020) showed that milk production of multiparous and primiparous cows with high daily yield and long MI was between 35% and 45% higher than that of cows with low daily yield and short MI. From all lactations analysed, the incidence of animals with high daily yield and long MI across farms was 7.5%. Thus, this study identified and quantified a new, AMS-specific phenotype (the combination of a relatively higher daily yield with a relatively longer milking interval) that could lead to the selection of more efficient animals for AMS. This has important implications for the stocking density of an AMS. The milking management regime being considered for an AMS needs to take cognisance of a number of factors such as type of cow (average yield), system of production (indoor or grass-based) and herd size; these parameters need to be aligned in order to optimise the performance and output of the AMS.

Initially, in the early years of AM (e.g. 2000–2002) it was considered that milk quality was negatively impacted, with effects such as increased free fatty acid levels (FFA), total bacteria count (TBC) and somatic cell count (SCC) being observed. Such effects could have significant negative consequences for the farmer and for the dairy industry. Elevated FFA content is undesirable in milk because it causes rancid flavours in dairy products. Increased levels of milk FFA have been reported as a result of increased MF (Klungel et al., 2000; Wiking et al., 2006; De Marchi et al., 2017). The study of Wiking et al. (2006) examined the effect of MF by collecting milk from udder halves that were milked two versus four times daily. The FFA content in the two kinds of milk was similar directly after milking but FFA was significantly higher in the milk from 4 times daily milking after 24 h of storage at 5°C. The authors indicated the weakness of the fat globule membrane as being the likely cause of the higher FFA in milk obtained from high-frequency milking. Increased FFA is also reported to result from excessive handling of small quantities of milk, for example air entrainment in the milking system, pumping of milk and cooling and stirring in the bulk tank (Wiking, 2022). However, improved management of cow MF and technical solutions such as correct cooling and gentle mechanical treatment are now implemented in the modern AMS.

Elevated milk TBC and milk SCC observed in the early years of AM may have been due to either herd management or equipment issues. The absence of a human operator during the teat cleaning process and/ or no visual inspection of the foremilk may have contributed to some of the milk quality issues. Contradictory data have been reported on the effect of AM on the microbiological quality of milk. Most of the earlier studies refer to increases in mean TBC with the advent of AM (Klungel et al., 2000; Rasmussen et al., 2002; de Koning et al., 2003). An increase in milk TBC may be attributable to the cleaning process of the teats and/or the cleaning process of the equipment. The teats represent an important source of bacteria, however, that is very much influenced by the level of bacteria in the cow's environment, for example faeces and bedding material, and this is particularly relevant in terms of thermoduric bacteria. Thus, the hygiene level of the environment and of the cow, together with the effectiveness of the teat cleaning process, is of crucial importance. The AM equipment cleaning protocols are generally devised by the manufacturing company and generally do not cause microbiological problems in the milk if the critical parameters of the protocols are maintained on the farm, for example sufficient quantities of water at the required temperature.

Milk quality in terms of milk SCC and mastitis is addressed under 'Health and welfare management of cows' in section 9 of this chapter.

4 Feeding concentrate supplementation in automatic milking systems

The initial commercialisation of AM focused predominantly on dairy production in intensive indoor housing systems. This was primarily due to AM systems being originally developed for use in such production systems, dominated by high costs of production and high-yielding cows (Lind et al., 2000). In such confinement systems, both Bach et al. (2007) and Halachmi et al. (2005) found no difference in MF when offering 3 kg/cow per day and 8 kg/cow per day and 1.2 kg/cow per day and 7 kg/cow per day, respectively.

In pasture-based AM systems, the use of supplementary feed can be a tactical decision to either cover pasture feed gaps, achieve high milk production levels or encourage cow traffic from pasture to the dairy (Salomonsson and Sporndly, 2000). As grass growth is seasonal (being dependent on prevailing climatic conditions), grass deficits may occur in spring and autumn, generally corresponding to the early and late lactation periods, respectively, in a spring-calving system which predominates in pasture-based systems. Thus, concentrate supplements may be introduced when grass growth levels are sub-optimal and not sufficient to meet herd demand; in this scenario, concentrate supplementation is used to maintain energy balance and increase the total dry matter intake (DMI) of dairy cows (Delaby et al., 2001) (particularly in early lactation), thus assisting in avoiding negative energy balance (Berry et al., 2006). The level of supplement offered normally depends on the deficit in grass quantity. When grass growth exceeds the demand of the herd, as may occur during the summer months in Ireland and some other countries with pasture-based systems, supplementing with concentrate is questionable as the milk response is limited (Kellaway and Harrington, 2004) and the substitution rate is increased (Bargo et al., 2002).

The successful operation of a pasture-based AM system relies on cows voluntarily trafficking from pasture to the milking yard and subsequently back to pasture again. Without voluntary traffic, milking events are not distributed evenly throughout the day, and failure to achieve this voluntary and distributed milking regime daily may have a negative effect on the uptake and adaption of AM technology at the farm level (Lyons et al., 2013a). The success of using supplementation to entice cows to an AM unit in a pasture-based system has been mixed. Jago et al. (2007) found that offering 1 kg of supplement/cow per day did increase the visits to the selection unit but did not result in an increased MF. It is possible that the provision of such a small amount of incentive combined with the longer walking distances associated with the farm layout was insufficient to significantly motivate the cows. A study by Lessire et al. (2017) found that supplementing with 4 kg/cow per day as opposed to 2 kg/cow per day increased visits to the milking unit but did not translate into more milkings. A further study by Shortall et al. (2018b) examined the effect of different concentrate supplementation levels in the early and late periods of lactation on milk production and cow traffic parameters in a seasonal calving pasture-based AM system. The experimental periods for the early (19 days in milk (DIM)) and late lactation trials (208 DIM) were 49 days in each trial period. The early lactation supplement levels were 2.3 kg/cow per day and 4.4 kg/cow per day for low concentrate (LC) and high concentrate (HC) levels, respectively, while the late lactation supplementation levels were 0.5 kg/cow per day and 2.7 kg/cow per day for LC and HC, respectively. This resulted in an approximate 2 kg differential in concentrate offered between the LC and HC treatments in both early and late lactation.

Supplementing with LC or HC in early lactation had no effect on milk production, milking characteristics or cow traffic variables. The similar milk yield observed for the different treatments may be a direct result of cows on the LC treatment mobilising more body reserves in the early lactation period (Bargo et al., 2002). Furthermore, Baudracco et al. (2010) considered that the high energy content of spring grass can result in milk response to concentrate being at its lowest during this period. There was an effect of treatment in late lactation, with HC cows having a greater milk yield/day than LC cows (12.4 kg/cow per day and 10.9 kg/cow, respectively). This may have been due to the quality of grass consumed by the cow at this time; a poorer quality base feed would result in a greater milk response (to concentrate). The milk response of 0.75 kg of milk per kg of concentrate was lower than that of Reid et al. (2015), who found a milk response of 0.96 kg of milk per kg of concentrate when moving from a grass diet with no concentrate supplementation to one with 3 kg of concentrate, in late lactation. However, that study was conducted in the context of a CM system, where cows were milked twice daily. The HC treatment also increased milking duration; but this longer milking duration/day was not detrimental to the optimisation of the AM system, as there can be a substantial surplus of capacity on the AM unit (depending on herd size) at that period of lactation. The Shortall et al. (2018b) study showed that offering an additional 2 kg/cow per day of concentrate in late lactation had a positive impact on cow traffic, with those cows returning from pasture 1.6 h sooner than the LC cows, while no effect on return time was observed in the early lactation period. It is likely that the motivation of cows to milk is high in early lactation due to a relatively high level of milk production and associated potential udder discomfort; so any further shortening of return time would be unlikely; alternatively, the positive impact of supplementation on milk yield in late lactation may have at least partially impacted on the reduced return time. This reduced return time, in turn, impacted on a significantly shorter MI (the HC treatment had a 9% shorter MI than the LC treatment, at 16.5 h and 18.2 h, respectively). In intensive indoor AM systems, concentrate supplement is offered not only to increase milk yield but also to increase MF (Prescott et al., 1998).

The successful operation of an AM system, irrespective of the production system, is dependent on cows presenting themselves at the milking unit on a voluntary and continuous basis; concentrate supplementation may assist this at different stages of lactation. However, any decision regarding the supplementing of dairy cows with concentrates needs to be examined from an economic perspective, to establish if the milk production and cow traffic benefits observed in late lactation outweigh the cost of the concentrate, thereby ensuring that the decision to supplement is financially prudent.

5 Grazing and grassland management for automatic milking systems

In pasture-based milk production systems, grassland and grazing management are key to maximising milk production from grazed pasture. Pasture-based AM systems are those where cows are outdoors and obtain more than 50% of their annual feed requirements from grazed pasture or forages (Lyons et al., 2014). Getting the grass right in terms of supply and quality ensures cows are fed high-quality feed resulting in high milk yields and high MFs, as well as ensuring good cow flow around the farm; this is particularly important in AM systems.

5.1 Farm infrastructure and layout

Feed is a primary motivator of voluntary movement of dairy cows in any AM system, and in pasture-based AM, that feed is grass. Pasture biomass is likely to be one of the main influences on cow behaviour in AM systems and is likely to significantly influence when cows decide to leave an allocated area (Lyons et al., 2014) in search of more biomass. How the grass is allocated across the farm depends on a number of factors including the shape and size of the farm, location of the milking unit, infrastructure and farmer preference. Farm infrastructure is very important in pasture-based AM systems; this includes good road surfaces that minimise discomfort as cows walk voluntarily to and from the milking unit, as poor surfaces may reduce the number of cow visits to the unit and restrict movement around the farm. While it is important to maintain good MF in pasture-based systems (average MF is generally between 1.5 milking/cow per day and 2.5 milking/cow per day), this MF should not affect the FFA level in milk to the same extent as has been observed in indoor systems, where increasing MF to three or four times a day can result in a high concentration of FFAs in milk, resulting in a rancid flavour in subsequent dairy products (Wiking, 2022).

Pasture allocation is critical in all grazing systems. Getting the pasture allocation correct ensures that grass is not wasted (Fulkerson et al., 2005) if there is over allocation and that cows are not hungry if there is under allocation. In a pasture-based AM system, allocating too much pasture means that cows will remain in the paddock longer, likely reducing visits to the robot; allocating too little grass means cows are in the collecting yard waiting for the gates to open for the next section of grass. An AM farm should be set up in paddock blocks to encourage cow movement. Generally, pasture-based AM farms are set up with two, three or four blocks (AB, ABC or ABCD, respectively). Lyons et al. (2013b) reported that a three-block system resulted in increased milk yield, greater MF and reduced MI compared to a two-block system. Three blocks motivated the cows to move around the farm for pasture, resulting in them visiting the milking facilities more frequently than in the two-block system. Each block can have a similar grass allocation, or the allocation can vary depending on the duration of cow residency in that block. For example, allocating a small evening grazing area encourages cows to leave the paddock during the night/early morning hours when cows are normally less active, thereby increasing traffic through the AM system and helping to ensure a more evenly distributed usage of the robot. If it is necessary that cows receive the supplementary feed, for example in periods of low grass growth or if inclement weather inhibits grazing, the supplementary feed area can represent a block. In terms of silage making, silage can be made from a proportion of each block, or alternatively a whole block could be used for silage.

5.2 Grazing management

There are many well-established grazing guidelines and principles used in conventional grazing systems; these can also be incorporated into pasture-based AM systems (Clark et al., 2015). Good grazing management and feed budgeting skills are vital to ensure the correct allocation of high-quality herbage on a pasture-based AM farm. Measuring herbage mass on a weekly basis provides a summary of the grass available on the farm and aids decision-making around managing surpluses and deficits in available herbage.

As grass growth and availability are significantly impacted by climate, different management guidelines may be appropriate in different geographical locations. The Irish context may be taken as an example, where, in spring, a tool such as the spring rotation planner (Teagasc, 2011) may be used. The spring rotation planner allows the farmer to allocate a fixed area of the farm to the herd on a daily basis. It is based on the start date of grazing in spring and the planned end date of the first rotation. It allows the farmer to allocate grass every day to the cows while ensuring that the first rotation does not extend beyond 'magic day' (the day when grass supply equals grass demand) or indeed end before 'magic day'. Combining farm herbage measurement with the spring rotation planner helps the farmer to decide how much supplementary feed, if any, is required in addition to grazed grass. In a pasture-based AM system, the area to be grazed per day can be divided across each block. For example, if 6% of the farm area should be grazed per day in an ABC farm set up with equal areas in A, B and C, then 6% each of the A, B, C blocks should be grazed.

In summer, the grass wedge is used (Teagasc, 2009, 2011). Ideally, a grass wedge (Fig. 2) should be developed for each block. To develop a grass wedge, the quantity of herbage present in the paddocks must be measured/estimated on a weekly basis at least (O'Donovan et al., 2002). The paddocks are arranged in a wedge shape from the highest quantity of herbage mass (on left) to the lowest (on right). A feed demand line is fitted over this wedge, from the ideal/target pre-grazing herbage mass to the target post-grazing herbage mass. The next paddocks to be grazed are those with the highest herbage mass. If there is excess grass on the farm (i.e. more than is required to meet the current feed demand), this will be represented by some paddocks having pre-grazing herbage mass greater than the feed demand line; then the herbage in these paddocks should be removed as surplus silage or the stocking rate should be increased. If paddocks have herbage mass levels below the feed demand line then there is a deficit in the feed budget and supplementation is required.

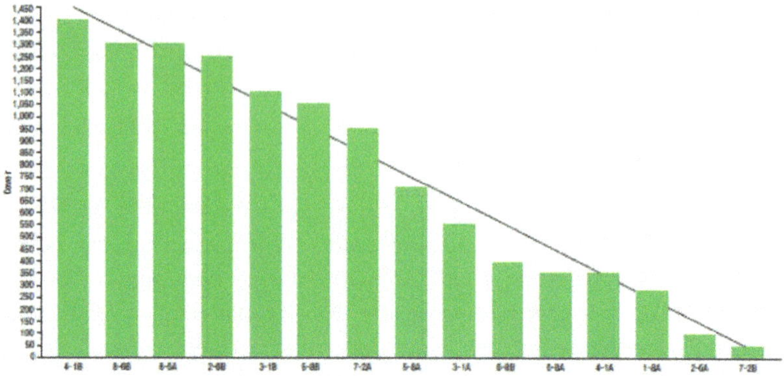

Figure 2 Grass wedge showing adequate grass supply. Source: Teagasc (2016).

Walking the farm on a weekly basis while completing the grass measurements also allows the farmer to evaluate the quality of the pasture and help make informed decisions about surpluses/deficits and supplement required.

Two main objectives for managing grass in the autumn in pasture-based systems are (i) to maintain grazed pasture in the dairy cow diet for as long as possible before housing for winter or drying off and (ii) to use good autumn grazing management to ensure good grass production and availability at cow turn-out in the following spring. An autumn feed budget is important. Guidelines such as the '60:40' rule can be used in the final grazing rotation to ensure grass growth over winter and in early spring; this rule targets having 60% of the grazing area grazed in the first 70–75% of the final rotation (i.e. if the rotation is 50 days, have 60% grazed in the first 35–37 days) and the remaining area grazed in the final part of the rotation. It is necessary to apply this system to each block on the grazing platform. Ideally, the paddocks to be grazed first in spring should be those closed first in the autumn. Swards should be tightly grazed in autumn to ensure that old material does not remain in the sward over winter leading to decay of herbage and tiller death. Grazing swards tightly at closing ensures that light can penetrate to the base of the sward to promote tiller production over winter, thus ensuring a productive sward in spring.

5.3 Integrating grazing and automatic milking in different geographical regions in Europe

While indoor feeding systems have traditionally been well adapted to AM, grazing systems have not, and this has been leading to a decrease in grazing on AM farms in different areas in Europe (Van den Pol-van Dasselaar et al., 2011). Grazing is considered to have many advantages for the economy, environment, animal welfare and product quality (O'Brien and Hennessey, 2017; Van den

Pol-van Dasselaar et al., 2020), thus, research has been undertaken to develop technological solutions to integrate grazing and AM in different geographical regions in Europe. Much of this research was conducted under an EU-funded project 'Autograssmilk'(AUTOGRASSMILK, 2017). Various strategies were developed to maximise milk output from integrated grazing and AM systems in a wide range of production conditions, from pasture-based extensive systems with low feed costs to intensive systems aimed at high milk yield per cow but with only small amounts of pasture in the diet. Recommended feeding strategies for periods of grass inadequacy as well as the performance of different cow breeds were investigated. A new automated grass measurement tool was used to ensure the precision required in grass allocation, particularly important in integrated grazing and AM systems. The construction of a web-based sustainability tool allowed farmers (in conjunction with their advisors) to understand and increase the sustainability of their farms. A further web-based decision support tool was developed to assist farmers and advisors in the decision-making process around grazing and AM systems from an economic perspective. This allowed more informed decisions to be taken by farmers. AUTOGRASSMILK provided a knowledge base for long-term impacts such as increased productivity in AM herds with grazing and increased numbers of AM system herds considering grazing as a realistic opportunity.

6 Benchmarking and optimising performance using key performance indicators

Milking is recognised as a core task on dairy farms, and AM represents a significantly different mode of conducting this task to CM and particularly so in pasture-based systems. Installation of an AM system represents a very significant economic investment. Therefore, it is important that information on system utilisation and performance is available to farmers considering different options. Conventional milking farmers need to understand what levels of performance and productivity are achievable under commercial AM operations, while AM farmers need to benchmark their performance against the potential of their system and other commercial AM farms at specific time points and over time, to promote optimisation of the system by informed decision-making. To initiate such a process, it was necessary that (i) KPIs be identified that can be used to monitor and benchmark performance, (ii) the relevant data on those KPIs is assembled and interpreted regularly and (iii) farm operators need to understand how to improve/manipulate those KPIs. Such information/data would put results into perspective, show the relationship between variables/KPIs, provide targets for KPIs and basically develop a KPI tool to assist operators to optimise their system at the farm level, while also providing scientific data and reports to advance scientific thinking and research in AM milking and promote appropriate developments by milking equipment manufacturers. Such a study was put in place in Australia and has been reported by Lyons and Kerrisk (2017).

Eight Australian commercial AM farms were monitored on a monthly basis for a 12-month period between July 2015 and June 2016. Data from the participating farms were subsequently summarised and collated. The participating farms were milking predominantly either Holstein cows or crossbreeds (Holstein and Jersey or Brown Swiss). Three, two and three farms had seasonal calving, split calving and year-round calving patterns, respectively. All farms except one were pasture based but generally had concentrate feeding stations as well.

6.1 Actual average, potential average and maximum potential average values for the key performance indicators

First, the actual system performance was addressed. The number of milking events (average number of milkings/AM unit per hour) was calculated for every hour of the day, on a monthly basis for each of the participating farms. The average number of milkings/AM unit per day was also calculated, together with the number of cows/AM unit (herd size divided by number of AM units) and milk yield/AM unit per day for every month on every farm.

Second, the potential average system performance was considered. The potential average extra milkings/unit per day that could be conducted if all of the hourly average number of milkings/unit per hour were at least as high as the daily average number of milkings/unit per hour was calculated for every month, that is the potential average milkings/unit per day was calculated as the value when all hourly averages below the daily average were brought up to the daily average. Then, using the actual average MF (number of milkings/cow per day) and daily milk yield (kg milk/cow per day), the potential extra cows (cows/unit above the actual average) that could be managed and extra milk yield (kg milk/ unit per day above the actual average) that could be harvested was calculated.

Finally, the potential maximum system performance was addressed. The potential maximum extra milkings/unit per day that could be conducted if all of the hourly average numbers of milkings/unit per hour were as high as the maximum value of the average number of milkings/unit per hour reached over the 24-h day were calculated for every month. Then, using the actual average MF (number of milkings/cow per day) and daily milk yield (kg milk/ cow per day), the potential extra cows (cows/unit above the actual average) that could be managed and extra milk yield (kg milk/unit per day above the actual average) that could be harvested, was calculated. The total potential maximum milking, cows and yield were then estimated by adding the actual and potential maximum values for each parameter, to give an indication of potential maximum improvement.

These authors found a wide range in farm performance. The distribution of milking events, expressed in milkings/AM unit per hour, varied across time of day, which was not unexpected as lower milking unit utilisation levels have been observed previously during the early hours of the day (John et al., 2016). The milking event values and distribution (average of 4.9 milkings/unit per hour, ranging between 0 milkings/unit per hour and 9.7 milkings/unit per hour) and the milkings/unit per day are highly related to cow, system and equipment factors. Cow factors include aspects related to cow traffic, daily milk yield and yield per milking. System factors include aspects related to herd size, management strategies and cow traffic; this may be partially related to the diurnal grazing behaviour of cows but feed is the main factor influencing cow traffic and therefore utilisation levels of robots (Lyons et al., 2013b,c). The distribution of milking events is also related to the MF and daily milk yield of the herd and, therefore, the amount of milk harvested per milking event, which in turn affects box duration time (time spent per cow in the milking unit). Equipment factors include aspects related to robot performance, cleaning cycles, number of washes and rinses per day and time of day the washes take place. The best pasture-based farms in the Lyons and Kerrisk (2017) study achieved quite consistent robot utilisation throughout the day mainly through manipulation of type, frequency, timing, size and location of feed (John et al., 2013). An increase in the number of milkings can be achieved by either milking cows in the herd more often or by milking more cows. Milking existing cows more often can increase daily milk yield (Stockdale, 2006), but only if genetic potential or nutrition are not limiting, and it may reduce efficiency, due to extra walking required in pasture-based systems. Milking more cows per AM unit is not feasible on all farms and is dependent on current stocking rates and feed availability. In the Lyons and Kerrisk (2017) study, the potential to increase the number of cows by a maximum of ~60% was observed; this has significant potential implications as it may be possible to manage the same number of cows with a reduced number of robots. This would reduce the capital investment required to install an AM system and thus could make the investment more attractive to a wider proportion of the industry. However, a higher ratio of cows to AM units increases the pressure on the system and requires the system to be very well managed. Maximising yield is the main aim within AM systems (Sonck and Donkers, 1995); this is influenced by adequate herd size and a management system that promotes highly efficient cow traffic (Lyons et al., 2014). Management of milking distribution, cows and individual milking as well as activities such as teat cleaning or equipment cleaning cycles all have an impact on the utilisation time of the AM unit. The levels of AM unit utilisation reported in the Lyons and Kerrisk (2017) study reflected the potential to increase the milking time on Australian AM farms by a maximum of ~60%; this would mean having the system operating up to 21–22 h/day.

Pasture-based dairy systems usually try to match herd nutritional requirements and pasture growth. This often dictates a seasonal calving pattern where all cows calve at a single time period each year; cows then enter peak lactation together during the maximum grass growth period. In such systems, peak demand for AM unit time is short-lived and so the AM unit is run at lower than potential utilisation levels for the remainder of the year; alternatively, this can also lead to the view that AM is more suitable for farms with year-round calving. However, according to the findings of the Lyons and Kerrisk (2017) study, farms with different calving patterns showed no difference in the current or potential maximum number of milking, number of cows or yield. However, the core issue is an understanding of how to manipulate the different factors to incentivise cow traffic (Lyons et al., 2014) and, therefore, system utilisation in a profitable and sustainable way. Even though the study results suggested that Australian AM farmers could increase the number of milking, number of cows, yield and time, their systems operate by up to 60% to achieve maximum potential, some farmers might find it either challenging or a high-risk decision to push the system to that extreme. To achieve that level of operation would require the adoption of a variety of strategies that might be quite farm specific. The rationale behind the potential average performance is, therefore, to show what can be achieved and opportunities for those farmers who wish to operate at a level somewhere between the current and potential maxima. Optimisation and benchmarking may help farmers who are considering investing in AM technology, while also challenging existing AM farmers who wish to push their systems to a higher level of performance.

A more recent study by Gargiulo et al. (2020a) compared the physical and economic performance of pasture-based AM with CM systems to identify opportunities for improving AM productivity and profitability. Data from 14 AM and 100 CM Australian farms were used over three years. Opportunities for improving pasture utilisation, labour efficiency and robot utilisation on the AM farms were identified, which if addressed could improve the productivity and profitability of these systems. The study also showed that physical performance indicators such as milk production per cow and per ha were similar between the AM and CM systems. This information should allow balanced consideration of the different systems by those replacing milking equipment or by new entrants to dairying.

7 Cow behaviour

Cows in AM systems largely have the potential to set their own milking schedule, they have more freedom to control their daily activities and rhythms and have more opportunities to interact with their environment than cows milked in CM systems. The main difference between AM and CM is the reliance on the cows to go voluntarily, and individually, to the milking unit a number of times daily to be milked. However, for this to operate efficiently in terms of performance and cow well-being, the cow traffic system has to operate well and facilitate an appropriate frequency of successful milking visits, adequate time for feeding (at pasture or in the barn) and adequate access to rest areas (Hermans et al., 2003; Bach et al., 2009). So it is important to understand the motivations and mechanisms that can induce cows, particularly those with access to pasture, to return to the AM unit to maintain the desired number of visits daily. Within such a system, cows may develop their own consistent feeding and drinking patterns (Melin et al., 2005), and they may also be influenced by other factors. Diurnal patterns of feeding and lying behaviour persist in AM systems, with fewer cows feeding and more cows lying down overnight, generally between 22:00 and 07:00 (Wagner-Storch and Palmer, 2003; De Vries et al., 2011; Munksgaard et al., 2011). Furthermore, low-ranking cows commonly visit the AM system during these low occupancy nighttime hours (Hopster et al., 2002) due to social competition in the herd. As long MIs can impair milk production, the potential exists for a low-ranked cow's milk yield to be dependent on other cows' schedules. This situation could be assisted by the presence of a sufficiently large waiting area in front of the AM unit (Melin et al., 2006) to alleviate stress on lower-ranking cows.

Sporndly and Wredle (2005) suggested that voluntary MF decreased to some extent on farms that combined AM and grazing; experimental studies have reported ranges of 1.4 milking per day to 2.3 milking per day for grazing cows, with higher rates for cows receiving forage in the barn compared with 100% grazing systems (Davis et al., 2008). Meanwhile, Jago et al. (2007) suggested that it might be more effective to reduce the number of milkings expected per cow on pasture farms and increase the number of cows per AM system. But a well-functioning cow traffic system becomes essential in this scenario of increased cow number, increased distances between the AM unit and feed source (pasture) and the fact that cow behaviour is more synchronised on pasture compared with indoor housing systems (Tucker et al., 2008). When grazing on pasture, cows often perform certain behaviours such as drinking (Jago et al., 2005) and grazing (Gregorini, 2012) in bouts or in a more synchronised way compared to cows fed indoors; this is a challenge since a number of cows returning to the milking unit simultaneously would create a large queue of cows standing in the waiting area. This behaviour is undesirable in an AM system which is designed to consistently have small numbers of cows present for milking at any one time. To reduce the synchrony of these behaviours and to encourage distributed milking visitation throughout the day and night, farm infrastructure and management of incentives (e.g. new grazing plots or feeding in milking stalls) can be adapted.

For cows at pasture, the grazing season is characterised by constant changes in weather, pasture supply, pasture quality and day length, and it is important to understand how cow behaviour responds to these conditions.

Cow behaviour parameters have been investigated (in terms of stress responses) as potential indicators in animal well-being assessment. The time budget of the cows, that is how much time they spend on different basic activities is of particular relevance in AM systems, for example increased standing time may indicate stress or discomfort. In such milking systems, time budgets may be influenced by the cow traffic system and by social rank; cows of low social rank spend more time standing, because they have to wait longer to enter the milking unit. A CM herringbone parlour and an AM system (both incorporating cow grazing) were investigated by Speroni et al. (2004); they reported that total time spent lying was higher, and time spent standing, including milking, was lower in the AM group compared to the CM group. However, the management system in place, as well as the number of cows per robot, can also influence behaviour characteristics.

The process of AM has been successfully integrated into pasture-based commercial and research farms in Europe, Australia and New Zealand. As outlined previously, a high voluntary cow flow is crucial for the success of AM in pasture-based systems. Previously, it has been shown that providing three rather than two fresh grass allocations during the day (a three-way grazing system) improved the cow flow through the system (Lyons et al., 2013b). Subsequently, it was considered that a four-way grazing system may improve cow flow further with a larger herd of >80 cows per milking unit. But it was necessary to assess the suitability of this system with regard to the natural behavioural pattern of cows. Thus, Werner et al. (2018) set up a study to examine the effect of a four-way system on cow behaviour and activity. A total of 18 cows in a herd of 84 were monitored to establish their behaviour patterns using RumiWatch sensors over a 9-day period. Analysis of the daily behavioural pattern was conducted in 3-h summaries. The behaviour data indicated that cows spent average times of 468 ± 166 min/day and 419 ± 93 min/day grazing and ruminating, respectively. Furthermore, cows spent an average of 93 ± 42 min/day walking while taking 2719 ± 1240 strides/day. Kennedy et al. (2011) reported average rumination times of 406 min/day and Werner et al. (2017) reported a walking time of 85 min/day in CM systems. However, a study by O'Driscoll et al. (2010) reported that cows in a CM system with a once-a-day or twice-a-day milking regime had longer lying times compared to the Werner et al. (2018) study (620 ± 15 and 627 ± 14 min/day, respectively, compared to 504 ± 218 min/day); but this might be explained by the different stage of lactation or treatment. There were two distinguishable periods of grazing 06:00 to 09:00 and 18:00 to 21:00, which were followed by intense rumination periods. These results show that the previously observed pattern of cows having two main feeding periods around dawn and dusk, a likely natural behaviour, is not impacted in a four-way grazing system.

So the four-way grazing system may be a novel approach to allow more cows to be milked per milking unit in a pasture-based system. This work by Werner et al. (2018) was a preliminary study; to the authors' knowledge, this was the first study to record such data. Further studies are required to report on cow location and behaviour data extended over a full lactation period and to confirm that cows can easily express natural behaviour in a four-way grazing AM system.

8 Training of cows and transition to automatic milking

Training is an important aspect of transitioning to AM for both the operator/farmer and the cows. Some of the main challenges for the operator include breaking the cows' routine, issues with nutrition (balancing feed and the cost of feed), trusting the AM system and changing the mind-set (accepting that the farm requires a different style of management and is now more computer-reliant and having to change the person's own routine), having an extremely demanding first few days/weeks at start-up (requiring an intense amount of physical labour to get cows milked in the AM system initially) and having to change health management (to deal with feet/leg, heat detection, reproduction and mastitis issues and to manage cow health on a more individual animal basis). Additional challenges for the operator include general maintenance of the AM unit, managing cow traffic to the AM unit and trying to become familiar with the technology. Common solutions to most of these challenges are time and patience, being proactive and being open to asking for assistance (e.g. from the equipment dealer and other AM system farmers). Alternatively, the cows must learn to operate within the framework of the new AM system, irrespective of whether it is on a pasture-based or indoor farm system. The cow must learn to move voluntarily to the AM unit, move into the unit once it becomes free, allow milking to take place and then move out and return to either the pasture or indoor environment through a series of gates. The AM companies often recommend a training program to help cows adapt to the AM system. Training may involve bringing cows to the AM unit from 1 to 4 times/day (without milking but with a high level of concentrate feed) for 3 to 14 days before start-up (DeLaval International AB, 2008; Hulsen and Rodenburg, 2008). It may take approximately 7–8 days to train cows or heifers to adapt to the AM system (Jacobs and Siegford, 2012; Spolders et al., 2004). Farmers sometimes choose not to train cows (to the AM system) because of the extra time and effort it requires; however, the pre-training of heifers is particularly important to save time in the long-term; efficient training for AM is important to reduce labour and minimise subsequent effects on production. Introducing heifers to the AM system before calving has been shown to have a positive effect on MIs and milk production after calving (Widegren, 2014). Training of heifers can be assisted by having the pre-trained cows (from previous lactations) calved and milking on the system first; the heifers can then follow the routines of the cows.

A study conducted by Jago and Kerrisk (2011) examined the effects of three levels of training of cows and heifers (pre-calving) to AM operated within a pasture-based dairy system. Animals received either no training, training which included handling in the AM unit and associated gate system or training in the AM unit including exposure to typical noises and mechanical movements as well as the gate system. Overall, heifers adapted to the new method of milking more quickly than cows. The majority of animals milked without assistance within seven days of calving, regardless of training method. Pre-calving training in the AM system improved entry at the first milking, indicating that habituation to the environment occurred during the training process. But additional exposure to the sounds of the working AM system and movement of the automatic arm did not improve AM unit entry beyond the basic training; this may be due to the cows making negative associations between noises/movement and the robot, as these associations can form readily in cattle (de Passille et al., 1996). Therefore, there does not appear to be any advantage in including exposure to the noises and mechanical movements of the AM system during training. However, providing a concentrated feed during AM milking provides an immediate positive re-enforcement for entering the milking unit, thus increasing the likelihood that the cow will repeat that behaviour. Cows can learn and quickly adapt to new concentrated feeding routines (Livshin et al., 1995). Both heifers and cows are able to learn a series of complex tasks, but training should be designed differentially for heifers and cows. The Jago and Kerrisk (2011) study showed that a longer period of training in the AM unit crate would be beneficial for heifers, while an increased focus on learning the farm layout would be beneficial for older cows; this may be a consequence of their prior experience of batch milking.

A study by Tse et al. (2018) surveyed 217 Canadian milk producers and found that 42% of producers trained their heifers, cows or both, prior to their first milking with the AM unit; the remaining 58% of producers allowed milking to occur during the cows' first experience with the AM unit. When training did take place, feed was often provided in the AM unit, which was presumably intended to encourage and/or distract cows during cluster application and during milk removal. Small training groups of <20 cows were most common (63% of respondents). On farms where mature cows were trained to AM, cows were required to visit the AM unit approximately twice per day during the first and second week of training, and it took approximately 7 days to train a cow or heifer.

9 Health and welfare management of cows

Animal well-being is a critical consideration to the dairy industry for both ethical and economic reasons. The term well-being may incorporate the health and welfare of the animal. In general, the well-being of cows milked conventionally or automatically appears to be comparable, as reported by a number of behavioural, physiological and health indicators. But careful farm management is vital to ensure a high standard of cow well-being. With an AM system, individual farmers may or may not be challenged to maintain cow well-being depending on their management strategy, farm design and cow handling procedures. Some of the most common issues considered in relation to AM are udder health and lameness.

Many studies reported an increase in milk SCC and a worsening of the udder health status after the introduction of AMS (Rasmussen et al., 2002; Kruip et al., 2002; de Koning et al., 2003). Maintenance of appropriate cow and milking management associated with mastitis and milk SCC was challenging. However, in more recent times, with the use of improved devices and sensor technology incorporated into AMS equipment, the mastitis and/or SCC status may be improved with AM compared to many CM scenarios, as observed by O'Brien (2016). The consistent pre-milking cleaning and post-milking disinfection of teats together with cluster cleaning between cows and very importantly the use and interpretation (by the farmer) of milk SCC data generated, all represent steps that can improve udder health and milk quality in AM systems. However, several studies comparing udder health between AM and CM or comparing udder health before and after the introduction of AM in the same herds have shown deterioration in udder health during the first year or more after the introduction of AM.

Properly designed yards are critical to maintain clean cows and efficient flow of cow traffic; this is also important to maintaining a reasonable MF per day, thus minimising the occurrence of long MIs. On one hand, increased MF promotes the removal of bacteria from the udder, reducing the time for bacteria to colonise the quarters. On the other hand, frequent milking provides greater opportunities for bacterial invasion during milking. These parameters may be particularly challenging when cows are at pasture. Experts on animal health and welfare should be involved in designing cow entry and exit routes to the milking unit, the waiting yard and the milking area to ensure comfortable and efficient cow flow. Milking by the AM unit should occur smoothly so that it does not cause distress through a long waiting time or restlessness through unpleasant or painful milking because of incorrect machine settings. The fact that AM is quarter-based means that it may be less irritating to the physiological characteristics of the udder by preventing overmilking of low yielding quarters (Bava et al., 2005) and reducing the risk of cross-contamination among quarters of the same cow through the milking claw. Uneven or inconsistent MIs inherent in an AM system can be considered a potential cause of increased SCC. Mollenhorst et al. (2011) examined this issue, that is if MI or its variation was related to changes in SCC, using data from 151 Dutch farms. The effect of MI was, although significant, not very strong; the effect of MI on SCC was marginal when corrected for other variables, such as production level, lactation stage, and parity. The study showed that variation in MI had only a limited association with SCC but variation in the interval was more important than the interval itself.

It is generally accepted that the different stakeholders (within the dairy industry) have a duty to ensure good cow udder health and quality milk and a study by Hovinen and Pyörälä (2011) has outlined guidelines for some specific stakeholders. The AM machine manufacturers should aim to develop more effective and precise methods for teat cleaning and disinfection, techniques for detection of subclinical mastitis as well as improving the design to reduce the spread of mastitis bacteria between cows. Farmers should become skilful in the observation and management of cows and knowledgeable on how to use the data captured by the system. Researchers should address questions concerning the effects of infrequent milking and systematic failures in teat cup attachment on udder health and the reasons for poor teat condition on some AM farms.

Cow lameness in an AM system can mean that less milk is produced during fewer milkings per day. Minimising lameness in a herd with an AM system is crucial because of the voluntary component of milking activity. A Canadian study (King et al., 2017) examined cow lameness issues on 41 farms with AM. Lame cows produced 1.6 kg/day less milk in 0.3 fewer milkings/day. Lame cows were 2.2 times more likely to be fetched to the milking unit on more than one occasion during the 6-day period (of the experiment) and spent 38 min/day more time lying down in bouts that were 3.5 min longer compared to non-lame cows. There was no difference in daily rumination or activity between lame and non-lame cows or in night:day rumination time, but lame cows had greater night:day activity ratios. That study supports the knowledge that lameness has negative effects on milk production, voluntary milking behaviour, and lying behaviour of cows in AM herds. These results highlight the negative effects of even moderate cases of lameness and that milk producers should be motivated to improve their lameness monitoring and treatment protocols.

A further aspect of cow well-being was examined by Wildridge (2017), that is the farmer-cow relationship after transitioning from CM to AM and the changes in farmer routine that may affect the relationship. Findings suggested that the farmer-cow relationship did change after transition to AM, as seen by the reduced fear responses expressed by cows when approached and handled. The near elimination of milk harvesting related tasks and the increase in time spent near the cows and in close contact with them in the AM system could have contributed to this change in the farmer-cow relationship. Other aspects potentially impacting the well-being of cows in a dairy system were also explored by Wildridge (2017), such as heat stress when cows were on a pasture-based AM system, and how these conditions could be mitigated. Milking frequency and milk yield of cows was negatively affected by high-temperature humidity index conditions. But, the study showed that various initiatives could be used to attract cows to the milking unit in these conditions, thus improving MI, which could reduce the risk of udder health problems and improve milk yield in the long term. These initiatives included encouraging cows to the milking unit at night when the temperature was cooler and provision of intermittent shade sources along the laneway between the feeding area and the milking facility, which increased travel by cows towards the milking facility during the day. Provision of shade in the pre-milking area also reduced cow respiration rate and body temperature, and also improved milk yield and concentrate intake (Wildridge et al., 2017).

Sensor technology provides a significant opportunity to positively manage the health and welfare status of cows in an AM system. But, it is important that the farmer engages positively with this technology; if used properly, multiple sensors and data analysis programs used in association with AM can be beneficial to the manager and cow alike. But alternatively, care must be taken not to allow too much dependency on sensors to detect animal health issues; the role of the stock person is still key.

10 People and the automatic milking system

The investment in AM may in some instances be looked on as part of a lifestyle plan. However, the investment needs to stack up in terms of the objectives and best option for both the dairy farming personnel and operation. Some positive aspects include the simplicity of the system and how relaxed and animal friendly the routine is, where cows can use their time of their own choice. Additionally, the system provides a considerable amount of information that allows farmers to have greater knowledge about their cows, and therefore the opportunity to make adjustments where appropriate. A further positive aspect of the system is that it allows people the opportunity of flexibility with regard to time and the opportunity for lifestyle benefits while managing and milking a dairy cow herd (Kerrisk and Ravenhill, 2010). A further key issue for some farmers is that AM could promote the continuation of the dairy farming tradition by making that occupation more attractive to the next generation through a better lifestyle and because technology generally appeals to young people (Butler et al., 2012). However, the focus must be placed on the economic investment required by AM systems (Shortall et al., 2016) as well as the weighting given by the farmer/potential investor to future profitability versus a better lifestyle; obviously, the achievement of both is the optimum scenario.

Findings regarding improvements to producers' lives with transition to AM were shown in a number of studies in Australia, Canada, New Zealand and Norway. Transitioning to AM meant more flexible working hours, which improved farmers' family and social life (Molfino et al., 2014); improved the health of farmers with the reduced physical workload (Meskens et al., 2001); as well as increased job satisfaction and less routine activities (Hansen, 2015; Woodford et al., 2015). In addition to the time saved, farmers particularly appreciated the additional flexibility in organising their work because they were able to personalise the schedule of tasks throughout the day and adapt it to their family life (Butler et al., 2012; Schewe and Stuart, 2015). Specifically in Canada, when Tse et al. (2018) surveyed AM milk producers (n = 217) from 8 Canadian provinces, the milk producers reported that their expectations had been largely met by the AM technology, the work was less stressful and physically demanding, employee management was easier and herd health had improved. The majority (86%) of producers indicated that they would recommend others to transition to AM.

But AM may not be suitable for all farm situations. When it is decided that AM could be an option for a specific farm, then much of the degree of success of the installation may be determined by the level of 'homework' undertaken in evaluating it for that particular farm situation. An important first step for the farmer considering AM is to engage with existing AM farmers to understand the overall challenges as well as the transition challenges, how they have adapted their farming systems and how AM has changed their work routines and lifestyles.

But it is also critically important to examine a range of aspects of AM which have not been as widely studied, such as the use of automatic and information technologies on-farm and the relationship between humans, cows and machines. Such studies require a mixture of social and scientific methodological approaches. Relevant questions may include the impact of AM on different dimensions of work, such as changes in tasks and routines, changes in the work profile or role of the dairy operator or stockperson, the farmer's mental workload and the human-animal relationship. The Butler et al. (2012) study showed that technology, including AM, can be viewed very differently, for example some farmers felt more restricted by AM, as they considered themselves on-call all of the time; when contacted by the robot (phone call), they felt compelled to check why the robot had called them. Also, some farmers considered that the setting of times and tasks was more difficult with AM, as the farmer now had to make his/her own choice on this matter, rather than previously having the profile of the day structured by conventional morning and evening milking times. The Tse et al. (2018) study did show that 'quality of life' was improved to a lesser extent on AM farms with larger herd sizes and that expectations (with AM) were also met to a lesser extent on such farms.

10.1 Person-animal interaction

The impact that AM has on the human-animal relationship has remained largely unexplored. At the early stages of AM, Seabrook (1992) suggested that AM could impact farmers' relationships with their cows and farmers' identities as stock persons. It may be argued that the role of the stock person is more important with AM, as the routine interaction with the cows may not occur on a twice daily basis (as with CM); interviewees in the Butler et al. (2012) study suggested that a better level of stockmanship was necessary with AM for early and precise identification of health and welfare issues. AM could involve a risk of damaging human-animal relationships, for example shorter duration of farmer-animal interactions, with that time representing negative interactions such as vaccinations, and trimming; these may be the only occasions when farmers are in direct contact with their animals in an AM system (Cornou, 2009). Alternatively, positive interactions for human-animal contact can occur in AM systems when farmers move among the herd to check the equipment or to bring forward cows that have not come to be milked; some farmers may use the time saved by AM to observe and be present with their animals. These interactions can have a beneficial effect on the animals and be a source of satisfaction for the farmers themselves (Fleuret and Marlet, 2014). A more recent study by Wildridge et al. (2020) examined the relationship between farmers and their animals on five Australian dairy farms. The study indicated that four of the five farms had less total farmer-cow interaction time with AM compared to CM systems, but cows were less fearful around people as seen by reduced avoidance distances and reduced stress responses to close handling. Tse et al. (2018) suggested that farmers needed time to adapt to the challenges of changing routines and having less contact with cows. For some farmers, physical contact with their cows plays a significant role in job satisfaction (Meskens et al., 2001); having reduced contact with cows was an important reason for not installing an AM system for some Swedish farmers (Bergman and Rabinowicz, 2013). So the optimum scenario is where the role of the stockperson is integral and complementary to the AM unit and system.

10.2 Person-data interaction

The presence of technology with the ability to conduct tasks, for example the milking unit, or the generation and access to data can represent a positive or a negative impact on farmers depending on the person and the situation. The information provided can reduce the mental workload of a farm manager/

operator due to the early knowledge and potential management of such animal events, for example onset of oestrus or health issues. However, the mental workload can sometimes be increased due to the complexity of the information involved in managing alarms, alerts and equipment failures. New tasks can emerge as a result of new technology, for example maintenance of new tools and analysis of the data generated by the technology; such operations can, in some cases, reduce time savings (Schewe and Stuart, 2015).

A considerable amount of information may be generated by some sensors, making it difficult to select which information is key for decision-making (Schewe and Stuart, 2015). The farmer must manage the data collected, organise data bases and sort and select the relevant information. The management of alarm warnings is highlighted by farmers as a source of stress. With regard to mastitis alerts, a study by Hogeveen et al. (2013) has shown that dairy farmers react to only 3% of mastitis warnings generated by AM units; this is generally a consequence of very frequent alarms arising due to settings being too precise. Farmers need to manage the information from the AM unit, analyse and cross-check data in order to set objective alerts (Hostiou et al., 2017). The Butler et al. (2012) study also indicated that owners of AM systems felt subjected to 'information overload' and had neither the skills nor time to explore possibilities that might arise from analysing the data. Many people just made use of a very small proportion of the data available. This may be addressed by providing more training for farmers in the use of AM-generated data, but the farmer has to be interested in this to be successful.

So the AM system can have a positive impact on the duration of work, the content and the nature of the tasks carried out by farmers, their mental workload and the relationship between farmers and their animals. But the adoption of such technology by the farmer must be facilitated. The AM system has the potential to make the profession of farming more attractive; it can give farm work a more modern image and encourage young farmers to continue in the business. But the tool must be adapted to farmers' skills and needs and then it will have a positive impact on both farmers and animals.

11 Impact on labour requirements

The decrease in labour requirement associated with AM is frequently an influencing factor in the decision to install this type of milking system by dairy farmers. The AM system can replace human labour in the milking unit and can reduce overall labour input on a dairy farm. It allows farmers to take more time off at weekends (27%), have more time available for their families (84%) and have a better quality of life (69%) (Fleuret and Marlet, 2014). An Irish study conducted by Shortall et al. (2016) quantified the difference in labour requirements (in terms of time) between CM and AM operating in pasture-based dairy systems. The on-farm labour study was conducted on 17 farms, 10 CM and 7 AM systems, respectively, over a 12-month period. The average herd size of the AM and CM farms were 105 (range 69–205) and 120 (range 70–160) cows, respectively. This study involved farm operators recording the duration of labour input associated with performing a range of different defined farm tasks throughout the day. Farm tasks defined for AM and CM groups focused on the 'Milking process' and tasks 'other than milking'. The 'milking process' tasks on AM farms included checking AM system data, fetching cows indoors, fetching cows outdoors, robot cleaning, maintenance and alarms and on CM farms included herding of cows pre- and post-milking, milking, milking plant and yard cleaning. The tasks 'other than milking' were similar for both systems and included grass allocation, other dairy tasks, other enterprise tasks and nonfarm activity. Data were recorded on a specially designed Smart Phone App. Records were compiled on three consecutive days on the third week of each month.

The total dairy labour input and the labour input for specific dairy-related tasks (hours/cow per year) averaged across AM and CM farms are shown in Table 1. Total dairy labour input was significantly less on AM compared to CM farms, with AM farmers requiring 15.8 h/cow per year and CM farmers working 25 h/cow per year. The average daily time spent at total dairy labour input was 5.2 and 8.9 h/day throughout the 12-m recording period, for AM and CM farms, respectively. Reduced labour with AM can be attributed to significantly less time spent on the daily milking process. On average, CM farmers spent 3 h/day (range 0.5–4.5) in the process of milking. As milking with an automatic system does not require the farmer to be present at milking time, the AM process

Table 1 Average dairy labour input (h/cow per year) for combined and specific dairy tasks on farms milking with automatic milking (AM) and conventional milking (CM) systems

	Milking system			
Item	AM	CM	SE	*P* value
Total dairy labour[1]	15.8	25.0	3.00	<.05
Milking process[2]	2.1	8.1	0.83	<.001
Grass allocation[3]	1.2	0.3	0.11	<.001
Other dairy tasks[4]	12.5	16.6	2.55	NS

[1]Sum of all tasks associated with the running of the dairy enterprise.
[2]Milking process on AM farms was the sum of time spent on robot cleaning and maintenance, checking AM data, attending to alarms and fetching any overdue cows from paddocks; milking tasks on CM farms were the sum of time spent herding cows to and from milking, milking and yard and machine cleaning.
[3]Task refers specifically to daily grass allocation to cows in the grazing paddocks.
[4]Tasks associated with the operation of the dairy enterprise (excluding milking and grass allocation) such as maintenance of buildings and machinery, machinery and office work, calf and replacement heifer rearing, veterinary treatment of animals, heat detection and AI, drying off cows and training of cows to milking system.

only consumed 0.7 h/day (range 0.2–1.1). This saving in labour associated with the milking process was partially counteracted by significantly more time being spent at grass allocation on farms with an AM system. The AM farmers spent, on average, 0.4 h/day ensuring their cows were allocated the correct amount of grass in each grazing area. However, as the CM farmers were not dependent on the allocation of grass for optimum cow traffic (and had only one grass allocation to administer, rather than three), they spent just 0.1 h/day attending to grass allocation. Despite labour being reduced by 9.2 h/cow per year, daily end of work times were similar for each milking system, at 18 h 32 min. However, daily start times were significantly different with AM farms starting work 50 min later than CM farms, at 07:55 and 07:05, respectively. The adoption of AM also changed the type of labour undertaken, with AM farmers spending more time managing and overseeing the system than carrying out manual labour.

11.1 Labour usage

Previous studies are in agreement that the adoption of AM technology leads to a reduction in the requirement for labour, although the level of labour reduction reported is variable; it varies between 11% and 36% (Mathijs, 2004; Bijl et al., 2007; Heikkila et al., 2010). Mathijs (2004) found that the reduction in labour, although averaging 19%, varied hugely by country, with Belgian farmers reducing labour by 28%, whereas Danish farmers only reduced labour by 11%. The evaluation of labour in the majority of these studies was based on retrospective survey responses from farmers estimating their time input rather than through the capture of real-time data as in the study of Shortall et al. (2016). Deming et al. (2018) showed that the milking process accounted for 33% of total dairy tasks on Irish CM dairy farms; this was in in line with the Shortall et al. (2016) study showing a similar proportion of labour associated with CM (32%), but the proportion associated with AM was just 13% of total labour input. The 36% reduction in labour associated with AM in the Shortall et al. (2016) study largely represented the reduction in time associated with the milking process from 3 h/day with CM to 40 min/day with AM. This reduction of labour is a key motivator for farmers to adopt AM. A survey of early AM adopters in Australia by Kerrisk and Ravenhill (2010) found that farmers had been quite frustrated with the quality of labour available and cited a history of on-going staffing problems as a contributory reason for investing in AM.

12 Use of resources

A significant limitation associated with the adoption of AM is the reduced profitability of that technology relative to CM technologies (Shortall et al., 2016).

Although the large capital cost associated with AM is one of the main factors contributing to this, the increased consumption of electricity associated with AM may also be considered a contributing factor (Bijl et al., 2007). Electricity consumption is one of the most relevant and uncontrollable operational costs of AM, ranging between 35% and 40% of the total annual operational cost. Dairy farmers have to focus on the efficient use of energy in order to improve cost competitiveness, and quantification of energy consumption is an essential component of this. Studies by Artmann and Bohlsen (2000) and Rasmussen and Pedersen (2004) have examined electricity consumption of AM systems and CM parlours. A more recent study by Calcante et al. (2016) showed that electric consumption by AM was mainly influenced by farm management (e.g. use of a steam cleaning system for teat-cups or choice of an efficient and correctly sized air compressor for opening/closing the entrance and exit gates of the milking stall and moving the automatic arm towards the udder) rather than machine characteristics/architectures. Further work by Upton and O'Brien (2013) analysed the energy consumption of an AM system as operated within a grass-based, seasonally calved dairy production system in Ireland. That work highlighted that the largest energy-demanding processes associated with milk harvesting in the AM system were heating water, compressing air and cooling milk. However, the daily and seasonal trends of electricity consumption and costs in a pasture-based AM system remained undocumented; this information was captured by Shortall et al. (2018c) over a year-long period. Water consumption on AM farms may also be significant. Information relating to on-farm water usage on AM pasture-based systems and particularly the daily and seasonal trends is minimal. This issue was also addressed in the work of Shortall et al. (2018c).

12.1 Daily and seasonal trends of consumption and cost analysis

The Shortall et al. (2018c) study was conducted on 7 pasture-based AM farms with a spring-calving system across Ireland; cumulative energy used (kWh) was calculated for each on-farm electricity consuming process at 15-min intervals. Average total electricity consumption (on the 7 study farms) was 62.6 Wh of electricity/L of milk produced (range = 47–84 Wh/L) or 336 kWh/cow (range = 246–422 kWh/cow). In total, 53% of all electricity consumed by the study farms occurred during the period of the higher-cost day rate tariff. The average cost of electricity on the study farms was 0.91 cents/L (range = 0.67–1.22 cents/L) or €49/cow (range = €36–57/cow) over the 12-mo period. Electricity used in the dairy milking shed accounted for 85% of the total electricity consumed on-farm. Within the dairy milking shed, the major processes of electricity consumption were milking (33%), air compressing

(26%), milk cooling (18%), auxiliary water heating (8%), water pumping (4%) and other (11%).

The profile of consumption followed a similar profile to the milk production curve of a seasonal production system. Total monthly farm consumption of electricity was at its lowest in January, at 1798 kWh, before rising steadily to 3538 kWh in March. Consumption remained consistent between March and July, peaking slightly in May at 3579 kWh. From July, consumption reduced gradually to 2493 kWh in December. There was a considerably greater demand for electric power to cool the greater volume of milk produced in the mid-lactation period, when cows were producing peak milk volumes. However, when electricity consumption was examined on a per litre basis, the months of greatest milk production resulted in the lowest electricity consumption per litre of milk. The daily electricity consumption profiles were quite consistent and were represented by multiple small peaks in consumption throughout each day; maximum consumption per hour represented just 6.5% of the daily total. This indicates a relatively constant demand for electricity, which in turn may make AM systems more suitable than CM systems for operation in association with renewable energy technologies.

The single milking unit configurations used more electricity (gross kWh) per AM unit on an annual and daily basis than the double-unit configurations. However, both configurations used a similar quantity of electricity per milking. This would indicate that the differences in electricity consumption per AM unit are a consequence of differing stocking densities per AM unit, with single-unit configurations tending to have a higher ratio of cows per unit than double-unit configurations (84 vs. 54 cows/AM unit), resulting in a higher number of milkings/AM unit per day.

The total farm electricity consumption and costs of the AM farms in the Shortall et al. (2018c) study were greater than those outlined by Upton et al. (2013) for an average Irish CM pasture-based system. Both Bijl et al. (2007) and Steeneveld et al. (2012) outlined greater electricity costs for AM systems compared with CM systems on Dutch dairy farms. The average AM farm, in the Shortall et al. (2018c) study, hot washed the milking unit(s) 1.85 times/day, whereas a typical CM farm may only hot wash on alternate days to coincide with the washing of the milk storage tank upon milk collection (Upton et al., 2015). The requirement for compressed air for the cleaning of milk lines and the opening and closing of entry and exit gates represented a main difference in electricity consumption between AM and CM systems in the Upton and O'Brien (2013) study. This component alone accounted for >25% of the total farm electricity consumption in the Shortall et al. (2018c) study. These findings have the potential to assist in developing future strategies that may improve the competitiveness of the AM system.

12.2 Water consumption

The Shortall et al. (2018c) study showed that total water consumption was 3.7 L of water/L of milk produced. This was less than the figure reported by Murphy et al. (2014) at 6.4 L of water/L of milk. Water consumption associated with the milking process was 1.5 L of water/L of milk, as water use was split between 42% and 58% for the milking process and livestock or miscellaneous, respectively. This 1.5 L of water/L of milk usage amounted to 957 693 L/year and included water for cleaning the AM unit, precooling of milk and washing down of the milking shed. Precooling of milk represented the largest consumer of water within the milking process, as was also observed by Murphy et al. (2014) with CM systems. However, the consumption was 0.5 L of water/L of milk less in the average AM precooling system of the current study compared with the CM system of Murphy et al. (2014). Water usage trends (as with electricity consumption) largely followed the seasonal milk production curve and generally followed AM unit utilisation. The average AM unit consumed 445 L/ per milking day. The different AM systems have their own washing protocols which are normally recommended to operate on three occasions per 24 h. The three times rather than twice per day washing recommendation may require more chemical detergents which could add to the cost.

13 Economics

AM is generally regarded as a system that requires two to three times more capital, initially, than the CM herringbone parlour (Rotz et al., 2003). Previous studies examining the economics of AM have suggested that AM is not cost-effective when examined solely on a financial return basis, because of higher capital costs combined with higher running costs. Also, cash flow and profitability can represent major concerns for farmers when transitioning to AM systems.

A study by Bijl et al. (2007) examined the profitability of Dutch dairy farms operating AM and CM systems, based on real accounting data. A total of 62 farms (31 using AM and 31 using CM) were examined. Dairy farmers with CM systems had larger revenues than those with AM systems, but AM farms used on average 29% less labour than farms with CM systems. The CM farms had more money available for rent, depreciation, interest, labour and profit than the AM farms; this was due to larger fixed costs (excluding labour), larger contractor costs and larger costs for gas, water and electricity associated with the AM farms.

A study by Steeneveld et al. (2012) again used actual farm accounting data from dairy farms in the Netherlands to examine capital costs and determine if the technical efficiency of AM farms differed from that of CM farms. The

63 AM and the 337 CM farms in the data set did not differ in general farm characteristics, for example cow number. The results indicated that an automatic or conventional system did not affect technical/farm efficiency, and the economic performance of AM and CM farms was similar. However, farms with AM systems had significantly higher capital costs (€12.71 per 100 kg of milk) compared to CM farms (€10.10 per 100 kg of milk). But, care must be exercised in considering technical efficiency in different milk production systems; technical efficiency challenges are different in indoor and outdoor production systems, for example capability in grassland management skills being critical for success in the pasture-based system. Furthermore, farm owners/operators should decide whether the reduced labour requirement/extra time made available by AM balances against the extra costs associated with such a system.

While previous studies have largely examined the economics of AM relative to conventional parlours that had little automation, it is crucially important to establish the economic parameters of AM relative to CM herringbone parlours with different levels of automation and technology. A robust economic evaluation of AM and CM was conducted in Ireland by Shortall et al. (2016). Three different parlour types were assessed: (1) an AM system, (2) CM system of medium specification (MS) and (3) CM system of high specification (HS). These systems were evaluated as relevant to 2 different farm sizes: a medium-sized farm milking 70 cows and a large-sized farm milking 140 cows. The six scenarios examined included (i) an AM system single unit (AMS-SU), (ii) an AM system double-unit (AMS-DU), (iii) a 12-unit MS CM parlour (12MS), (iv) a 12-unit HS CM parlour (12HS), (v) a 20-unit MS CM herringbone parlour (20MS) and (vi) a 20-unit HS CM herringbone parlour (20HS). The initial machine and infrastructural investment costs, annual machine maintenance, energy use, machine running costs and on-farm dairy labour input were all incorporated into the Moorepark Dairy Systems Model (MDSM) (Shalloo et al., 2004). Financial and biological model assumptions were based on historical Irish data.

13.1 Effect of milking system on profitability

The Shortall et al. (2016) study showed that the AMS-SU and AMS-DU milking equipment were 3.5 and 3.0 times more expensive than the 12MS and 20MS milking equipment scenarios, respectively. However, when the smaller area required to accommodate an AM system relative to a herringbone parlour was taken into consideration, the overall investment in the AMS-SU and AMS-DU (equipment and infrastructure) required 41% more capital than MS CM systems. This was a consequence of the greater interest, depreciation, energy and maintenance costs of AM, even though labour costs were significantly lower. When compared with the HS CM parlours, the AMS-SU and AMS-DU still required considerably more capital investment than the 12HS and 20HS, at 21% and 44%, respectively. Given that the milking process on CM farms accounted for 32% of all dairy labour, the investment in HS CM parlours over AM may be questionable, particularly at the medium farm size, as it (12 HS) generated the least profit of the investment scenarios examined at this farm size while still maintaining a considerable labour requirement (Fig. 3). Furthermore, despite the potential benefits, it is difficult to establish the monetary value of individual cow data at each milking, as its use in supporting decision-making on the farm is dependent on and at the discretion of the operator.

13.2 Effect of capital costs and labour on profitability

Lightfoot and Mulvany (2002) noted that AM could become economically viable in the context of the Australian dairy industry, if the price of units decreased by 25% and if the cost of labour increased to AUS$20 per hour. When the investment cost was reduced in the Shortall et al. (2016) study, it had some positive effects at the medium farm size but had little effect on the difference in profitability between the scenarios at the large farm size. When the cost of labour was increased to €20/h, AM increased in competitiveness and profitability relative to MS technology, particularly at larger herd sizes. But it is important to consider whole-farm profitability, particularly for farms that do not employ labour; the AM system is associated with reduced labour requirement, but if this labour was not hired before the introduction of AM, then the farmer does not benefit in terms of reduced labour costs through the introduction of an AM system. In these situations, no monetary gains are available to the farmer from reducing the requirements for labour and, thus, the farmer's disposable income may be reduced. This diminishes the competitiveness of AM, as the CM farmer is working more hours but, in real terms the financial return is greater,

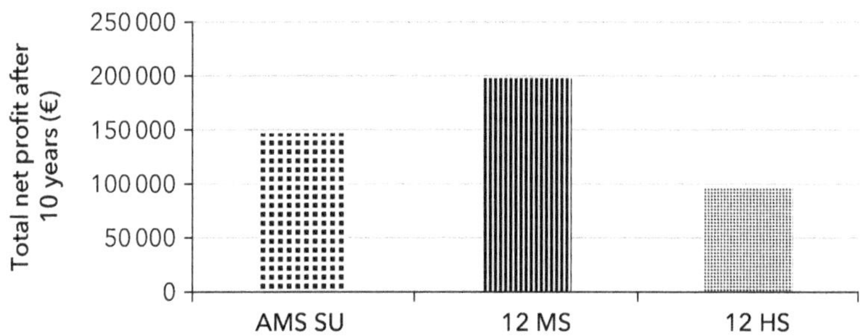

Figure 3 Total net profit (€) after a 10-year period for a herd size of 70 cows with automatic milking (AM) and conventional milking (CM) systems (AMS-SU = an AM system single unit; 12 MS and 12 HS = a 12-unit medium specification and a 12-unit high specification CM parlour, respectively).

with the AM farmer losing the difference in labour costs as well as profit. However, if the incentive in investing in an AM system is to reduce the labour requirement of the farmer, then it will achieve that objective; then the investor prioritises reduced labour and improved lifestyle choices over maximising financial return.

Running and maintenance costs can also be higher for the AM system, which can be expected given the magnitude of electronics and mechanics associated with AM compared with MS milking technology. A Danish report (Sorensen et al., 2013) on the maintenance and service costs of 52 AM farms (units installed between 1999 and 2011) found a large disparity in service and maintenance costs between farms, and a correlation between the maintenance costs and the number of milkings per AM unit was not observed. Studies based on high-input confinement AM systems have in the past shown milk production increases of 5–10% with AM (de Koning, 2010) with potential to increase profitability, but such increases are not expected in the context of a grass-based system. A study to identify the physical factors (i.e. production outputs, physical inputs and productivity measures) associated with productivity and to quantify their effect on profitability in pasture-based AM was conducted by Gargiulo et al. (2020b). Two datasets incorporating farm business and economic data as well as detailed AM system performance data were collected from AM pasture-based farms (mainly in Australia) over three years. The study showed that if pasture-based AM farms managed to increase the number of cows per robot from 55 (current average) to 71 (top 25% average), they would increase their chances of achieving more than 1500 kg of milk harvested (from 24% to 58%) and almost double the overall farm profitability.

Although in the Shortall et al. (2016) study the AM system was associated with greater interest and capital repayments, depreciation, maintenance and running costs and lower profitability, the reduced labour associated with AM can still make it an attractive lifestyle choice for some farmers. The increased labour requirement of CM is a social cost to be considered, particularly as herds expand. It may be advised that profitability should not be the only reason for investing in AM technologies; any decision to invest in AM should consider several factors, such as the availability of skilled labour, lifestyle sought by the farmer, interest in technology, the initial capital investment requirement by the milking system and the overall economic goals of the farm.

14 A place for mobile automatic milking systems?

The majority of AM systems involve fixed milking units but in recent years mobile AM systems have been developed. In the past, combining AM with grazing was almost impossible if the pasture area was located on the other side of a road from the shed/barn or if it was located at some distance from the barn.

However, a mobile AM unit may now allow cow grazing when the farm land is at a significant distance from the shed/unit. This has largely been designed for systems in which a flat milk production curve prevails, with milk produced by cows grazing on pasture during summer months and indoors during the remainder of the year, as operated in for example France or Belgium. In this scenario, the mobile AM unit is installed on a trailer (Figs 4, 5a) and is usually moved between two sites/locations (e.g. pasture [summer] and barn [winter] sites) during the year. The unit may be transported to the pasture area (at the start of the season) and connected up to the services (e.g. bulk tank) (Fig. 5b). The unit remains on the trailer during the period at pasture. It is then transported back to the barn site in the autumn. The trailer with the AM unit is parked within the shed/barn with connections to the bulk tank outside.

Research has been conducted on mobile AM systems (Lessire et al., 2015; Brocard et al., 2015). Guidelines on the combination of mobile AM and grazing

Figure 4 Mobile AM unit being transported to the new site on a farm in Belgium.

Figure 5 (a) The mobile AM unit is being transported to and operated at its (b) summer milking site in France.

have been produced based on experiments at the Trévarez site (Instutute de l'Elevage, France) and at the Sart Tilmant site (University of Liège, Belgium). The experimental site at Trévarez was equipped with a mobile AM unit (VMS Delaval on a Rolland chassis) and involved a herd of dairy cows managed in an organic farming set-up; the main aim was to test the AM system while operating on plots of land that were distant from the main farm site. A similar experiment was conducted at the Sart Tilmant experimental site which was equipped with a Lely mobile AM unit.

14.1 Evaluation of the requirements of mobile automatic milking units

An evaluation was conducted on how these new mobile AM units could operate and contribute a technical solution to having cows grazing distant from the barn during summer time and indoors during winter time. When at the pasture site, it is necessary that the mobile AM system is located close to a roadway for easy access. Planning for the passage of milk collection tankers needs to be carried out at an early stage and long and complex circuits between the AM unit and the milk tank should be avoided. The management of non-marketable milk (e.g. subsequent to cow antibiotic treatment) should also be considered. In order to facilitate cow access particularly when wet weather prevails, some light fixings are necessary around the mobile AM system (e.g. drainage in wet soils, some slatted floors, etc.). Electricity is required at the pasture site and a minimum power of 25 kVA should be provided. The site should have connections to at least two three-phase electric sockets to power the AM unit and its related items. The emergency power generator should also be movable. A good water supply is necessary for the internal washing of the milk line system and the teat cup liners and the external washing of the installation. Cow feed, washing solutions and animal treatments must be stored securely near the AM unit and protected from pests and wildlife. A communication network is necessary to provide for emergency calls from mobile phones, computer data transfer or cameras and so on. A wi-fi connection is necessary for alarm systems, monitoring and maintenance of the system. From an environmental impact perspective, wash waters have to be collected and stored in a specific container. The system must allow for the management of contaminated water runoff from uncovered exercise yards used by animals, water from washing the stall and holding area and water from the internal cleaning of the AM unit and milk tank.

14.2 Economics of mobile automatic milking units

The cost of mobile and fixed AM systems as relevant to different situations need to be compared and the recovery period of the extra costs of the most expensive system versus the expected working life of that system needs to be established. For countries where milk production from cows largely maintained indoors is normal, for example France or Belgium, the cost associated with operating a mobile AM system could be partially offset due to the lower cost of milk production arising from grazed grass compared to the more costly indoor feed (Van den Pol-van Dasselaar et al., 2008). There are also other benefits associated with a mobile AM system through its association with grazing (O'Brien and Hennessey, 2017), for example improved animal health, welfare and sustainability, and while these impacts may be difficult to measure, they are important. Furthermore, a mobile AM system represents a solution for organic farming, if the pastures are distant from the farm, since organic regulations require a specific duration of cows being outdoors at pasture. Alternatively, there may be a place for mobile AM to be used on rented/leased farms, or where the farm land base is fragmented. However, the cost associated with such options needs to be weighed carefully. The mobile AM unit does not require any specific installations in the barn, so the building expenses are therefore reduced compared to a fixed AM system. But other questions need to be addressed, for example what is the economic gain of introducing a mobile AM system, will the money saved on cow feed costs (pasture normally being cheaper than concentrate feed) fund the mobile AM, are there any other solutions such as land plot exchanges, plot re-organisation and so on? An evaluation of the additional investment costs related to the mobility of AM systems was conducted at Trévarez, France. It was estimated that mobility of the AM unit and additional related items (purchase of trailers and amenities) would cost an additional €40,000 (over a fixed installation). Development of the pasture site (earthwork, waste management and storage facilities) would cost an additional €55 000. Thus, the total additional estimated cost of a mobile AM unit would be €95 000.

Mobile AM units appear to be technically feasible, and they can add value to pastures that are distant from the main farm. However, the installation of a mobile AM system requires considerable planning by the farmer at a very early stage. The motivations and objectives behind investment in a mobile AM unit must be clear to all relevant stakeholders, that is the farmer, the suppliers, the advisors and so on. The decision-making must take into account the choice to milk and graze cows on distant pastures and ensure that all other options have been considered. There are both positive and negative aspects to mobile AM. Current knowledge indicates that mobile AM technology is being used in a number of European countries, for example Germany, Norway, Denmark, France and Belgium.

15 Current and future focus of automatic milking research

An increase in adoption rates of AM systems, in different areas globally, in recent years suggests that this technology will play a large role in the future of the dairy industry. Much research on AM has been conducted, the focus of which has mainly been on milk production and sustainability issues, but further in-depth research is needed to elucidate some farm, machine, animal and person management variables and their influence on efficiency and economic performance of the AM system. Current work is addressing some issues of key importance for the future if AM is to be acceptable to the wider farming groups. This includes addressing precision milking management, for example sources of variability in milk yield, milking time and MI of the cow; the optimum projected milk yield for milking permission; milking the slowest quarter faster; reducing udder preparation and attachment time; improving the economics of individual cow concentrate feeding; and the use of economic models to identify less profitable cows as candidates for dry-off and culling.

But AM systems can be quite variable, depending on the context that they are operated in; three main scenarios include milk production systems where (a) cows are indoors full time, (b) the AM system is integrated with different levels of cow grazing and (c) cows are outdoors either partially or full time, sometimes in climates with relatively high temperatures where heat stress is an issue. Each scenario has different but related challenges. In pasture-based AM systems average MF is lower (MI is higher) compared to indoor-based AM. Increased MF (or lower MI) is associated with greater daily milk yields. The persistency of MF throughout lactation of cows in pasture-based, seasonal AM systems would be an important focus together with the effect of different MIs on milk yield. Also, a greater understanding of management strategies for hot climates would help alleviate a decline in AM unit utilisation during the hottest period of the day. Likewise, the management of extended walking distances would become more important as AM is adopted into larger pasture-based farms.

The number of cows that can be milked by a single AM unit per day is often based on the manufacturer's guidelines rather than empirically derived recommendations. This suggests a need to determine the relationship between the number of cows milked daily by a single AM unit and milk production per unit. Strategies for optimising AM unit use through changing the number of cows per AM unit are being examined; for example on pasture it may be more feasible to increase milk production by increasing the number of cows rather than the number of milkings per cow. Stocking density may be a significant factor affecting MI and milk production variables. Some research concerns identification of opportunities to improve AM efficiency through strategic management of stocking rates and milking permission and through the selection of cows for milking speed and teat placement. Though several researchers have examined the effect of cow traffic systems, questions still remain about how best to optimise cow flow through the system. Changes in cow energy balance and anticipated milk yield suggest that higher stocking rates, different motivations to encourage cows to milk and effects of weather in AM pasture systems are topics requiring further attention; a grazing season is characterised by constant changes in weather, pasture supply, pasture quality and day length and therefore is associated with greater variation than systems where cows are housed indoors and fed a total mixed ration diet.

A higher synchronisation of activities typically occurs with grazing dairy cows, and cows tend to concentrate grazing activity during daylight, leading to a low level of cow traffic and visits to the robot during the early morning hours. Any study of, or research into management practices that could increase cow traffic at this time is likely to have a positive impact on MF and system utilisation. However, care must be taken not to negatively impact any aspect of cow welfare, for example rest time. A more complete understanding of diurnal patterns of feeding, lying and milking behaviour in AM systems is needed. Future research should focus on developing strategies to achieve a more distributed visitation and milking pattern. These strategies may involve the availability of an additional allocation of fresh pasture during the early morning period; the offering of three and four grass allocations per 24 h has been investigated, with the 4 allocation system having both advantages and disadvantages. Thus, work examining cow traffic needs to be continued, particularly with regard to how individual cow and group behaviour affect the use of the AM system.

Even where the physical operation of the AM system is optimised, inter-related parameters of cow number, MF and milk yield may be affected by different less obvious parameters such as grassland management, cow lameness issues and social hierarchy of cows in the herd. Perhaps the least understood area to date is a social hierarchy within AM herds and how this impacts feed and robot access in both indoor and pasture-based AM systems. Particularly in a large herd, it may be difficult for lower social ranking cows to access resources, so it would be important to quantify the impact of infrastructure such as a priority pre-milking yard on these cows. All of these parameters are more critical in AM systems than in CM systems and sometimes represent the reason for disparity between performance results from the two systems rather than the direct effect of the milking systems.

It is also important to develop an understanding of the circumstances surrounding the adoption of this technology by farmers and to examine the way in which the information generated by the AM system is used subsequently by the farmer. It would also be important to establish if farmers are satisfied with their decision post-adoption. Such studies require both social and scientific methodological approaches. This would encourage greater decision-making around the consideration, purchase and implementation of AM systems, minimising risk and maximising the likelihood of successful adoption. Research to date has mostly focused on the technology itself and collection of data and not on collation of that data, its use and integration into the farming system.

Addressing this issue may assist in moving the focus from 'does the technology work' to 'how the information generated by the technology can be best used'; that is from the collection of data to use of information. Finally, the sensors associated with AM should be studied more fully to elucidate their ability to accurately detect important changes in cow health, behaviour and welfare, such as oestrus, mastitis, lameness and feeding and how to convert the output data to useful information for decision-making.

16 Conclusion

The international dairy industry has undergone significant growth in recent times. Globally, farms are undergoing a profound structural change, the speed of which is often underestimated. A dramatic shift has occurred in the number of dairy farms with a decrease of 18% and 60% in Australia and New Zealand, respectively, since 1990, and herd size increasing by 160% (Dairy Australia, 2015; DairyNZ and LIC, 2015). According to the IFCN Dairy Report (2019), the average farm in the EU and the USA has experienced an annual rise of 8% in milk production between 2013 and 2018. It is well-documented that the milking task accounts for the majority of labour required on dairy farms (40–50% of total farm labour in Australia and 33% in Ireland). Thus, it is not surprising that AM is increasing in popularity, and developing knowledge with regard to the efficient operation of AM systems together with integration of the system into pasture-based milk production systems is a critically important field of research for the future. Much work has been cited in the previous sections on various aspects of successful AM operation, such as performance of the system, training of cows, management of cow traffic, nutritional aspects, grass allocation and the health and well-being of the cow. While much advancement has been made and AM systems generally work very well, both as indoor and pasture systems, questions relating to farm infrastructure, cow health, pasture management and capital investment still remain. It is very important to ensure that the operator is 'in-tune' with and understands the management required by an AM system, both at a technical and data level, as well as having the more traditional skills as an animal operator. This is fundamental, as one of the key aspects of the system is that it allows the operator to focus on management issues and not spend all the time on operational issues. Concerns do surround the profitability of the system, and this is a potential reason for the technology not being adopted more widely. Previous comparisons of AM with CM parlours have indicated that the findings back up these concerns. The reduced profitability of AM relative to MS CM parlours is a consequence of increased costs associated with the system, such as electricity, maintenance, depreciation and interest. However, while the MS CM parlour performs the basics of milk harvesting, the technology of an AM parlour has the potential to provide significant additional data/information that may be used to make better management decisions. Thus, decision-making regarding the adoption of new AM milking technology needs to encompass the likely use of data/information generated on the farm for improved efficiency, desired workload of the individual, cost and availability of skilled labour and economic goals of the farm. Awareness of AM is present but not prominent among rural professionals; some are still unsure if the technology is suited for intensive dairy systems. However, technology is playing an increasing role in agriculture generally, and, specifically, AM focuses on using technology to monitor and collect cow and milking data on a continuous and automated basis, with the ultimate goal of providing a mechanism, whereby this data can be used to inform the resource management and decision-making process, ultimately increasing the precision of the operation. From both business and societal viewpoints, AM can assist the dairy sector to focus on performance efficiency, the well-being of the livestock unit and of the manager as well as economic performance, which together help to ensure the sustainability of farm and food production in the future.

17 Where to look for further information

The Milking Edge project is a training and extension program for automatic milking in Australia. It is designed to support dairy farmers in Australia and elsewhere. It is a joint initiative of the NSW Department of Primary Industries, Dairy Australia and DeLaval, to 'Support industry to invest and operate AMS successfully'. It is based on a number of years of research, as well as the experience of almost 50 commercial farms around Australia who have invested in the technology. It represents a source of information to assist decision-making around the consideration, purchase and implementation of AMS.

https://www.dpi.nsw.gov.au/animals-and-livestock/dairy/dairy-technologies/robotic-milking-systems/ams-projects/milking-edge. The DairyNZ and Dairy Australia websites (www.dairynz.co.nz; www.dairyaustralia.com.au) also include information on automatic milking. The integration of automatic milking with pasture grazing represents a significant departure from the more predominant indoor operation observed in much of Europe and North America. However, integrated milking and grazing systems are very important in some locations, such as Ireland, New Zealand and Australia. The general principles of grazing are consistently relevant in the context of AM and may be found in the book: *Grass: Its production and utilisation,* Hopkins, A. (Ed.) (2000), Third edition, Published for the British Grassland Society by Blackwell Science; a review paper by Chapman, D. F. (2016) Using Ecophysiology to Improve Farm Efficiency: Application in Temperate Dairy Grazing Systems. *Agriculture* 6: 17; the Grazing Notebook as described in http://www.agresearch.teagasc.ie/moorepark/Publications/pdfs/Open%20Day%20Moorepark%202009%20Grazing%20Manual.pdf; and the

European Grassland Federation Working Group whose work is outlined in http://www.europeangrassland.org/working-groups/grazing.html.

18 References

Artmann, R. and Bohlsen, E. (2000). Results from the implementation of automatic milking system (AMS)-multi-box facilities. In: *Robotic Milking: Proceedings of the International Symposium*, Hogeveen, H. and Meijering, A. (Eds). Wageningen Academic Publishers, Wageningen, the Netherlands, pp. 221–231.

Auldist, M. J., Pyman, M. F. S., Grainger, C. and Macmillan, K. L. (2007). Comparative reproductive performance and early lactation productivity of Jersey×Holstein cows in predominantly Holstein herds in a pasture-based dairying system. *J. Dairy Sci.* 90(10), 4856–4862.

AUTOGRASSMILK Final Report Summary (2017). *Innovative and Sustainable Systems Combining Automatic Milking and Precision Grazing*. https://cordis.europa.eu/project/id/314879/reporting.

Bach, A., Devant, M., Igleasias, C. and Ferrer, A. (2009). Forced traffic in automatic milking systems effectively reduces the need to get cows, but alters eating behavior and does not improve milk yield of dairy cattle. *J. Dairy Sci.* 92(3), 1272–1280.

Bach, A., Iglesias, C., Calsamiglia, S. and Devant, M. (2007). Effect of amount of concentrate offered in automatic milking systems on milking frequency, feeding behavior, and milk production of dairy cattle consuming high amounts of corn silage. *J. Dairy Sci.* 90(11), 5049–5055.

Bargo, F., Muller, L. D., Delahoy, J. E. and Cassidy, T. W. (2002). Milk response to concentrate supplementation of high producing dairy cows grazing at two pasture allowances. *J. Dairy Sci.* 85(7), 1777–1792.

Baudracco, J., Lopez-Villalobos, N., Holmes, C. W. and Macdonald, K. A. (2010). Effects of stocking rate, supplementation, genotype and their interactions on grazing dairy systems: A review. *N. Z. J. Agric. Res.* 53(2), 109–133.

Bava, L., Sandrucci, A., Tamburini, A. and Succi, G. (2005). Milk ejection during automatic milking in dairy cows. *Ital. J. Anim. Sci.* 4 (Suppl. 2), 218–220.

Begley, N. (2008). An evaluation of Norwegian Red and Norwegian Red x Holstein friesian dairy cows under Irish production conditions (Ph.D. thesis). School of Agriculture, Food Science and Veterinary Medicine. University College Dublin, Dublin, Ireland.

Bergman, K. and Rabinowicz, E. (2013). Adoption of the automatic milking system by Swedish milk producers. AgriFoods Economics Centre. http://www.agrifood.se/Files/AgriFoodWP20137.pdf (accessed 26 February 2018).

Berry, D. P., Veerkamp, R. F. and Dillon, P. (2006). Phenotypic profiles for body weight, body condition score, energy intake, and energy balance across different parities and concentrate feeding levels. *Livest. Sci.* 104(1-2), 1-12.

Bijl, R., Kooistra, S. R. and Hogeveen, H. (2007). The profitability of automatic milking on Dutch dairy farms. *J. Dairy Sci.* 90(1), 239-248.

Bogucki, M., Sawa, A. and Kuropatwińska, I. (2017). Association of automatic milking systems milking frequency in primiparous and multiparous cows with their yield and milkability. *Acta Agric. Scand. A* 67(1-2), 66-70, 66-70.

Brocard, V., Dufrasne, I., Lessire, F. and Francois, J. (2015). Challenging land fragmentation thanks to a mobile milking robot: Statement of two cases of implementation. In: Proceedings of the European Association of Animal Production, 31 August–4 September 2015. Abstract, p. 330.

Butler, D., Holloway, L. and Bear, C. (2012). The impact of technological change in dairy farming: Robotic milking systems and the changing role of the stockperson. *J. R. Agric. Soc. Engl.* 173, 1–6.

Calcante, A., Tangorra, F. M. and Oberti, R. (2016). Analysis of electric energy consumption of automatic milking systems in different configurations and operative conditions. *J. Dairy Sci.* 99(5), 4043–4047.

Clark, C. E. F., Farina, S. R., Garcia, S. C., Islam, M. R., Kerrisk, K. L. and Fulkerson, W. J. (2015). A comparison of conventional and automatic milking system pasture utilization and pre- and post-grazing pasture mass. *Grass Forage Sci.* 71(1), 153–159.

Clark, D. A., Phyn, C. V. C., Tong, M. J., Collis, S. J. and Dalley, D. E. (2006). A systems comparison of once- versus twice-daily milking of pastured dairy cows. *J. Dairy Sci.* 89(5), 1854–1862.

Cogato, A., Bršćić, M., Guo, H., Marinello, F. and Pezzuolo, A. (2021). Challenges and tendencies of automatic milking systems (AMS): A 20-years systematic review of literature and patents. *Animals (Basel)* 11(2), 356. doi:10.3390/ani11020356.

Cornou, C. (2009). Automation systems for farm animals: Potential impacts on the human-animal relationship and on animal welfare. *Anthrozoös* 22(3), 213–220.

Dairy Australia (2015). *Australian Dairy Industry in Focus 2015*. Dairy Australia, Southbank, VIC. Dairy Australia Limited ABN 60 105 227 987 Level 5, IBM Centre 60 City Road, Southbank VIC 3006 Australia.

DairyNZ and Livestock Improvement Corporation (LIC) (2015). *New Zealand Dairy Statistics 2014–15*. LIC and DairyNZ, Hamilton, New Zealand.

Davis, K. L., Fulkerson, W. J., Garcia, S. C., Dickeson, D. and Barchia, I. M. (2008). Pre-milking teat preparation for Australian pasture- based cows milked by an automatic milking system. *J. Dairy Sci.* 91(7), 2604–2609.

de Koning, C. J. A. M. (2011). *Milking Machines | Robotic Milking. Encyclopedia of Dairy Sciences* (2nd edn.). Reference Module in Food Science, pp. 952–958. Available at: https://doi.org/10.1016/B978-0-12-374407-4.00360-5

de Koning, K. (2010). Automatic milking–Common practice on dairy farms. In: *Proceedings of the Second North American Conference on Robotic Milking*, Toronto, Canada. Precision Dairy Operators, Elora, Canada, pp. V59–V63.

de Koning, K., Slaghuis, B. and van der Vorst, Y. (2003). Robotic milking and milk quality: Effects on bacterial counts, somatic cell counts, freezing point and free fatty acids. *Ital. J. Anim. Sci.* 2(4), 291–299.

de Passille, A. M., Rushen, J., Ladewig, J. and Petherick, C. (1996). Dairy calves' discrimination of people based on previous handling. *J. Anim. Sci.* 74(5), 969–974.

Delaby, L., Peyraud, J. L. and Delagarde, R. (2001). Effect of the level of concentrate supplementation, herbage allowance and milk yield at turn-out on the performance of dairy cows in mid lactation at grazing. *Anim. Sci.* 73(1), 171–181.

DeLaval International, A. B. (2008). *Instruction Book VMS Best Practices [Brochure]*. DeLaval International AB, Tumba, Sweden.

De Marchi, M., Penasa, M. and Cassandro, M. (2017). Comparison between automatic and conventional milking systems for milk coagulation properties and fatty acid composition in commercial dairy herds. *Ital. J. Anim. Sci.* 16(3), 363–370.

Deming, J., Gleeson, D., O'Dwyer, T., Kinsella, J. and O'Brien, B. (2018). Measuring labor input on pasture-based dairy farms using a smartphone. *J. Dairy Sci.* 101(10), 9527–9543.

De Vries, T. J., Deming, J. A., Rodenburg, J., Seguin, G., Leslie, K. E. and Barkema, H. W. (2011). Association of standing and lying behaviour patterns and incidence of intra-mammary infection in dairy cows milked with an automatic milking system. *J. Dairy Sci.* 94(8), 3845–3855.

Erdman, R. A. and Varner, M. (1995). Fixed yield responses to increased milking frequency. *J. Dairy Sci.* 78(5), 1199–1203.

Evans, R. D., Dillon, P., Buckley, F., Berry, D. P., Wallace, M., Ducrocq, V. and Garrick, D. J. (2006). Trends in milk production, calving rate and survival of cows in 14 Irish dairy herds as a result of the introgression of Holstein-Friesian genes. *Anim. Sci.* 82(4), 423–433.

Fleuret, M. and Marlet, A. (2014). De la sale de traite au robot, Quels impacts? *Terra* 449, 23–29.

Friggens, N. C. and Rasmussen, M. D. (2001). Milk quality assessment in automatic milking systems: Accounting for the effects of variable intervals between milkings on milk composition. *Livest. Prod. Sci.* 73(1), 45–54.

Fulkerson, W. J., McKean, K., Nandra, K. S. and Barchia, I. M. (2005). Benefits of accurately allocating feed on a daily basis to dairy cows grazing pasture. *Aust. J. Exp. Agric.* 45(4), 331–336.

Gargiulo, J. I., Lyons, N. A., Kempton, K., Armstrong, D. A. and Garcia, S. C. (2020a). Physical and economic comparison of pasture-based automatic and conventional milking systems. *J. Dairy Sci.* 103(9), 8231–8240.

Gargiulo, J. I., Lyons, N. A. and Garcia, S. C. (2020b). Factors affecting productivity and profitability in pasture-based automatic milking systems. In: *Proceedings of the Dairy Research Foundation 2020 Symposium*, University of Sydney.

Goddard, M. E. and Grainger, C. (2004). A review of the effects of dairy breed on feed conversion efficiency–an opportunity lost? In: *Proceedings of the 25th Biennial Conference of the Australian Society of Animal Production*, University of Melbourne, Melbourne, VIC, pp. 77–80.

Gregorini, P. (2012). Diurnal grazing pattern: Its physiological basis and strategic management. *Anim. Prod. Sci.* 52(7), 416–430.

Halachmi, I., Ofir, S. and Miron, J. (2005). Comparing two concentrate allowances in an automatic milking system. *Anim. Sci.* 80(3), 339–343.

Hansen, B. G. (2015). Robotic milking-farmer experiences and adoption rate in Jaeren, Norway. *J. Rural Stud.* 41, 109–117.

Harris, B. L. and Kolver, E. S. (2001). Review of Holsteinization on intensive pastoral dairy farming in New Zealand. *J. Dairy Sci.* 84, E56–E61.

Heikkila, A. M., Vanninen, L. and Manninen, E. (2010). Economics of small-scale dairy farms having robotic milking. In: *Proceedings of the First North American Conference on Precision Dairy Management*, Toronto, Canada. www.precisiondairy.com/proceedings/s3heikkila.pdf (accessed 19 May 2017).

Heins, B. J., Hansen, L. B., Seykora, A. J., Johnson, D. G., Linn, J. G., Romano, J. E. and Hazel, A. R. (2008). Crossbreds of Jersey × Holstein compared with pure Holsteins for production, fertility, and body and udder measurements during first lactation. *J. Dairy Sci.* 91(3), 1270–1278.

Hermans, G. G. N., Ipema, A. H., Stefanowska, J. and Metz, J. H. M. (2003). The effect of two traffic situations on the behaviour and performance of cows in an automatic milking system. *J. Dairy Sci.* 86(6), 1997–2004.

Hogenboom, J. A., Pellegrino, L., Sandrucci, A., Rosi, V. and D'Incecco, P. (2019). Invited review: Hygienic quality, composition, and technological performance of raw milk obtained by robotic milking of cows. *J. Dairy Sci.* 102(9), 7640–7654.

Hogeveen, H., Buma, K. J. and Jorritsma, R. (2013). Use and interpretation of mastitis alerts by farmers. In: *Proceedings of the Precision Livestock Farming and Farmers' Work of the 6th European Conference on Precision Livestock Farming*, Berckmans, D. and Vandermeulen, J. (Eds), Leuven, Belgium, 10–12 September, pp. 313–319.

Hopster, H., Bruckmaier, R. M., Van der Werf, J. T. N., Korte, S. M., Macuhova, J., Korte-Bouws, G. and van Reenen, C. G. (2002). Stress responses during milking; comparing conventional and automatic milking in primiparous dairy cows. *J. Dairy Sci.* 85(12), 3206–3216.

Hostiou, N., Fagon, J., Chauvat, S., Turlot, A., Kling-Eveillard, F., Boivin, X. and Allain, C. (2017). Impact of precision livestock farming on work and human-animal interactions on dairy farms. A review. *Biotechnol. Agron. Soc. Environ.* 21(4), 268-275.

Hovinen, M. and Pyörälä, S. (2011). Invited review: Udder health of dairy cows in automatic milking. *J. Dairy Sci.* 94(2), 547–562.

Hulsen, J. and Rodenburg, J. (2008). *Robotic Milking [Brochure]*. Roodbont Publishers, Zutphen, the Netherlands.

IFCN (2019). IFCN Dairy Report for 2019. Hemme, T. Global Dairy Trends and Drivers 2019 - IFCN Dairy Research Network, 20th Anniversary Edition, Kiel, Germany. October 2019. Page 1.

Jacobs, J. A. and Siegford, J. M. (2012). Invited review: The impact of automatic milking systems on dairy cow management, behavior, health, and welfare. *J. Dairy Sci.* 95(5), 2227–2247.

Jago, J., Jackson, A. and Woolford, M. (2003). Dominance effects on the time budget and milking behaviour of cows managed on pasture and milked in an automated milking system. *N. Z. Soc. Anim. Prod.* 63, 120–123.

Jago, J. and Kerrisk, K. (2011). Training methods for introducing cows to a pasture-based automatic milking system. *Appl. Anim. Behav. Sci.* 131(3–4), 79–85.

Jago, J. G., Davis, K. L., Copeman, P. J., Ohnstad, I. and Woolford, M. M. (2007). Supplementary feeding at milking and minimum milking interval effects on cow traffic and milking performance in a pasture-based automatic milking system. *J. Dairy Res.* 74(4), 492–499.

Jago, J. G., Roche, J. R., Kolver, E. S. and Woolford, M. W. (2005). The drinking behaviour of dairy cows in late lactation. *Proc. N. Z. Soc. Anim. Prod.* 65, 209–214.

John, A. J., Clark, C. E., Freeman, M. J., Kerrisk, K. L., Garcia, S. C. and Halachmi, I. (2016). Review: Milking robot utilization, a successful precision livestock farming evolution. *Animal* 10(9), 1484–1492.

John, A. J., Clark, C. E. F., Freeman, M. J., Kerrisk, K. and Rawnsley, R. P. (2013). Pasture management in two high performing automatic milking systems. In: *Proceedings of the Dairy Research Foundation Symposium*, Kiama, NSW, Kerrisk, K., Garcia, Y., Catt, S. and Heward, M. (Eds). The University of Sydney, Camden, NSW, pp. 99–103.

Kellaway, R. and Harrington, T. (2004). *Feeding Concentrates: Supplements for Dairy Cows*. Landlinks Press, Collingwood, VIC.

Kennedy, E., Curran, J., Mayes, B., Mcevoy, M., Murphy, J. P. and O'Donovan, M. (2011). Restricting dairy cow access time to pasture in early lactation: The effects on milk production, grazing behaviour and dry matter intake. *Animal* 5(11), 1805-1813.

Kerrisk, K. and Ravenhill, B. (2010). Learnings around automatic milking system adoption on-farm. In: *Proceedings of the Australian Dairy Conference*. Dairy Australia, Southbank, VIC, pp. 45-49.

Ketelaar-de Lauwere, C. C. K., Devir, S. and Metz, J. H. M. (1996). The influence of social hierarchy on the time budget of cows and their visits to an automatic milking system. *Appl. Anim. Behav. Sci.* 49(2), 199-211.

King, M. T. M., LeBlanc, S. J., Pajor, E. A. and DeVries, T. J. (2017). Cow-level associations of lameness, behavior, and milk yield of cows milked in automated systems. *J. Dairy Sci.* 100(6), 4818-4828.

Klungel, G. H., Slaghuis, B. A. and Hogeveen, H. (2000). The effect of the introduction of automatic milking systems on milk quality. *J. Dairy Sci.* 83(9), 1998-2003.

Kruip, T. A. M., Morice, H., Robert, M. and Ouweltjes, W. (2002). Robotic milking and its effect on fertility and cell counts. *J. Dairy Sci.* 85(10), 2576-2581.

Lessire, F., Froidmont, E., Shortall, J., Hornick, J. L. and Dufrasne, I. (2017). The effect of concentrate allocation on traffic and milk production of pasture-based cows milked by an automatic milking system. *Animal*. 11(11), 2061-2069. doi:10.1017/S1751731117000659.

Lessire, F., Hornick, J. L., Minet, J. and Dufrasne, I. (2015). Rumination time, milk yield, milking frequency of grazing dairy cows milked by a mobile automatic system during mild heat stress. *Adv. Anim. Biosci.* 6(1), 12-14.

Lightfoot, C. and Mulvany, J. (2002). *Farming Systems for Automatic Milking in Australia*. Occasional Report Prepared for Dairy Australia, Melbourne, VIC.

Lind, O., Ipema, A., de Koning, C., Mottram, T. and Hermann, H. (2000). Automatic milking: Reality, challenges and opportunities. In: *Robotic Milking: Proceedings of the International Symposium*, Lelystad, the Netherlands, 17-19 August 2000, pp. 19-31.

Livshin, N., Maltz, E. and Edan, Y. (1995). Regularity of dairy cow feeding behaviour with computer-controlled feeders. *J. Dairy Sci.* 78(2), 296-304.

Lopez-Villalobos, N., Garrick, D. J., Blair, H. T. and Holmes, C. W. (2000). Possible effects of 25 years of selection and crossbreeding on the genetic merit and productivity of New Zealand dairy cattle. *J. Dairy Sci.* 83(1), 154-163.

Lyons, N. A. and Kerrisk, K. L. (2017). Current and potential system performance on commercial automatic milking farms. *Anim. Prod. Sci.* 57(7), 1550-1556.

Lyons, N. A., Kerrisk, K. L. and Garcia, S. C. (2013a). Effect of pre-versus postmilking supplementation on traffic and performance of cows milked in a pasture-based automatic milking system. *J. Dairy Sci.* 96(7), 4397-4405.

Lyons, N. A., Kerrisk, K. L. and Garcia, S. C. (2013b). Comparison of two systems of pasture allocation on milking intervals and total daily milk yield of dairy cows in a pasture-based automatic milking system. *J. Dairy Sci.* 96(7), 4494-4504.

Lyons, N. A., Kerrisk, K. L., Dhand, N. K. and Garcia, S. C. (2013c). Factors associated with extended milking intervals in a pasture-based automatic milking system. *Livest. Sci.* 158(1-3), 179-188.

Lyons, N. A., Kerrisk, K. L. and Garcia, S. C. (2014). Milking frequency management in pasture-based automatic milking systems: A review. *Livest. Sci.* 159, 102-116.

Masía, F. M., Lyons, N. A., Piccardi, M., Balzarini, M., Hovey, R. C. and Garcia, S. C. (2020). Modeling variability of the lactation curves of cows in automated milking systems. *J. Dairy Sci.* 103(9), 8189-8196.

Mathijs, E. (2004). Socio-economic aspects of automatic milking. In: *Proceedings of the Automatic Milking: A Better Understanding*, Meijering, A., Hogeveen, H. and de Koning, C. J. A. M. (Eds). Wageningen Academic Publishers, Wageningen, The Netherlands, pp. 46-55.

McNamara, S., Murphy, J. J., O'Mara, F. P., Rath, M. and Mee, J. F. (2008). Effect of milking frequency in early lactation on energy metabolism, milk production and reproductive performance of dairy cows. *Livest. Sci.* 117(1), 70-78.

Melin, M., Hermans, G. G. N., Pettersson, G. and Wiktorsson, H. (2006). Cow traffic in relation to social rank and motivation of cows in an automatic milking system with control gates and an open waiting area. *Appl. Anim. Behav. Sci.* 96(3-4), 201-214.

Melin, M., Wiktorsson, H. and Norell, L. (2005). Analysis of feeding and drinking patterns of dairy cows in two cow traffic situations in automatic milking systems. *J. Dairy Sci.* 88(1), 71-85.

Meskens, L., Vandermersch, M. and Mathijs, E. (2001). Implication of the introduction of automatic milking on dairy farms. Deliverable D2 from EU project 'Implications of the introduction of automatic milking on dairy farms' (QLK5 2000-31006). http://www.automaticmilking.nl (accessed 8 June 2018).

Molfino, J., Kerrisk, K. and Garcia, S. C. (2014). Investigation into the labour and lifestyle impacts of automatic milking systems (AMS) on commercial farms in Australia. In: *Proceedings of the 5th Australian Dairy Science Symposium*, Melbourne, VIC, pp. 339-342. http:// futuredairy.com.au/wp-content/uploads/2016/02/63MolfinoADSS 2014.pdf (accessed 8 June 2018).

Mollenhorst, H., Hidayat, M. M., van den Broek, J., Neijenhuis, F. and Hogeveen, H. (2011). The relationship between milking interval and somatic cell count in automatic milking systems. *J. Dairy Sci.* 94(9), 4531-4537.

Munksgaard, L., Rushen, J., de Passille, A. M. and Krohn, C. C. (2011). Forced versus free traffic in an automated milking system. *Livest. Sci.* 138(1-3), 244-250.

Murphy, E., Upton, J., Holden, N. M., Curran, T. P., Cummins, E. J. and Curran, T. P. (Eds.) (2014). Direct water use on Irish dairy farms. In: *Biosystems Engineering Research Review 19, Biosystems Engineering Research Review. 19th Annual Research Review.* University College Dublin. Research Repository University College Dublin, Dublin Ireland, pp. 146-148.

O'Brien, B. (2016). CellCheck Newsletter, July 2016, Animal Health Ireland. *Guest Contribution on 'Mastitis Control in AMS'*, pp. 2-3.

O'Brien, B. and Hennessey, D. (2017). Scientific appraisal of the Irish grass-based milk production system as a sustainable source of premium quality milk and dairy products. *Ir. J. Agric. Food Res.* 56(1), 120-129.

O'Donovan, M., Dillon, P., Rath, M. and Stakelum, G. (2002). A comparison of four methods of herbage mass estimation. *Ir. J. Agric. Food Res.* 41, 17-27.

O'Driscoll, K., O'Brien, B., Gleeson, D. and Boyle, L. (2010). Milking frequency and nutritional level affect grazing behaviour of dairy cows: A case study. *Appl. Anim. Behav. Sci.* 122(2-4), 77-83.

Penno, J. (1998). Principles of profitable dairying. In: *Proceedings of the Ruakura Farmers Conference*, Ruakura, New Zealand, pp. 1-14.

Pettersson, G., Svennersten-Sjaunja, K. and Knight, C. H. (2011). Relationships between milking frequency, lactation persistency and milk yield in Swedish Red heifers and cows milked in a voluntary attendance automatic milking system. *J. Dairy Res.* 78(3), 379-384.

Prendiville, R., Pierce, K. M. and Buckley, F. (2009). An evaluation of production efficiencies among lactating Holstein-Friesian, Jersey, and Jersey × Holstein-Friesian cows at pasture. *J. Dairy Sci.* 92(12), 6176-6185.

Prendiville, R., Shalloo, L., Pierce, K. and Buckley, F. (2011). Comparative performance and economic appraisal of Holstein-Friesian, Jersey and Jersey x Holstein-Friesian cows under seasonal pasture-based management. *Ir. J. Agric. Food Res.* 50, 123-140.

Prescott, N. B., Mottram, T. T. and Webster, A. J. F. (1998). Relative motivations of dairy cows to be milked or fed in a Y-maze and an automatic milking system. *Appl. Anim. Behav. Sci.* 57(1-2), 23-33.

Rasmussen, J. B. and Pedersen, J. (2004). Electricity and water consumption at milking. Farm Test-Cattle nr. 17, Danish Agricultural Advisory Service, National Centre, Building and Technique, Udkaersvej 15, Skejby DK-8200 Aarhus. www.farmtest.dk ISSN: 1601-6785, pp. 1-42.

Rasmussen, M. D., Bjerring, M., Justesen, P. and Jepsen, L. (2002). Milk quality on Danish farms with automatic milking systems. *J. Dairy Sci.* 85(11), 2869-2878.

Reid, M., O'Donovan, M., Murphy, J. P., Fleming, C., Kennedy, E. and Lewis, E. (2015). The effect of high and low levels of supplementation on milk production, nitrogen utilization efficiency, and milk protein fractions in late-lactation dairy cows. *J. Dairy Sci.* 98(8), 5529-5544.

Rotz, C. A., Coiner, C. U. and Soder, K. J. (2003). Automatic milking systems, farm size, and milk production. *J. Dairy Sci.* 86(12), 4167-4177.

Salomonsson, M. and Sporndly, E. (2000). Cow behaviour at pasture with or without supplementary roughage in automatic milking systems. In: *Proceedings of the Robotic Milking Report*, Lelystad, the Neatherlands, pp. 192-193.

Schewe, R. L. and Stuart, D. (2015). Diversity in agricultural technology adoption: How are automatic milking systems used and to what end? *Agric. Hum. Values* 32(2), 199-213.

Seabrook, M. (1992). The perception by stockpersons of the effect on their esteem, self-concept and satisfaction of the incorporation of automatic milking into their herds. In: Proceedings of the International Symposium on Prospects for Automatic Milking, Ipema, A., Lippus, A., Metz, J. and Rossing, W. (Eds), Wageningen, The Netherlands, 23-25 November 1992. European Association for Animal Production, Rome, publication no 65, pp. 409-413.

Shalloo, L., Dillon, P., Rath, M. and Wallace, M. (2004). Description and validation of the Moorepark Dairy System Model. *J. Dairy Sci.* 87(6), 1945-1959.

Shalloo, L., O'Donnell, S. and Horan, B. (2007). Profitable dairying in an increased EU milk quota scenario. In: *Proceedings of the Nat Dairy Conference*, Kilkenny, Ireland. Teagasc, Moorepark, Fermoy, Co., Cork, Ireland, pp. 20-42.

Shortall, J., Foley, C., Sleator, R. D. and O'Brien, B. (2018a). The effect of dairy cow breed on milk production, cow traffic and milking characteristics in a pasture-based automatic milking system. *Livest. Sci.* 209, 1-7.

Shortall, J., Foley, C., Sleator, R. D. and O'Brien, B. (2018b). The effect of concentrate supplementation on milk production and cow traffic in early and late lactation in a pasture-based automatic milking system. *Animal* 12(4), 853-863.

Shortall, J., O'Brien, B., Sleator, R. D. and Upton, J. (2018c). Daily and seasonal trends of electricity and water use on pasture-based automatic milking dairy farms. *J. Dairy Sci.* 101(2), 1565–1578.

Shortall, J., Shalloo, L., Foley, C., Sleator, R. D. and O'Brien, B. (2016). Investment appraisal of automatic milking and conventional milking technologies in a pasture-based dairy system. *J. Dairy Sci.* 99(9), 7700–7713.

Smith, J. W., Ely, L. O., Graves, W. M. and Gilson, W. D. (2002). Effect of milking frequency on DHI performance measures. *J. Dairy Sci.* 85(12), 3526–3533.

Sonck, B. R. and Donkers, H. W. J. (1995). The milking capacity of a milking robot. *J. Agric. Eng. Res.* 62(1), 25–37.

Sorensen, M., Lund, O. and Krog, J. (2013). *Kvaeginfo 2381: Omkostninger til vedligehold ved malkerobotter*. https:. www.landbrugsinfo.dk/byggeri/stalde/kvaegstalde/kostalde/ams-automatiske-malkesystemer/sider/2381-omkostninger-tilvedligehold-ved-malkerobotter.aspx (accessed 9 December 2015).

Speroni, M., Lolli, S. and Pirlo, G. (2004). Introduction of AMS in Italian herds: Effects on animal behaviour. In: *Automatic Milking - A Better Understanding*, Meijering, A., Hogeveen, H. and de Koning, C. J. A. M. (Eds). Wageningen Academic Publishers, Wageningen, The Netherlands, pp. 429–443.

Spolders, M., Meyer, U., Flachowsky, G. and Coenen, M. (2004). Differences between primiparous and multiparous cows in voluntary milking frequency in an automatic milking system. *Ital. J. Anim. Sci.* 3(2), 167–175.

Sporndly, E. and Wredle, E. (2005). Automatic milking and grazing - Effects of location of drinking water on water intake, milk yield, and cow behavior. *J. Dairy Sci.* 88(5), 1711–1722.

Steeneveld, W., Tauer, L. W., Hogeveen, H. and Oude Lansink, A. G. J. M. (2012). Comparing technical efficiency of farms with an automatic milking system and a conventional milking system. *J. Dairy Sci.* 95(12), 7391–7398.

Stockdale, C. R. (2006). Influence of milking frequency on the productivity of dairy cows. *Aust. J. Exp. Agric.* 46(7), 965–974.

Svennersten-Sjaunja, K. M. and Pettersson, G. (2008). Pros and cons of automatic milking in Europe. *J. Anim. Sci.* 86(13) (Suppl.), 37–46.

Teagasc (2009). *Grazing Notebook*. https://www.opt-in.ie/static/GrazingNotebook.pdf.

Teagasc (2011). *Grazing Guide*. https://www.teagasc.ie/media/website/crops/grassland/Grazing-Guide-Book-2011_0-21.pdf (accessed 16 July 2020).

Teagasc (2016). *Teagasc Dairy Manual*. Teagasc, Carlow, Ireland.

Tse, C., Barkema, H. W., DeVries, T. J., Rushen, J., Vasseur, E. and Pajor, E. A. (2018). Producer experience with transitioning to automatic milking: Cow training, challenges, and effect on quality of life. *J. Dairy Sci.* 101(10), 9599–9607.

Tucker, C. B., Rogers, A. R. and Schutz, K. E. (2008). Effect of solar tradiation on dairy cattle behaviour, use of shade and body temperature in a pasture-based system. *Appl. Anim. Behav. Sci.* 109(2–4), 141–154.

Upton, J., Humphreys, J., Groot Koerkamp, P. W. G., French, P., Dillon, P. and De Boer, I. J. M. (2013). Energy demand on dairy farms in Ireland. *J. Dairy Sci.* 96(10), 6489–6498.

Upton, J., Murphy, M., De Boer, I. J. M., Groot Koerkamp, P. W. G. G., Berentsen, P. B. M. and Shalloo, L. (2015). Investment appraisal of technology innovations on dairy farm electricity consumption. *J. Dairy Sci.* 98(2), 898–909.

Upton, J. and O'Brien, B. (2013). Analysis of energy consumption in robotic milking. In: *Proceedings of the Precision Livestock Farming '13*. Berckmans, D. and

Vandermeulen, J. (Eds). Katholieke Universiteit Leuven, Leuven, Belgium, pp. 465–470.

Vance, E. R., Ferris, C. P., Elliott, C. T., Hartley, H. M. and Kilpatrick, D. J. (2013). Comparison of the performance of Holstein-Friesian and Jersey×Holstein-Friesian crossbred dairy cows within three contrasting grassland-based systems of milk production. *Livest. Sci.* 151(1), 66–79.

Van den Pol-van Dasselaar, A., de Vliegher, A., Hennessy, D., Peyraud, J. L. and Pinxterhuis, J. B. (2011). Research methodology of grazing. In: *Proceedings of the EGF Working Group Grazing*. Report 405. Wageningen UR Center Livestock Research, Lelystad, the Netherlands, 19 pp.

Van den Pol-van Dasselaar, A., Hennessy, D. and Isselstein, J. (2020). Grazing of dairy cows in Europe – An in-depth analysis based on the perception of grassland experts. *Sustainability* 12(3), 1098.

Van den Pol-van Dasselaar, A., Vellinga, T. V., Johansen, A. and Kennedy, E. (2008). To graze or not to graze, that's the question. Biodiversity and animal Feed Future Challenges for grassland Production. In: *Proceedings of the 22nd General Meeting of the European Grassland Federation*, Hopkins, A., Gustafsson, T., Bertilsson, J., Dalin, G., Nilsdotter-Linde, N. and Spörndly, E. (Eds). Uppsala, Sweden, pp. 706–716.

Vijayakumar, M., Park, J. H., Ki, K. S., Lim, D. H., Kim, S. B., Park, S. M., Jeong, H. Y., Park, B. Y. and Kim, T. I. (2017). The effect of lactation number, stage, length, and milking frequency on milk yield in Korean Holstein dairy cows using automatic milking system. *Anim Biosci. Asian-Australas J. Anim. Sci.* 30(8), 1093–1098.

Wagner-Storch, A. M. and Palmer, R. W. (2003). Feeding behavior, milking behavior, and milk yields of cows milked in a parlor versus and automatic milking system. *J. Dairy Sci.* 86(4), 1494–1502.

Walsh, S., Buckley, F., Berry, D. P., Rath, M., Pierce, K., Byrne, N. and Dillon, P. (2007). Effects of breed, feeding system, and parity on udder health and milking characteristics. *J. Dairy Sci.* 90(12), 5767–5779.

Werner, J., Leso, L., Umstätter, C., Kennedy, E., O'Leary, N., Schick, M. and O' Brien, B. (2017). Effect of restricted feeding conditions on cow's feeding behaviour and activity on pasture based milk production systems. In: *Precision Livestock Farming `17*, Berckmans, D. and Keita, A. (Eds), Nantes, France, pp. 443–445.

Werner, J., O'Leary, N., Umstatter, C. and O'Brien, B. (2018). Behavioural pattern of dairy cows in an automatic milking system with a 4-way grazing setup. In: *Sustainable Meat and Milk Production from Grasslands* (vol. 23), Horan, B., Hennessy, D., O'Donovan, M., Kennedy, E., McCarthy, B., Finn, J. A. and O'Brien, B. (Eds), Grassland Science in Europe. European Grassland Federation, Cork, Ireland, pp. 500–502.

Widegren, S. (2014). Introduction of heifers to an automatic milking system. BS Thesis. Department of Animal Nutrition and Management, Swedish University of Agricultural Sciences, Uppsala, Sweden.

Wiking, L. (2022). Milking and handling of raw milk: Influence on free fatty acids. In: *Encyclopedia of Dairy Sciences* (3rd edn.), McSweeney, P. L. H. and McNamara, J. P. (Eds). Elsevier Ltd., pp. 979–983. ISBN 978-0-12-818767-8

Wiking, L., Nielsen, J. H., Båvius, A. K., Edvardssont, A. E. and Svennersten-Sjaunja, K. (2006). Impact of milking frequencies on the level of free fatty acids in milk, fat globule size, and fatty acid composition. *J. Dairy Sci.* 89(3), 1004–1009.

Wildridge, A. (2017). Management and infrastructure considerations and their impact on cow behaviour and welfare in a pasture-based automatic milking system. PhD

Thesis. University of Sydney, Faculty of Science, School of Life and Environmental Science. The University of Sydney, NSW 2006, Australia.
Wildridge, A. M., Garcia, S. C., Thomson, P. C., Jongman, E. C., Clark, C. E. F. and Kerrisk, K. L. (2017). The impact of a shaded pre-milking yard on a pasture-based automatic milking system. *Anim. Prod. Sci.* 57(7), 1219–1225.
Wildridge, A. M., Thomson, P. C., Garcia, S. C., Jongman, E. C. and Kerrisk, K. L. (2020). Transitioning from conventional to automatic milking: Effects on the human-animal relationship. *J. Dairy Sci.* 103(2), 1608–1619.
Woodford, K. B., Brakenrig, M. H. and Pangborn, M. C. (2015). New Zealand case studies of automatic-milking-systems adoption. In: *Proceedings of the New Zealand Society of Animal Production*, Dunedin, New Zealand, pp. 127–131. http://www.sciquest.org.nz/node/142523 (accessed 8 June 2018).

Chapter 13

Developments in monitoring grazing behaviour and automated grazing management in extensive systems

Dana L. M. Campbell, Gregory J. Bishop-Hurley, Caroline Lee and Ed Charmley, CSIRO, Australia

1 Introduction

With the global human population predicted to rise to 9 billion by 2050 it has been estimated that food production will have to be doubled to meet demand (Thornton, 2010). A growing population coupled with improving living standards will increase the demand for meat and dairy products, thus increasing the demand for the land needed to raise livestock. Ruminants are unique among domestic livestock as their complex rumen microbiota enables them to produce meat and milk from cellulosic feeds grown on marginal land (Van Soest, 1982), unlocking resources that would otherwise be unusable in the human food supply chain. Grazing livestock form essential contributions to the human population through the provision of food and other animal-based products such as wool, hides and transport and have been part of human life for centuries (Butzer and Helgren, 2008; Dahlström et al., 2006; Thomson and Simpson, 2006). In developing countries, livestock play an integral role in daily survival where more sustainable management practices could help ease poverty, hunger and malnutrition in these regions (FAO, 2018).

http://dx.doi.org/10.19103/AS.2021.0090.14

The modern farming world is rapidly changing in response to multiple drivers such as maximising productivity, optimising resource-use, reducing environmental impacts, improving animal welfare and health and the overall need for sustainable systems to carry the population and the planet into the future. Technology that enables improved management of the animal/pasture interface can contribute to lifting the efficiency of ruminant production from grazing lands. Efficiency can be defined in many ways, such as the efficient use of feedstuffs, natural resources and human capital, and economic efficiency (Holmes, 2015). Thus, technology applications for management (and genetic) improvements that reduce labour needs, operational costs and animal losses can have clear economic benefits and ensure the long-term sustainability of farming families and communities (Bell et al., 2011). While technology is continually modifying and/or advancing aspects of farming life, its impact on the grazing animal is relatively recent (Anderson, 2007; Rutter, 2014).

This chapter will provide an overview of the development and use of information technology to improve the management and welfare of grazing livestock under extensive systems. It will discuss the evolution of grazing management through the animal/pasture interface, and the development and application of grazing management technologies, including, on-animal managements through virtual fencing technology and on-animal monitoring of pasture intake, location, health and activity. Technological examples will focus on the Australian beef industry as a potential early adopter of precision livestock farming (PLF) solutions. A case study will also be presented on the commercial implementation of the eShepherd® virtual fencing system for protection of an environmentally sensitive area. The chapter concludes with a summary of developments to date, discussion of future research trends and where to look for additional information.

2 The animal/pasture interface

The management of livestock on grazing lands is known to have existed for at least 10000 years (Clutton-Brock, 1999). The interaction between the grazing animal and the sward is central not only to the welfare and performance of the animal (O'Reagain et al., 2011; Rutter, 2014) but also to the productivity and resilience of the sward (Ash et al., 1995, 1997; Hunt et al., 2014). An imbalance between feed availability and animal numbers can lead to overgrazing, poor animal performance, changes in sward botanical characteristics and long-term run-down in condition and productivity of the pasture (Ash et al., 1995; McKeon et al., 2004). Despite the critical importance of maintaining an optimal balance between sward production and animal requirements, the management of this interaction has been limited by our ability to control the grazing animal. Historically, transhumance was, and still is, in some locations, employed to match the feed requirements of nomadic herds and flocks with the seasonal ebb and flow of pasture growth (Schlolz and Schlee, 2015). With intensification and the popularity of enclosures, the scale and frequency of livestock movement across the landscape changed (Chambers and Mingay, 1966). However, with intensification came increased risk of mismanagement, often due to the vagaries of weather and climate. Concurrently, the human-animal connection diminished. From a situation of almost constant contact between herd and herder, in some production systems, livestock can be left for months without human oversight. Combined, intensification and lack of human oversight can lead to failures in the sustainability of the animal/sward interface.

Information technology presents new opportunities for understanding and managing grazing behaviour. The supplanting of the animal/human interface with an animal/digital interface presents opportunities to refine management of both sward and animal to optimise persistence and productivity of the sward with welfare and performance of the animal. Monitoring of the sward has evolved from periodic visual assessment to near real-time remote sensing and prediction of future patterns based on historical data, sensors and weather forecasting. The use of precision technology to monitor animals is a more recent advent. Within the last 20 years, advances in electronics have allowed on-animal devices to become smaller, smarter and more energy self-sufficient (Anderson, 2007; Rutter, 2014).

Monitoring of the pasture and animal allows for timely management intervention. Today, technological solutions now allow us to take the final step from just monitoring the grazing animal to remotely managing the animal by understanding the animal's response to cues, coming not from the herder but from a device programmed to emulate the animal/human interface.

2.1 The development of precision livestock farming solutions for the grazing animal

Successful animal production off pasture relies on a synergy between the animal and the sward being grazed. The successful application of technology, therefore, presupposes that a similar level of technological development can be applied to the pasture as well as the animal grazing that pasture. Information technologies applied to pastures have benefitted from developments in cropping agriculture (Schellberg et al., 2008) and have, until recently, not been matched by developments in animal-based technologies.

2.2 Pasture monitoring

One of the earliest examples of pasture monitoring was at the Jornada Experimental Range in the USA (Rango et al., 2011). Over time, physical recording of pasture on the ground has been replaced by near-ground image analysis (Gillan et al., 2019) and satellite technology (Edirisinghe et al.,

2011). These technologies continue to improve with increased spatial (pixel size) and spectral (e.g. hyperspectral) resolution. Species identification and biomass estimation are potentially possible (Schellberg et al., 2008). Near-ground imaging has the advantage of higher resolution and higher frequency (Handcock et al., 2016), but static nodes have low spatial coverage. Unmanned aerial vehicles can overcome this disadvantage (Gillan et al., 2019). While the ultimate goal of estimation of biomass, species composition and potentially nutritive value remains elusive, progress to that endpoint continues.

2.3 Livestock grazing behaviour

The interaction between the grazing animal and the sward is a complex and dynamic process influenced by both the animal and the grazing environment. Species of domesticated ruminants differ in their grazing behaviour (Van Soest, 1982) dependent on preferences and energy needs, thus affecting sward botanical composition and rate of pasture utilisation (PISC, 2007). Stocking rates can be modified by management and these are optimised to control the rate of biomass disappearance over time or the duration of grazing on a particular area of pasture (McIvor and Gardener, 1995). Livestock can be stocked at a constant rate relative to available pasture biomass or can be periodically rotated between pastures under a more invasive management approach (Hunt et al., 2014). Rotational grazing involves moving livestock approximately every 3-4 weeks, whereas cell grazing involves the daily movement of livestock to fresh grazing paddocks (cells) but has high capital costs for water, fencing and labour.

In small homogeneous paddocks the effect of the grazing environment on grazing behaviour is small and livestock tend to utilise the sward evenly across the grazing area (Tomkins et al., 2009). However, as paddock size increases, the behaviour of the animal has an increasingly important impact. The animal responds to attractants and repellents in the landscape (Henkin et al., 2012; Thornley et al., 1994), resulting in patchy grazing patterns (Tomkins and O'Reagain, 2007; Tomkins et al., 2009). The distance to water points can also create a piosphere grazing effect in larger paddocks (Hunt et al., 2014). Other drivers in the landscape influence the location of cattle and these can change in priority over time. For example, terrain, abundance of herbage biomass and variations in herbage nutritive value will influence the distribution of cattle in large paddocks (Henkin et al., 2012).

2.4 Environmental impacts of the grazing animal

Livestock production systems can have detrimental effects on native flora and fauna, through clearance of land for grazing, and destruction of natural habitats (Dettenmaier et al., 2017; Forrester et al., 2017; Howell et al., 2019). Livestock can also pollute waterways through faecal contamination (Collins et al., 2007) and exacerbate soil erosion. This leads to loss of topsoil and nutrients to aquatic systems where they can have negative effects on marine and freshwater ecosystems (Thorburn and Wilkinson, 2013). Livestock are also contributors to climate change through greenhouse gas emissions in the form of methane (Ripple et al., 2014; Thorburn et al., 2011). These environmental impacts are a key element of social license to maintain grazing herds. Conversely, carefully managed grazing of livestock can control weeds or overgrowth and can promote environmental improvements such as increased species diversity (Howell et al., 2019; Marty, 2015).

2.5 Animal welfare impacts on grazing systems

Similar to societal concerns for environmental impacts, there is also unease over how livestock may fare under human management and whether current practices sufficiently cater to the animal's wellbeing and health (Cornish et al., 2016; Grandin, 2014). There are increasing obligations for farmers to provide quality assurance of their products to consumers, legislators and welfare certification bodies. This type of assurance can be particularly challenging in extensive grazing systems where farmers may have limited ability to know where their animals are at a given point in time, and if they may be suffering from illness or injury. Even for animals in intensive systems with more regular human contact, there may be limited ability to monitor individuals continuously when housed in groups of hundreds or thousands of animals.

3 Active precision livestock farming management of pasture and animal

3.1 Virtual fencing

Technology such as virtual fencing that enables continual animal monitoring and measuring behaviours of individuals provides a tool for guiding necessary system improvements. The concept of virtual fencing has been around for several decades now with many researchers entertaining the idea of being able to have fences that contain animals, without the same effort and infrastructure as standard physical (electric) fences. The appeal of being able to fence almost any type of area and avoid laying down and maintaining miles of fence line is easy to understand. However, although multiple different types of virtual fencing systems have been proposed or tested with experimental prototypes (reviewed in Umstatter, 2011), taking a product to full-scale commercialisation is yet to be achieved. Approximately 15 years ago, a research team at the Commonwealth Scientific and Industrial Research Organisation (CSIRO) in Australia patented an algorithm for virtually fencing cattle using audio cues and electrical pulses (Lee, 2006; Lee et al., 2010) and tested a range of cue and control options (Bishop-Hurley et al., 2007). This algorithm was proposed to train cattle to associate a benign audio warning tone with an aversive electrical pulse delivered via a collar device. The cues would be delivered based on the animal's behaviour resulting in the animals learning to appropriately turn away from a virtual fence based on the audio warning tone only. Initial trials with manually operated experimental prototype collars demonstrated that animals were able to successfully learn this association and be kept away from a feed trough attractant and conspecifics using only these collar signals (Lee et al., 2007; Lee et al., 2009). In addition, bulls also learnt to associate the visual cue of another bull with the electrical pulse to maintain spatial separation and prevent fighting (Lee et al., 2008a). Further trials were undertaken to investigate the feasibility of reducing the amount of time animals spent in a desirable area over an extended period of time (Ruiz-Mirazo et al., 2011). In 2015, the start-up company Agersens (www.agersens.com), based in Melbourne, Victoria, Australia, became world-wide exclusive licensees for the patented IP and set on the path to developing a virtual fencing device for commercial distribution.

The virtual fencing eShepherd® system has been previously described (Campbell et al., 2019a,b), but briefly, it consists of a neckband worn by individual cattle with a device that sits on top of the neck (Fig. 1). A base station (or multiple base stations for extensive systems spanning thousands of hectares) is installed on site to communicate with the devices with all fence

Figure 1 A steer fitted with the pre-commercial prototype eShepherd® neckband. An IceQube R® is also fitted on the front leg to measure behavioural time budgets during experimental trials.

activations and animal monitoring occurring through an online user interface. A virtual fence boundary (separating inclusion vs. exclusion zones) is specified through the interface using GPS coordinates and transmitted to the unit using a radio frequency link. As an animal approaches the virtual fence boundary, a single non-aversive audio tone within the animal's hearing range is emitted. If the animal stands still or turns away after the tone, no electrical stimulus is applied. However, if the animal continues to move through the virtual fence boundary into the exclusion zone, the unit delivers a short, sharp electrical pulse sequence. This audio/pulse sequence is repeated if the animal continues walking through the fence line into the exclusion zone, but all stimuli cease if the animal turns around to re-enter the inclusion zone. Animal locations can be viewed online in real-time and all GPS and cue data in the pre-commercial prototypes are stored on the neckband devices, enabling download for analysis when the neckbands are removed.

The commercialisation journey is long and challenging when developing a technology that will be reliable, affordable, robust enough to work outdoors on livestock and does not cause any welfare issues for the animals. To comply with these requirements, research must occur at every step to ensure the safe and effective development of the product. A multi-institutional research project funded by industry and the Australian Government has supported the conduct of multiple trials with varying iterations of pre-commercial prototypes of the neckband devices to understand how the animals respond to the technology as well as the applications and limitations of the system to grazing livestock. The majority of the work has occurred with beef cattle, with some studies being conducted on dairy cattle, and the feasibility of application to sheep has also been explored. The results of the beef cattle and sheep studies are presented henceforth.

3.2 Application of the technology - cattle trials

To successfully develop an automated device that will fence and move animals, it is necessary to first understand how animals respond to the technology so that it aligns with their behaviour and learning abilities. The first trials conducted with the automated neckband prototypes were aimed at assessing whether the virtual fence could keep cattle away from a preferred food source (hay), what their behavioural reactions were to the audio and electrical stimuli, and whether they would learn the association between the two signals and start responding to the audio cue alone. A group of 12 cattle were individually tested across a series of 3 small paddocks (~64 mL × 35 mW) each with a bale of hay at the end and a virtual fence placed approximately one third into the paddock (Campbell et al., 2018). The cattle were led into the paddock one by one to document behaviour and learning at the level of the individual. The results showed that each animal varied in their behavioural reactions towards the stimuli, with some animals running forward when first receiving a pulse, whereas other animals did exhibit the desired reaction of turning back around when they received the pulse. The cattle were kept away from the hay and were able to learn to stop or turn around in response to the audio cue, with a mean of 6 interactions until 50% of the animals were turning around at the audio cue alone. However, individual variation in learning and success rate was high (Campbell et al., 2018), emphasising the importance of algorithms that respond specifically to the animal's behaviour. Additionally, a maximum of 9 paddock tests were able to be conducted per animal before the cattle started refusing to walk down the paddock. It appeared that the cattle had learned the location of the virtual fence and were avoiding any further interaction with it (Campbell et al., 2018). This learned avoidance was an undesirable outcome. One of the appeals of virtual fencing is its temporary application; if a fence is subsequently deactivated, a farmer may want cattle to re-access a previously excluded area. Ultimately, although individual testing is informative, it is an artificial situation as cattle on-farm would be exposed to the virtual fencing system as a group. The next study addressed virtual fencing in this context.

Some of the same cattle from the feed attractant trial were placed into a larger 6 ha paddock to be tested in a group situation (6 heifers) with a series of fences that shifted location. The cattle initially had access to the entire paddock to become familiarised with the area (Campbell et al., 2017) and were then restricted to only 40% of the paddock area by a single straight virtual fence line. Once it was clear the animals were trained to the cues and were staying within the 40% inclusion zone, the fence was shifted to include 60% of the paddock area. Although the cattle had been restricted to 40% of the paddock for eight days, it only took approximately 4 h until they showed interaction with the new fence line, indicating rapid movement into the previously excluded area rather than avoiding a learned fence location. This same behaviour was repeated when the fence was again shifted to include 80% of the available paddock area, and even when the fence was shifted to be placed lengthways down the paddock rather than across its width. The animals spent the majority of their time in the intended zones, and approximately 84% of the time, they were turning away from the fence line based on the audio cue alone (once the animals were trained). The trial was able to show the cattle were tuning into the audio signal to know where the virtual boundary was rather than avoiding a specific location (Campbell et al., 2017). This was an encouraging result for temporary applications of a virtual fence, which was further confirmed by a trial on a commercial property that included a riparian zone within the grazing paddocks.

Tumbarumba, NSW, was selected as the site of the first commercial test of the pre-commercial neckband prototypes, to determine if the technology could keep a group of 11 cattle out of a waterway running through the property

(Campbell et al., 2019a). The cattle were first given full access to a grazing paddock that included the riparian zone and further land across the other side of the river. A straight virtual fence line was then set along the front of the water area, blocking access to the river (a water trough provided drinking water) and additional land. The cattle were again able to rapidly learn to respond to the audio cue alone and remained outside the water course. However, their desire to access the water remained high. The animals interacted with the fence line every day of the 11-day exclusion period and 4 animals broke through the fence line for approximately 30 min on the third day before they were shepherded back by the neckband algorithm that continues to deliver signals as the animal moves farther into the exclusion zone. After deactivation of the temporary virtual fence, the cattle accessed the water within a few hours and were utilising all available land once more (Campbell et al., 2019a).

3.3 Application of the technology for sheep and goats

Development of virtual fencing for sheep and goats would enable improved management of livestock and better control of weeds, particularly in mixed farming systems. A cost-benefit analysis identified potential whole-farm income gains of 15% for a low rainfall crop-livestock farm in Australia, with further potential benefits from general grazing efficiency, labour savings and specific weed control savings (Llewellyn et al., 2018).

Research on virtual fencing of sheep and goats is in the early stages. Previous studies in both goats (Fay et al., 1989) and sheep (Jouven et al., 2012) have tested a pet containment system using a wire buried in the ground to implement the virtual fence with the cues administered via a collar worn by the animal. Both species learnt the virtual fence in this context; however, there are limitations with this system as it is not readily movable, and animals that cross the boundary are not able to return without receiving the aversive stimuli.

To overcome the limitations of a fixed boundary, a GPS-based virtual fencing system is recommended for commercial application. An automated GPS collar device, Nofence, has been developed and tested on sheep; however, large variation in animal responses were reported that limited the effectiveness of the fence (Brunberg et al., 2015).

More recently, manually controlled, electronic dog collars were used to determine key parameters relating to virtual fencing in sheep (Fig. 2), including minimum electrical stimulus levels and the speed of learning the association between the audio cue and electrical stimulus (Marini et al., 2018a). Sheep were tested individually using a feed attractant model and 50% of sheep learnt to respond to the audio cue within three interactions with the virtual fence. In agreement with the studies of Brunberg et al. (2015), large individual differences in learning were also apparent. The only study to date looking at individual

Figure 2 A sheep wearing a manually controlled electric dog collar approaching conspecifics to test the effectiveness of audio and electrical cues signalling a virtual boundary in sheep. Photo courtesy of Danila Marini.

differences in temperament and variation in learning the virtual fence in sheep did not find any relationship between animals classified as 'nervous' or 'calm' and their responses to the audio and electrical stimuli (Marini et al., 2019). However, this was with small numbers (20 animals) and further investigation using larger numbers is needed.

Associative learning of the virtual fence cues was more rapid when sheep were tested in a group of six, with only a 19% probability of sheep receiving an electrical stimulus following the audio cue (Marini et al., 2018b). With sheep being a highly gregarious species, it is likely that social learning increased the effectiveness of the virtual fence with observations of sheep naive to the cues responding to the behavioural reactions of others that received the cues. Importantly, when the virtual fence was removed, sheep rapidly entered the previous exclusion zone, indicating that they were responding to the audio and shock cues, rather than the location in which they received them (Marini et al., 2018b). This was not found in sheep using the Nofence collars, where moving the boundary resulted in sheep escaping, and was suggested to be related to inconsistent electric shock strength due to wool growth around the neck (Brunberg et al., 2015).

Development of an automated virtual fence device for sheep that utilises GPS and other platforms to apply cues based on the animal's behaviour is needed. There are several challenges with developing a commercial product for sheep as a collar is not feasible due to the insulating property of wool and the constant need to clip the neck region to allow contact of the electrodes with the skin. An alternative currently being investigated is whether application of electrical stimuli via an ear tag may provide a solution. Initial trials show promise and further development and testing is on-going. Other challenges associated with the development of virtual fencing for sheep include how to provide enough power to the devices while minimising size and weight due to the smaller size of sheep compared to cattle.

3.4 Animal welfare

When designing and implementing a new technology, animal welfare is a clear concern. Pressure comes from the public to allow social license to operate, and from legislators that are tasked with deciding on governing regulations (e.g., virtual fencing is not currently legal to use commercially in all Australian states and some states have made recent adjustments to their legislation to allow the product to be used for research purposes only). Additionally, producers that are searching to optimise their operations do not want to apply a product that is going to cause undue stress and decrease the wellbeing and productivity of their livestock. Exposure to the virtual fencing cues does elicit an acute stress response that is comparable to common handling procedures in both cattle (Lee et al., 2008b) and sheep (Kearton et al., 2019). However, as animals learn to respond to the audio cue to avoid receiving the electrical pulse typically within the first two days or less, the stress response is not on-going, and the animal has agency over how it chooses to engage with the virtual fence. A framework that discusses these concepts has been proposed to assess welfare outcomes based on predictability and controllability using virtual fencing as a case study (Lee et al., 2018).

To examine the welfare impacts of virtual fencing, a trial was conducted to compare behaviour and welfare of groups (n = 8/group) of beef cattle exposed to either a virtual fence or an electric tape fence across a 4-week duration (Campbell et al., 2019b). During this time, faecal samples were collected periodically to measure stress hormone metabolites, animals were weighed weekly, and automated behaviour monitoring devices collected data on standing/lying time and lying bouts to compare behavioural profiles between fence types. The results showed there were minimal differences between the groups. The virtually fenced animals showed a small decrease in lying time (less than 20 min/day) but no consistent differences in other measures (Campbell et al., 2019b). This study indicated that a virtual fencing system could be implemented with similar impacts to standard electric fences but further study across different groups of animals and situations would continue to inform potential users of the system and increase societal acceptance. Refining the device algorithms based on animal behaviour has been an integral part of commercial product development to ensure the technology is welfare friendly.

3.5 Summary

The commercialisation of the eShepherd® virtual fencing system could see the implementation of many changes across the grazing livestock industries. The ability for cattle to be remotely monitored and moved could see vast improvements for extensive beef cattle systems, particularly in Australia, where northern beef producers have herds of thousands of animals that roam across thousands of hectares of land and may not be seen until mustering time. Tracking of these animals will enable farmers to have better control over herd management, knowledge of animal numbers and potential for remote mustering – improving the safety of both the animals and personnel involved in this process. Virtual fencing systems could also potentially be implemented for easier strip grazing management on dairy cattle farms (Lomax et al., 2019). Temporary fences could be implemented for specific grazing control, and fences along difficult terrain such as waterways are made possible. The ability to erect fences immediately after natural disasters such as fire or flood is also a benefit of the virtual system. The pre-commercial products are continually being improved to ensure high success of the technology along with minimal animal impact. Further research will continue to look at factors that can influence the responses of animals. Pre-exposure to a standard electric fence improves associative learning of virtual fencing cues (Verdon et al., 2020), and cattle will socially learn to avoid a virtual fence from others in their group (Keshavarzi et al., 2020) but more studies will help to understand the complexities of virtual fence system applications to cattle herds to ensure optimal implementation of the technology. In sheep, the development of a commercial product will require an alternative design as a collar will not be effective in a commercial context due to issues with wool. The application of the electrical pulse via an ear tag requires overcoming challenges such as long-term attachment and management of battery power.

4 Alternative monitoring technologies

4.1 Ceres Tag

For many livestock producers, locating livestock and monitoring their behaviour is a time-consuming and often costly challenge, particularly in extensive grazing systems. Recent advances in technology including smaller components, reduced energy requirements and improvements in power generation have led to a proliferation of animal tracking devices. Over the last decade, CSIRO have been experimenting with such technologies (Bishop-Hurley et al., 2016) and in 2018 partnered with a small-medium-enterprise (SME), Ceres Tag (www.cerestag.com) to develop a commercial solar-powered smart ear tag for livestock (Fig. 3). It features geo-location for greater traceability and provenance of livestock, and an accelerometer for activity classification, and provides a platform for other applications in the future. Currently, there are a number of similar devices about to enter the market, but all require infrastructure to be deployed to relay data back to a central location. Ceres Tag utilises direct to low earth orbit satellite communications, making it easily deployed anywhere.

The data collected from the tags can be translated into knowledge to improve management of both livestock and the pastures. The system being developed using the ear tags will give producers greater control over grazing management, allowing them to locate livestock remotely, and will alert them to stock theft or illness. The research community will be able to develop algorithms for inclusion on the device to identify other events of interest to commercial producers, such as oestrous. The aim is to reduce operating costs and increase operational efficiency and additional financing opportunities through better management of livestock based on the remotely collected data. In addition, the system aspires to meet the Australian National Livestock Identification System (NLIS) tag requirements and the identification and traceability requirements of other countries.

Figure 3 Steer fitted with a prototype Ceres Tag smart ear tag. Photo courtesy of Ceres Tag.

4.2 eGrazor

Livestock producers are under increased pressure to produce more from a smaller environmental footprint. On-farm productivity gains by the grass-fed beef sector have been static at approximately 0.5% per annum for the past 15 years in Australia. To be competitive, the on-farm sector needs to lift productivity gain to around 2.5% per annum. Production efficiency in the grass-fed beef sector provides the most cost-effective way to grow production and profitability while reducing greenhouse gas emissions intensity.

Over the last 15 years, CSIRO in collaboration with the NSW Department of Primary Industries has developed a cattle location and behaviour monitoring collar that determines pasture intake by cattle based on behavioural information (Greenwood et al., 2014). The solar-powered devices are worn by grazing livestock and use accelerometers to compute real-time behaviour information (grazing, ruminating, resting, walking, drinking and other behaviours) that can be interpreted to inform management strategies for sustainable livestock production and accelerate the rate of genetic gain for feed efficiency in grazing cattle. Location can be determined as often as once per second, with the accelerometer recording raw values at 60 Hz. Operation of the system requires minimal labour, and once deployed the cattle are free to return to their grazing environment.

While all data are stored on the livestock-monitoring collars, a sub-set of engineering data including battery voltage, solar charge, data storage, and location and behaviour data are sent to the cloud via a LoRa gateway.

Figure 4 A group of heifers fitted with eGrazor collars.

This provides the researcher and soon the farmer or caretaker with summary information on how the device is functioning, the device's location and a summary of the animals' behaviours over the last five minutes. Producers can achieve sustainable production gains by optimising the production system through information on how the animals use the landscape. For example, the combination of location data and behaviours allows the user to understand how the livestock are using the paddock and can assist with the placement of fences, water points and other infrastructure to improve pasture utilisation and prevent overgrazing.

The measurement of pasture intake by grazing free-ranging livestock has been impractical in the past. However, this is changing with the development of the eGrazor (Greenwood et al., 2017). Producers can improve productivity through the accurate measurement of pasture intake and the application of advanced livestock management strategies based on accurate, real-time information on animal performance or the development of breeding selection strategies to identify animals with genetically superior feed efficiency (Fig. 4).

5 Case study: virtual fencing for sensitive environments

The flexibility of virtual fencing including minimal infrastructure requirements makes the technology suitable for protection of environmentally sensitive areas. If fences can be activated temporarily and with minimal labour requirement, they can aid farmers in protecting specific sections of a grazing property and maximise pasture utilisation without compromising sensitive areas. To test the application of virtual fences for environmental protection, a commercial grazing property in Eden Valley, South Australia, was selected with an area of regenerating Eucalyptus (*Eucalyptus camaldulensis*) saplings at one end of a 14-ha grazing paddock (Campbell et al., 2020). A group of 20 Santa Gertrudis heifers were fitted with pre-commercial eShepherd® neckbands and released into the paddock area. A series of training fences were first activated to ensure the animals were learning to respond to the audio cue and turn around before finally implementing a more complicated contoured fence line that followed the boundaries of the sapling section. This fence remained activated for a period of 44 days after which the trial was terminated as cattle needed to be moved to a new grazing paddock. During this trial period the animals were able to be monitored remotely using the eShepherd® system's online user interface to check their location within the paddock, including if they crossed over the boundary and went into the sapling area. Quadrat cuts of pasture were also collected at the start and conclusion of the trial to compare pasture quality within the inclusion (grazing) and exclusion (non-grazing) zones.

The results showed that the cattle learned the virtual fencing cues within the first day and were able to respond correctly to the audio cue for the majority of the time and avoid receiving electrical pulses (Campbell et al., 2020). Animals were also able to navigate the contoured fence line and remained within the inclusion zone for most of the trial duration. Although there were incursions into the exclusion zone, these were typically only for short durations of time with less than 20 min per animal spent in the excluded area across the 44-day period. The saplings remained protected and pasture analysis showed more and better pasture quality in the exclusion zone relative to the grazed area (Fig. 5). This trial demonstrated a practical application of the virtual

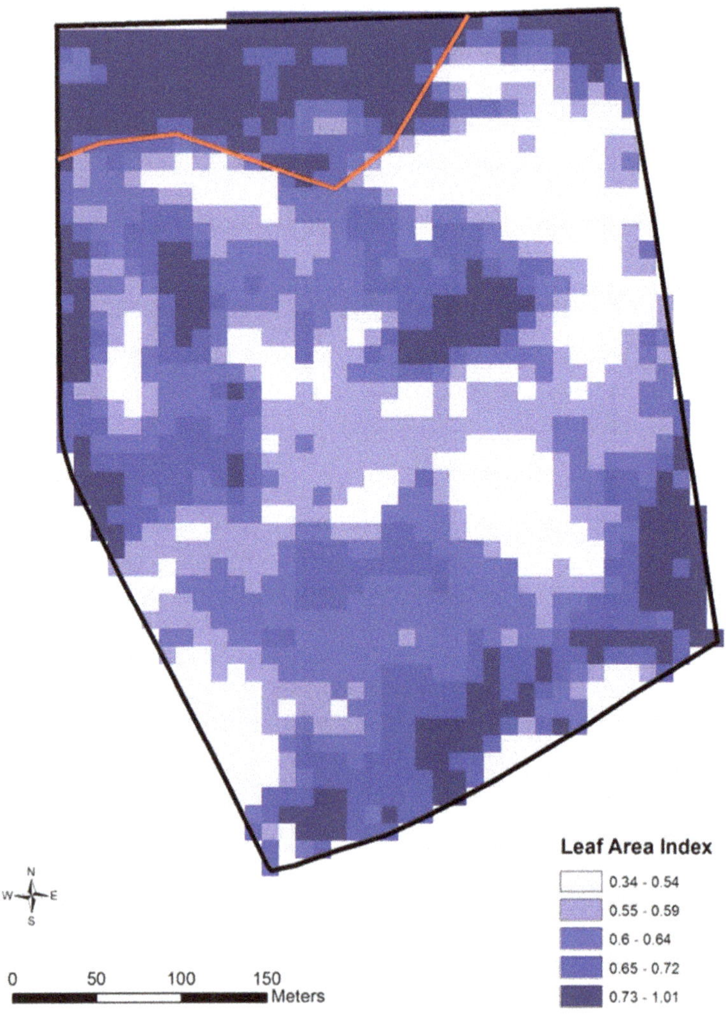

Figure 5 Leaf Area Index showing the virtual fence line and higher pasture vegetation within the exclusion zone at the conclusion of the trial. Reproduced from Campbell et al. (2020).

fencing technology in setting up a temporary fence around a set of developing trees within a grazing paddock. This could be achieved with minimal labour and highlights the benefits of this type of remote technology in optimising management of grazing and land.

6 Conclusion

Information technology has advanced exponentially over the last three decades or so as described by Moore's Law. Thirty years ago, our understanding of grazing systems relied on direct observation aided with devices that were expensive, unreliable, and unwieldy and recorded only the most trivial of information at very low spatial and temporal granularity. Compare that situation with today, where the challenge is not in the paucity of information but in discerning what is important and in interpreting that information in a meaningful manner. Our understanding of the relationship between the grazing animal and the pasture has grown with the development of new measurement and monitoring devices and our ability to handle large amounts of data. In turn, this has allowed for improved management interventions to optimise the health and productivity of both the sward and the grazing animal. The culmination of these developments has led to our ability to remotely manage the grazing animal through the use of on-animal devices that interact in real time with the animal in response to the animal's behaviour. Ongoing development and continued research will see the full realisation of artificial intelligence and the internet of things applied to grazing systems. Our ability will extend to control the sward at the individual plant scale and the individual animal at the organ level, and to predict their interactions into the future. However, there is the danger that the centuries-old relationship between humankind and their livestock will be diminished. In the face of new remote-monitoring technologies, the challenge will be to ensure that our instinctive empathy for the livestock we rely upon is not lost forever.

7 Future trends in research

Precision livestock farming technology for measuring the behaviour of individual animals has progressed considerably in recent years, but further research and technological developments are needed before it is integrated into daily management for the majority of commercial farms. With multiple different types of on-animal technologies available, or multiple companies rushing for first place in the commercialisation race, it can be overwhelming for a producer knowing which product to select. There is also a limit to the number of on-animal devices a single individual can sustain before behaviour and productivity are negatively impacted. Integration of wearable livestock devices such as eShepherd®, eGrazor and Ceres Tag into one platform would enable producers to purchase one product that can both monitor and control livestock. Eventual development of user-friendly platforms, software and devices that can enable most farmers to manage pasture and livestock remotely would transform the way livestock are managed and would substantially reduce labour requirements while maximising productivity.

While new digital technologies may benefit the livestock industries, recognition that animals are sentient, have agency and show individual differences is important to consider when designing new systems. Livestock are intelligent beings and new technologies such as virtual fencing can harness their learning capabilities to improve their life experiences. However, the complexity of the systems may be limited by the ability of animals to adapt without negative consequences. Thus, alignment of the expectations of the new technology with the cognitive and learning abilities of animals must be determined before widespread release. Smart algorithms that can adapt to individual variation in learning and behavioural responses may enhance the successful application of technology to all animals within a herd. Additionally, research on the effects of both individual variation coupled with social group dynamics is necessary to understand the impacts of new systems such as virtual fencing.

Further research applying virtual fencing devices on commercial farms across different scenarios, different breeds, different group sizes, for example, is needed to determine the cost-benefits of the technology in both cattle and sheep across varying contexts. In particular, understanding of longer-term impacts (>6 months) is still needed. Further studies testing the effectiveness of virtual fencing when livestock are highly motivated to access an area such as during strip grazing or when preventing access to environmentally sensitive areas would contribute to the current knowledge base.

Devices that monitor livestock location and behaviour can inform on animal welfare, but ultimately, the development of additional sensors that can measure physiological stress responses, such as heart rate and body temperature, may provide the opportunity for real-time assessment of welfare and health states of animals in the field. This knowledge could be utilised for animal welfare assurance to provide consumers with confidence that welfare standards have been met for the products they purchase.

8 Where to look for further information

In addition to the literature cited within this chapter, updated information on the discussed technologies can be found on different websites.

- eShepherd® Virtual fencing: www.agersens.com.
- Nofence: https://nofence.no/en/.
- Ceres Tag: www.cerestag.com.

- eGrazor: https://www.csiro.au/en/Research/AF/Areas/Livestock/eGrazor-measuring-cattle-pasture-intake.
- Information on the Australian Government-funded Virtual Herding project can be found at: https://www.dairyaustralia.com.au/farm/animal-management/technologies/virtual-herding-program.

Research centres for further information and collaboration within Australia include the Commonwealth Scientific and Industrial Research Organisation (CSIRO) with different teams located at Armidale (NSW), Brisbane (NSW) and Townsville (QLD).

9 References

Anderson, D. M. 2007. Virtual fencing–past, present and future. *Rangeland J.* 29, 65-78. https://doi.org/10.1071/rj06036.

Ash, A. J., McIvor, J. G., Corfield, J. P. and Winter, W. H. 1995. How land condition alters plant-animal relationships in Australia's tropical rangelands. *Agric. Ecosys. Environ.* 56(2), 77-92.

Ash, A. J., McIvor, J. G., Mott, J. J. and Andrew, M. H. 1997. Building grass castles: integrating ecology and management of Australia's tropical tallgrass rangelands. *Rangeland J.* 19, 123-144. https://doi:org/10.1071/RJ9970123.

Bell, A. W., Charmley, E., Hunter, R. A. and Archer, J. A. 2011. The Australasian beef industries–challenges and opportunities in the 21st century. *Anim. Front.* 1(2), 10-19 https://doi:org/10.2527/af.2011-00159.

Bishop-Hurley, G. J., Perry, J., Charmley, E. and Watson, I. 2016. Developing technologies to track cattle in remote areas. In: *Proceedings of the Northern Beef Research Update Conference*. Available at: https://www.nabrc.com.au/nbruc (vol. 2016).

Bishop-Hurley, G. J., Swain, D. L., Anderson, D. M., Sikka, P., Crossman, C. and Corke, P. I. 2007. Virtual fencing applications: implementing and testing an automated cattle control system. *Comput. Electron. Agric.* 56(1), 14-22. https://doi.org/10.1016/j.compag.2006.12.003.

Brunberg, E. I., Boe, K. E. and Sorheim, K. M. 2015. Testing a new virtual fencing system on sheep. *Acta Agr. Scand. A-AN.* 65(3-4), 168-175.

Butzer, K. W. and Helgren, D. M. 2008. Livestock, land cover, and environmental history: the tablelands of New South Wales, Australia, 1820-1920. *Ann. Am. Assoc. Geogr.* 95, 80-111. https://doi.org/10.1111/j.1467-8306.2005.00451.x.

Campbell, D. L. M., Lea, J. M., Farrer, W. J., Haynes, S. J. and Lee, C. 2017. Tech-savvy beef cattle? How heifers respond to moving virtual fence lines. *Animals* 7(9), 72. https://doi.org/10.3390/ani7090072.

Campbell, D. L. M., Lea, J. M., Haynes, S. J., Farrer, W. J., Leigh-Lancaster, C. J. and Lee, C. 2018. Virtual fencing of cattle using an automated collar in a feed attractant trial. *Appl. Anim. Behav. Sci.* 200, 71-77. https://doi.org/10.1016/j.applanim.2017.12.002.

Campbell, D. L. M., Haynes, S. J., Lea, J. M., Farrer, W. J. and Lee, C. 2019a. Temporary exclusion of cattle from a riparian zone using virtual fencing technology. *Animals* 9(1), 5. https://doi.org/10.3390/ani9010005.

Campbell, D. L. M., Lea, J. M., Keshavarzi, H. and Lee, C. 2019b. Virtual fencing is comparable to electric tape fencing for cattle behavior and welfare. *Front. Vet. Sci.* 6, 445. https://doi.org/10.3389/fvets.2019.00445.

Campbell, D. L. M., Ouzman, J., Mowat, D., Lea, J. M., Lee, C. and Llewellyn, R. S. 2020. Virtual fencing technology excludes beef cattle from an environmentally sensitive area. *Animals* 10(6), 1069; https://doi.org/10.3390/ani10061069.

Chambers, J. D. and Mingay, G. E. 1966. *The Agricultural Revolution 1750–1880*. Batsford, London, UK.

Clutton-Brock, J. 1999. *A Natural History of Domesticated Animals*. Cambridge University Press, Cambridge, UK.

Collins, R., Mcleod, M., Hefley, M., Donnison, A., Close, M., Hanly, J., Horne, D., Ross, C., Davies-Colley, R., Bagshaw, C. and Matthews, L. 2007. Best management practices to mitigate faecal contamination by livestock of New Zealand waters. *J. Agr. Res.* 50(2), 267–278. https://doi.org/10.1080/00288230709510294.

Cornish, A., Raubenheimer, D. and McGreevy, P. 2016. What we know about the public's level of concern for farm animal welfare in food production in developed countries. *Animals* 6(11), 74. https://doi.org/10.3390/ani6110074.

Dahlström, A., Cousins, S. A. O. and Eriksson, O. 2006. The history (1620–2003) of land use, people and livestock, and the relationship to present plant species diversity in a rural landscape in Sweden. *Environ. Hist.* 12(2), 191–212. https://doi.org/10.3197/096734006776680218.

Dettenmaier, S. J., Messmer, T. A., Hovick, T. J. and Dahlgren, D. K. 2017. Effects of livestock grazing on rangeland biodiversity: a meta-analysis of grouse populations. *Ecol. Evol.* 7, 7620–7627. https://doi.org/10.1002/ece3.328.

Edirisinghe, A., Hill, M. J., Donald, G. E.. and Hyder, M. 2011. Quantitative mapping of pasture biomass using satellite imagery. *Int. J. Remote Sens.* 32(10), 2699–2724. doi: 10.1080/01431161003743181.

FAO 2018. World livestock: transforming the livestock sector through the Sustainable Development Goals. Rome. 222 pp. Licence: CC BY-NC-SA 3.0 IGO.

Fay, P. K., McElligott, V. T. and Havstad, K. M. 1989. Containment of free-ranging goats using pulsed-radio-wave-activated shock collars. *Appl. Anim. Behav. Sci.* 23(1–2), 165–171. https://doi.org/10.1016/0168-1591(89)90016-6.

Forrester, T. R., Green, D. J., McKibbin, R. and Bishop, C. A. 2017. Evaluating the efficacy of seasonal grazing and livestock exclusion as restoration tools for birds in riparian habitat of the Okanagan Valley, British Columbia, Canada. *Restor. Ecol.* 25(5), 768–777. https://doi.org/10.1111/rec.12495.

Gillan, J. K., McClaran, M. P., Swetnam, T. L. and Heilman, P. 2019. Estimating forage utilization with drone-based photogrammetric point clouds. *Rangeland Ecol. Manag.* 72(4), 575–585. https://doi.org/10.1016/j.rama.2019.02.009.

Grandin, T. 2014. Animal welfare and society concerns finding the missing link. *Meat Sci.* 98(3), 461–469. https://doi.org/10.1016/j.meatsci.2014.05.011.

Greenwood, P. L., Paull, D. R., McNally, J., Kalinowski, T., Ebert, D., Little, B., Smith, D. V., Rahman, A., Valencia, P., Ingham, A. B. and Bishop-Hurley, G. J. 2017. Use of sensor-determined behaviours to develop algorithms for pasture intake by individual grazing cattle. *Crop Pasture Sci.* 68(12), 1091–1099. https://doi.org/10.1071/CP16383.

Greenwood, P. L., Valencia, P., Overs, L., Paull, D. R. and Purvis, I. W. 2014. New ways of measuring intake, efficiency and behaviour of grazing livestock. *Anim. Prod. Sci.* 54(10), 1796–1804. https://doi.org/10.1071/AN14409.

Handcock, R. N., Gobbett, D. L., Gonzalez, L. A., Bishop-Hurley, G. J. and McGavin, S. L. 2016. A pilot project combining multispectral proximal sensors and digital cameras for monitoring tropical pastures. *Biogeosciences* 13, 4673–4695. https://doi.org/10.5194/bg-13-4673-2016.

Henkin, Z., Ungar, E. D. and Dolev, A. 2012. Foraging behaviour of beef cattle in the hilly terrain of a Mediterranean grassland. *Rangel. J.* 34(2), 163–172. https://doi.org/10.1071/RJ11096.

Holmes, P. R. 2015. Rangeland pastoralism in northern Australia: change and sustainability. *Rangel. J.* 37(6), 609–616. https://doi.org/10.1071/RJ15051.

Howell, H. J., Mothes, C. C., Clements, S. L., Catania, S. V., Rothermel, B. B. and Searcy, C. A. 2019. Amphibian responses to livestock use of wetlands: new empirical data and a global review. *Ecol. Appl.* 29(8), e01976. https://doi.org/10.1002/eap.1976.

Hunt, L. P., McIvor, J. G., Grice, A. C. and Bray, S. G. 2014. Principles and guidelines for managing cattle in the grazing lands of northern Australia: stocking rates, pasture resting, prescribed fire, paddock size and water points–a review. *Rangeland J.* 36, 105–119. http://dx.doi.org/10.1071/rj13070.

Jouven, M., Leroy, H., Ickowicz, A. and Lapeyronie, P. 2012. Can virtual fences be used to control grazing sheep? *Rangel. J.* 34(1), 111–123. http://doi.org/10.1071/RJ11044.

Kearton, T., Marini, D., Cowley, F., Belson, S. and Lee, C. 2019. The effect of virtual fencing stimuli on stress responses and behavior in sheep. *Animals* 9(1), 30. https://doi.org/10.3390/ani9010030.

Keshavarzi, H., Lee, C., Lea, J. M. and Campbell, D. L. M. 2020. Virtual fence responses are socially facilitated in beef cattle. *Front. Vet. Sci.* 7, 543158. https://doi.org/10.3389/fvets.2020.543158.

Lee, C. 2006. An apparatus and method for the virtual fencing of an animal. In: *International Patent Application PCT/AUT2005/001056*. Published 26 Jan. International Publication No. Wo 2006/007643.

Lee, C., Colditz, I. G. and Campbell, D. L. M. 2018. A framework to assess the impact of new animal management technologies on welfare: a case study of virtual fencing. *Front. Vet. Sci.* 5, 187. https://doi.org/10.3389/fvets.2018.00187.

Lee, C., Prayaga, K. C., Fisher, A. D. and Henshall, J. M. 2008a. Behavioral aspects of electronic bull separation and mate allocation in multiple-sire mating paddocks. *J. Anim. Sci.* 86(7), 1690–1696. https://doi.org/10.2527/jas.2007-0647.

Lee, C., Fisher, A. D., Reed, M. T. and Henshall, J. M. 2008b. The effect of low energy electric shock on cortisol, beta-endorphin, heart rate and behaviour of cattle. *Appl. Anim. Behav. Sci.* 113(1–3), 32–42. https://doi.org/10.1016/j.applanim.2007.10.002.

Lee, C., Henshall, J. M., Wark, T. J., Crossman, C. C., Reed, M. T., Brewer, H. G., O'Grady, J. and Fisher, A. D. 2009. Associative learning by cattle to enable effective and ethical virtual fences. *Appl. Anim. Behav. Sci.* 119(1–2), 15–22. https://doi.org/10.1016/j.applanim.2009.03.010.

Lee, C., Prayaga, K., Reed, M. and Henshall, J. 2007. Methods of training cattle to avoid a location using electrical cues. *Appl. Anim. Behav. Sci.* 108(3–4), 229–238. https://doi.org/10.1016/j.applanim.2006.12.003.

Lee, C., Reed, M. T., Wark, T., Crossman, C. and Valencia, P. 2010. A control device, and method, for controlling the location of an animal. PCT/AU2009/000943. Published 28 Jan 2010. WIPO No. WO2010/009509.

Llewellyn, R., Marini, D., Lee, C., Belson, S., Ouzman, J. and Monjardino, M. 2018. Cost effective weed management using targeted sheep grazing technology. Final report for the Australian Government. Department of Agriculture and Water Resources.

Lomax, S., Colusso, P. and Clark, C. E. F. 2019. Does virtual fencing work for grazing dairy cattle? *Animals* 9(7), 7, 429. https://doi.org/10.3390/ani9070429.

Marini, D., Cowley, F., Belson, S. and Lee, C. 2019. The importance of an audio cue warning in training sheep to a virtual fence and differences in learning when tested individually or in small groups. *Appl. Anim. Behav. Sci.* 221. https://doi.org/10.1016/j.applanim.2019.104862, 104862.

Marini, D., Meuleman, M. D., Belson, S., Rodenburg, T. B., Llewellyn, R. and Lee, C. 2018a. Developing an ethically acceptable virtual fencing system for sheep. *Animals* 8(3), 33. https://doi.org/10.3390/ani8030033.

Marini, D., Llewellyn, R., Belson, S. and Lee, C. 2018b. Controlling within-field sheep movement using virtual fencing. *Animals* 8(3), 31. https://doi.org/10.3390/ani8030031.

Marty, J. T. 2015. Loss of biodiversity and hydrologic function in seasonal wetlands persists over 10 years of livestock grazing removal. *Restor. Ecol.* 23(5), 548–554. https://doi.org/10.1111/rec.12226.

McIvor, J. G. and Gardener, C. J. 1995. Pasture management in semi-arid tropical woodlands: effects on herbage yields and botanical composition. *Aust. J. Exp. Agric.* 35(6), 705–715. https://doi.org/10.1071/EA9950705.

McKeon, G. M., Hall, W. B., Henry, B. K., Stone, G. S. and Watson, I. W. 2004. *Pasture Degradation and Recovery in Australia's Rangelands: Learning from History.* Queensland Department of Natural Resources, Mines and Energy, Brisbane.

O'Reagain, P., Bushell, J. and Holmes, B. 2011. Managing for rainfall variability: long-term profitability of different grazing strategies in a northern Australian tropical savanna. *Anim. Prod. Sci.* 51(3), 210–224. https://doi.org/10.1071/AN10106.

Primary Industries Standing Committee, (PISC) 2007. *Nutrient Requirements of Domesticated Ruminants.* CSIRO Publishing, Melbourne.

Rango, A., Havstad, K. and Estell, R. 2011. The utilization of historical data and geospatial technology advances at the Jornada Experimental Range to support Western America ranching culture. *Remote Sens.* 3(9), 2089–2109. https://doi.org/10.3390/rs3092089.

Ripple, W. J., Smith, P., Haberl, H., Montzka, S. A., McAlpine, C. and Boucher, D. H. 2014. Ruminants, climate change and climate policy. *Nat. Clim. Change* 4(1), 2–5. https://doi.org/10.1038/nclimate2081.

Ruiz-Mirazo, J., Bishop-Hurley, G. J. and Swain, D. L. 2011. Automated animal control: can discontinuous monitoring and aversive stimulation modify cattle grazing behavior? *Rangel Ecol. Manag.* 64(3), 240–248. https://doi.org/10.2111/REM-D-10-00087.1.

Rutter, S. M. 2014. Smart technologies for detecting animal welfare status and delivering health remedies for rangeland systems. *Rev. Sci. Tech.* 33(1), 181–187. doi: 10.20506/rst.33.1.2274.

Schellberg, J., Hill, M. J., Gerhards, R., Rothmund, M. and Braun, M. 2008. Precision agriculture on grassland: applications, perspectives and constraints. *Eur. J. Agron.* 29(2–3), 59–71. https://doi.org/10.1016/j.eja.2008.05.005.

Schlolz, F. and Schlee, G. 2015. *Nomads and Nomadism in History. International Encyclopedia of the Social & Behavioural Sciences*, pp. 838–843. https://doi.org/10.1016/B978-0-08-097086-8.62018-4.

Thomson, A. M. and Simpson, I. A. 2006. Modelling historic rangeland management and grazing pressure in landscapes of settlement. *Hum. Ecol.* 35, 151–168. http://dx.doi.org/10.1007/s10745-006-9055-8.

Thornley, J. H. M., Parsons, A. J., Newman, J. and Penning, P. D. 1994. A cost-benefit model of grazing intake and diet selection in a two-species temperate grassland sward. *Funct. Ecol.* 8(1), 5–16. https://www.jstor.org/stable/2390105.

Thorburn, P. J., Robertson, M. J., Clothier, B. E., Snow, V. O., Charmley, E., Sanderman, J., Teixeira, E., Dynes, R. A., Hall, A., Brown, H., Howden, S. M. and Battaglia, M. 2011. Australia and New Zealand perspectives on climate change and agriculture. In: Hillel, D. and Rosenzweig, C. (Eds.), *ICP Series on Climate Change Impacts, Adaptation, and Mitigation–Vol. 2 Handbook of Climate Change and Agroecosystems. Global and Regional Aspects and Implications.* Joint Publication with the American Society of Agronomy, Crop Science Society of America, and Soil Science Society of America. 107–141. https://doi.org/10.1142/9781848169845_0007.

Thorburn, P. J. and Wilkinson, S. N. 2013. Conceptual frameworks for estimating the water quality benefits of improved agricultural management practices in large catchments. *Agric. Ecosys. Environ.* 180, 192–209. https://doi.org/10.1016/j.agee.2011.12.021.

Thornton, P. K. 2010. Livestock production: recent trends, future prospects. *Philos. T. R. Soc. B* 365(1554), 2853–2867. https://doi.org/10.1098/rstb.2010.0134.

Tomkins, N. W. and O'Reagain, P. J. O. 2007. Global positioning systems indicate landscape preferences of cattle in the subtropical savannas. *Rangel. J.* 29(2), 217–222. https://doi.org/10.1071/RJ07024.

Tomkins, N. W., O'Reagain, P. J. O., Swain, D., Bishop-Hurley, G. and Charmley, E. 2009. Determining the effect of stocking rate on the spatial distribution of cattle for the subtropical savannas. *Rangel. J.* 31(3), 267–276. https://doi.org/10.1071/RJ07070.

Umstatter, C. 2011. The evolution of virtual fences: a review. *Comput. Electron. Agric.* 75(1), 10–22. https://doi.org/10.1016/j.compag.2010.10.005.

Van Soest, P. J. 1982. *Nutritional Ecology of the Ruminant.* O&B Books, Inc., Corvallis, OR.

Verdon, M., Lee, C., Marini, D. and Rawnsley, R. 2020. Pre-exposure to an electrical stimulus primes associative pairing of audio and electrical stimuli for dairy heifers in a virtual fencing feed attractant trial. *Animals* 10(2), 217. https://doi.org/10.3390/ani10020217.

Index

www.ingramcontent.com/pod-product-compliance
Ingram Content Group UK Ltd.
Pitfield, Milton Keynes, MK11 3LW, UK
UKHW021825100926
8556IPUK00003B/159

9 781786 764713